# ANIMALS AND THE ENVIRONMENT

Contemporary earth and animal activists rarely collaborate, perhaps because environmentalists focus on species and ecosystems, while animal advocates look to the individual, and neither seems to have much respect for the other. This diverse collection of essays highlights common ground between earth and animal advocates, most notably the protection of wildlife and personal dietary choice.

If earth and animal advocates move beyond philosophical differences and resultant divergent priorities, turning attention to shared goals, both will be more effective—and both animals and the environment will benefit. Given the undeniable seriousness of the problems that we face, including climate change and species extinction, it is essential that activists join forces. Drawing on a wide range of issues and disciplines, ranging from wildlife management, hunting, and the work of NGOs, to ethics, ecofeminism, religion, and animal welfare, this volume provides a stimulating collection of ideas and challenges for anyone who cares about the environment or animals.

**Lisa Kemmerer** is a Professor of Philosophy and Religion at Montana State University, Billings, USA. She is the author or editor of a number of books on animal ethics, ecofeminism, animals and religion, and the environment.

# ANIMALS AND THE ENVIRONMENT

## Advocacy, Activism, and the Quest for Common Ground

*Edited by Lisa Kemmerer*

Routledge
Taylor & Francis Group
LONDON AND NEW YORK

earthscan
from Routledge

First published 2015
by Routledge
2 Park Square, Milton Park, Abingdon, Oxon OX14 4RN

and by Routledge
711 Third Avenue, New York, NY 10017

*Routledge is an imprint of the Taylor and Francis Group, an informa business*



*British Library Cataloguing-in-Publication Data*
A catalogue record for this book is available from the British Library

*Library of Congress Cataloging in Publication Data*
Animals and the environment : advocacy, activism, and the quest for common ground / edited by Lisa Kemmerer.
pages cm
Includes bibliographical references and index.
1. Animal welfare. 2. Animal rights. 3. Animal ecology. 4. Environmental protection. I. Kemmerer, Lisa.
HV4708.A5713 2015
363.7--dc23
2014027788

ISBN: 978-1-138-82587-1 (hbk)
ISBN: 978-1-138-82588-8 (pbk)
ISBN: 978-1-315-73966-3 (ebk)

Typeset in Bembo
by GreenGate Publishing Services, Tonbridge, Kent

In memory of Randall Gloege and other activists, largely unknown to the world, who commit significant amounts of energy to the difficult work of bringing change for earth and animals.

Any proceeds received by the editor from the sale of this book will be reinvested to bring change for earth and animals.

# CONTENTS

# CONTRIBUTORS

**Nathan Baillet** is a MBA student at the University of Montana who holds a BS in business administration and a BA in history with a minor in philosophy. He grew up in the mountains of Montana where he developed a strong connection with nature and the environment and, as a vegan, has developed a strong desire to be a voice to the voiceless. His goal is to meld the two disciplines into a career in community-based activism. He is currently serving as a trustee on the board for the Laurel Public Library, always looking for ways to improve and expand access to education.

**Haida Bolton** was born and raised in Nelson, British Columbia, Canada, obtained her Bachelor of Science from the University of British Columbia in 1990, then headed off to explore thirty countries throughout Africa, Asia, Australia, North America, and Europe. Her travels enhanced her appreciation of the earth's vast beauty. In 2006 she founded Camp Uganda (www.campuganda.org), a registered charity. Bolton remained the president of Camp Uganda Conservation Education Society until 2011, which aims to "empower youth to save endangered chimpanzees" in Uganda. Bolton, a proud advocate of children's camps, worked in Canadian children's camps for ten years before founding Camp Uganda.

**Valerie J. Chalcraft** is an experimental psychologist with an MA and a PhD from the University of Nevada, Reno where she studied chimpanzees who communicate with American Sign Language. Chalcraft is an independent researcher who is interested in applying the principles of behavior to improve the welfare and rights of humans and nonhumans. Currently, she practices applied animal behavior in Chicago, Illinois and is a professor at the Animal Behavior Institute.

**Cara Chamberlain** attended the University of Utah and Purdue University. She has published fiction and poetry in numerous journals. *The Devil's Party*, her novella about genetically modified organisms, was a 2004 finalist in the Low Fidelity Novella Contest, and her poetry collection, *Hidden Things*, was produced by FootHills Publishing in 2009. Another poetry collection, *The Divine Botany*, is forthcoming from WordTech Communications. She has volunteered with The Nature Conservancy and the Bureau of Land Management to survey scrub jays, gopher tortoises, eagles, and sage grouse. A member of Montana Conservation Voters, she teaches at Rocky Mountain College in Billings, Montana.

**Chris Darimont**, although trained as an evolutionary ecologist, has developed strong personal, scholarly, and practical interests in animal welfare. As Science Director for the Raincoast Conservation Foundation and Assistant Professor in the Geography Department at the University of Victoria, Canada, his research focuses on sensitive carnivores, like wolves and bears, who endure severe suffering because of humans, both through persecution and competition (for land and "huntable" species). Chris has received fellowships from the Natural Science and Engineering Research Council of Canada, and several animal welfare awards, including a Compassion in Science Award from the International Fund for Animal Welfare, a Christine Stevens Wildlife Award from the Animal Welfare Institute, and an Earth Day Canada Finalist Award.

**Josephine Donovan** co-edited (with Carol J. Adams) *The Feminist Care Tradition in Animal Ethics* (2007), which contains her articles "Animal Rights and Feminist Theory," "Attention to Suffering," "Caring to Dialogue," and *Animals and Women: Feminist Theoretical Explorations* (1995). She is also the author of *Feminist Theory: The Intellectual Traditions* (4th ed., 2012) and numerous other works, a complete list of which is available on her website: www.english.umaine.edu/faculty/josephine.donovan. She is Professor Emerita at the University of Maine.

**Bethany Dopp** holds a Bachelor of Arts in environmental studies from Montana State University–Billings. Born in a small town in Northeastern Montana, she advocates for social justice with special enthusiasm for animal and earth liberation. Dopp works at a Billings college library, and enjoys hiking, bird-watching, bicycling, writing, tie-dying, gardening, and music in her spare time. Her passion is writing fiction, and she ultimately hopes to put this interest to work for social justice.

**Debra Erenberg** is an organizer and activist for the environment, animal rights, human rights, and social justice in general. Erenberg is Director of State Affairs for Justice at Stake, and previously served as Midwest Regional Director for Amnesty International USA, Organizing Director for Rainforest Action Network, and as Director of Affiliate Development at NARAL Pro-Choice America. Erenberg has also worked with In Defense of Animals, the Great Ape Project, and is currently on the advisory board for the Food Empowerment Project. She received her Master's degree in public policy from the University of Michigan and a Juris Doctorate from the George Washington University National Law Center.

**Carrie P. Freeman** is a U.S. activist–scholar who studies the fundamental role that communication plays in creating a more just and sustainable world for all species. An Associate Professor of Communication at Georgia State University in Atlanta, she publishes on critical animal studies, environmental communication, and media ethics, with a specialty in vegan advocacy, which is the topic of her book *Framing Farming: Communication Strategies for Animal Rights* (Rodopi Press, 2014). A volunteer animal activist for two decades, Freeman led three grassroots animal rights organizations, and now co-hosts animal and environmental protection radio shows on indie station WRFG.org.

**Xylem T. Galadhon** has been active for animal rights, old-growth forests, Central American solidarity, the Obama 2008 campaign, and health care reform, and has campaigned against the Iraq invasion, globalization, mountain top removal coalmining, and the forces of climate change. As a physicist and astronomer he has analyzed data taken from particle physics colliders in two states, as well as images taken from telescopes both on Earth and in space.

**Chris Genovali** has served as Executive Director for the Raincoast Conservation Foundation for twelve years. His articles on Canadian wildlife and conservation have been printed by media throughout Canada and internationally (in *Globe and Mail, Vancouver Sun, Victoria Times Colonist, The Ecologist, Seattle Post-Intelligencer, Vancouver Province, Edmonton Journal,* etc), as well as by online publications (*Huffington Post, Common Dreams, Truthout, Counterpunch, The Tyee,* etc.). Genovali has also appeared as a radio and television spokesperson with CBC's *As It Happens*, CBC *Newsworld*, US National Public Radio, CKNW, CTV, Global TV, BBC radio, BBC television, Channel 4 UK, the Knowledge Network, and CBC News Vancouver.

**Carol L. Glasser** works professionally, academically, and at the grassroots for the liberation of human and nonhuman animals. Since 2009, she has focused on ending animal experimentation, and is a cofounder of and organizer with the anti-vivisection group Progress For Science (www.ProgressForScience.com). Glasser's academic work addresses social movements and the intersections of nonhuman and human animal exploitation. She is currently an assistant professor of sociology at the University of Minnesota, Mankato.

**Randall Gloege** began his writing life as a poet. Three of his poems may be found on the Pryors Coalition website (www.pryormountains.com). In the 1970s he became an activist in the wilderness movement. His work and that of many others led to the designation of the Absaroka-Beartooth Wilderness. As a college teacher, he focused on writing, American literature, philosophy, and environmental ethics. He understood himself as a work in progress. In retirement, he continued to study and work to protect wilderness and on behalf of other environmental causes. He died peacefully in his chair February 19, 2013.

**Jennifer Gross** is a community activist working at the intersection of feminism and animal liberation. She currently holds a BA in Environmental Studies and her goal is to teach nonviolence and compassion through diet. Gross has worked on a Montana ballot initiative for Trap-Free Public Lands and is currently working as a community organizer for a women's health organization. She is motivated by love for the earth and all creatures.

**John M. Halley** grew up on a farm in a farming community in Ireland. He graduated as an electronics engineer from University College Dublin in 1983, then earned his MSc and PhD degrees from University College London in 1989. He travelled extensively in China, took a job in England, then Scotland, and finally settled in Greece, where he is currently associate professor of ecology at the University of Ioannina. Animals and the environment have always interested Halley, but his areas of scientific expertise are in the application of mathematics and statistics to biodiversity, climate change, and extinction.

**Anja Heister,** born and raised in Germany, holds a Masters in biology from the Johann Wolfgang von Goethe University in Frankfurt, and is currently enrolled in an Interdisciplinary PhD program at the University of Montana in Missoula, where she is studying wildlife conservation and ethics. For five years Heister was Executive Director of Footloose Montana (www.footloosemontana.org), a nonprofit organization working to end recreational and commercial trapping on Montana's public lands, and is currently Director of the Wild and Free – Habitats Campaign for In Defense of Animals (IDA).

**Chris Hunt** is the food program director for GRACE Communications Foundation, an organization that raises public awareness about environmental and public health issues created by our food, water, and energy systems. At GRACE, Hunt oversees Sustainable Table, Eat Well Guide, and The Meatrix, and writes for publications such as *Civil Eats*, *Huffington Post*, and *Ecocentric*. He earned a BA in environmental economics and peace studies from Colgate University and currently focuses on industrial livestock production, food waste, and urban agriculture. Hunt served two terms on the board of directors of the Hudson River Sloop Clearwater, and is proud to be a New York City Master Composter.

**Lisa Kemmerer** is a philosopher–activist who has been working on behalf of the environment, nonhuman animals, and disempowered humans for more than thirty years. A graduate of Reed, Harvard, and Glasgow University, she is the author of half a dozen books, including *Eating Earth: Dietary Choice and Planetary Health*, *Animals and World Religions*, and *Sister Species: Women, Animals, and Social Justice*. You can learn more about her work at lisakemmerer.com.

**Daniel Kirjner** is a vegan–feminist activist and doctoral sociology student at University of Brasília, Brazil. His research focuses on links between the construction of masculinity and the glorification of violence against animals and women in contemporary capitalist societies. He continues to learn about, question, and deconstruct masculinity in his own life in order to heighten his own awareness of predatory male behavior in the hope of being himself part of bringing change to the world.

**Charlotte Laws**, PhD, a weekly NBC commentator, served four terms on the Greater Valley Glen Council in Southern California and was previously a Los Angeles city commissioner. She holds a doctorate in social ethics from USC and was the recipient of the 2006 LA Animal Humanitarian Award. Laws has authored *Armed for Ideological Warfare* (USC, 1999), and articles in *Call to Compassion* (Lantern, 2011) and *Igniting a Revolution* (AK Press, 2006). Additionally, she has appeared in dozens of television shows, including *Larry King Live*, *The Late Show*, and *Fox News*. She is the founder and president of the League for Earth and Animal Protection and the Los Angeles Directors of Animal Welfare.

**Phaik Kee Lim** has worked for Friends of the Earth Malaysia (Sahabat Alam Malaysia—SAM) for more than thirty years. She creates and distributes petitions, sends out action alerts, corresponds with the media, local authorities, and other NGOs on a range of issues on behalf of nonhuman animals and the environment, focusing on the exploitation and destruction of wildlife and species (via entertainment industries, laboratories, and the pet trade) and of domestic animals (factory farming, and mistreatment of "pets").

**Melanie J. Martin** holds a Master of Arts in English from Slippery Rock University of Pennsylvania, and is the author of "Myths of Loss, Myths of Power," published in *Of Mice and Men: Animals in Human Culture* (Cambridge Scholars Publishing, 2009). Her Master's thesis, "Why We Search for Animals," explores the American fascination with finding the rarest of nonhuman animals. Martin's work has been printed in publications as diverse as *Alternatives Journal* and *Professional Artist*. She writes essays and poetry exploring themes of environmental justice, cultural attitudes toward animals and nature, and what a liveable future might look like.

**Paul Paquet** is an internationally recognized authority on mammalian carnivores, especially wolves, with research experience in several regions of the world. He worked for many years as a biologist for the Canadian Wildlife Service. He now serves as Senior Scientist with the Raincoast Conservation Foundation, as an Adjunct Professor in the Geography Department at the University of Victoria, and as an international consultant and lecturer. Paquet was one of the architects of the World Wide Fund for Nature and the European Union's Large Carnivore Initiative for Europe. He also holds positions at the Universities of Calgary, Manitoba, and New Brunswick, where his current research focuses on the effects of human activities on the survival of large carnivores, and the conservation of these highly endangered species.

**Bernard Quetchenbach** was born in Rochester, New York, on Lake Ontario at the mouth of the Genesee River. He is a faculty member in the Department of English, Philosophy, and Modern Languages at Montana State University Billings, where his courses include Studies in Literature and the Environment. He has been involved in various writing and environmental organizations, and currently serves on the board of the Montana Wilderness Association's Eastern Wildlands Chapter. His poetry, essays, and critical articles have appeared in books, anthologies, and periodicals.

**Christina Service** is a biologist with Raincoast Conservation Foundation and a graduate student in the Geography Department at the University of Victoria. Her academic interests include wildlife conservation and ethics in wildlife research. She has received the Natural Science and Engineering Research Council of Canada Undergraduate Student Research Award, Canadian Millennium Excellence award, and the Tom Perry Award for Social Responsibility from the University of Victoria.

**Deric Shannon** is an Assistant Professor of Sociology at Oxford College of Emory University. He has co-written/edited a number of books, journal articles, and book chapters. His current research interests include political economy, crisis/austerity, food justice, and sustainability.

**Jon E. Swenson,** professor of ecology and natural resource management at the Norwegian University of Life Sciences in Ås, Norway, leads the Scandinavian Brown Bear Research Project (www.bearproject.info). He received a Bachelor and Master of Science in fish and wildlife management at Montana State University. After working as a wildlife management biologist with the Montana Department of Fish, Wildlife, and Parks for ten years, he earned a PhD from the University of Alberta. He has been awarded the International Bear Association President's Award for "Outstanding Service to Bear Conservation" and several Norwegian awards for communicating research results to the public.

# ACKNOWLEDGMENTS

Thank you to Randall Gloege, gone over the rainbow bridge, for commenting on an earlier draft of this manuscript. Thank you to Daniel Kirjner for comments on the final manuscript, and to Daniel Kirjner and Morgan Bernitt for feedback on the order of the essays. Thank you to Carol Adams for helpful advice on the literature review, and to Morgan Bernitt and Nathan Baillet for converting references from MLA to Harvard style on several essays at the last moment. Thanks to Tim Hardwick of Routledge, and to a handful of anonymous reviewers, for suggestions on Chapter 1. Finally, many thanks to all the contributors who patiently and diligently worked with me to create this anthology.

# INTRODUCTION

*Lisa Kemmerer*

Long after my last grade school dot-to-dot puzzle, in a classic "ah-hah!" moment, I connected those very visual little black dots on paper with the idea of "connecting the dots." Much later, I pondered how hard it would be, were I no bigger than a gnat, to connect those many little black dots—especially if they were on a multi-dimensional plane. What if a gnat-sized human were trying to connect figurative dots, and those dots were partially or wholly obscured (by social conventions)? This seems to be our predicament, and our task is to connect the otherwise seemingly disparate aspects (dots) that form our lives. If we are able to do so, we gain a richer and fuller understanding of existence—of life and our place in the universe. This is exactly what ecofeminists have set out to do.

In the last third of the twentieth century, ecofeminists exposed various connections between different forms of oppression, more precisely between sexism and environmental degradation. Eventually, ecofeminists also linked poverty, racism, and animal exploitation with environmental problems. Indeed, these thinkers and writers have demonstrated that we must simultaneously work against poverty, racism, environmental degradation, sexism, speciesism, and homophobia, because all are connected. Social justice is only possible if we understand connections between various forms of oppression. We have little hope of putting an end to racism or homophobia (or even understanding them) if we fail to recognize how these harmful oppressions are connected. Social justice requires that we dismantle *systems of oppression* undergirding a plethora of "isms." It is hard work to ferret out forces that have been purposefully obscured behind common, everyday aspects of our lives and our communities, but ecofeminist writings explain why we must connect the dots of oppression, and attack all of them simultaneously to liberate and protect, among other things, animals and our much-beleaguered earth.

## Philosophical and Practical Differences

Environmentalists and animal activists are often keenly aware of philosophical and practical differences played out in the arenas of animal and earth advocacy—including different points of focus and distinct goals that are likely approached via contrasting methods. Perhaps most noticeably, environmentalists strive to preserve species and ecosystems, focusing their energy on *wild* flora and fauna while ignoring (sometimes despising) domestic animals. Environmentalists tend to willingly protect endangered species at the expense of individuals from more plentiful species—including domestic animals. They sometimes even resort to killing plentiful species, invasive species, and domestic animals (burros and cats and pigs) on behalf of fragile ecosystems and/or endangered species. Conversely, animal activists tend to value the individual, viewing every life as precious, and generally value the preservation of individuals—whether labeled domestic or invasive—over the preservation of ecosystems.

These philosophical differences are often reflected in lifestyle choices, especially consumer choices. Serious environmentalists are likely to shop with cloth bags, select organic or local foods, cycle or walk when possible, use fluorescent bulbs, recycle almost all "garbage," and lobby against environmentally destructive products such as palm oil and genetically modified foods. Meanwhile, these same earth activists might buy leather, defend and/or align with hunters and hunting, accept animal experimentation, and frequent zoos or aquariums. Environmentalists are more likely than animal activists to purchase (rather than adopt) companion animals, and may or may not be committed to spay/neuter.

In contrast, serious animal advocates are less likely to shop organic, carry cloth bags, cycle or recycle, or be concerned about palm oil and other environmentally damaging products. Animal advocates are more apt to allow a beloved, adopted, spayed/neutered companion dog or cat to run free, viewing this as essential to their dog or cat's happiness. They are less likely to drive small cars, and more likely to lobby against the use of animals in science and for entertainment or education, including their exploitation in aquariums and zoos. Serious animal advocates generally tend toward a vegan lifestyle, avoiding animal products such as leather, flesh, eggs, and dairy.

But once the figurative dots have been connected, these consumer/lifestyle differences become unreasonable: Earth and animal activists, if serious about their cause, must adopt many of the lifestyle/consumer choices of the "other" camp. Environmentalists might become aware that leather processing and hunting policies are extremely damaging to ecosystems. Animal experimentation, when practiced on certain species, depletes and endangers wildlife and ecosystems because animals such as frogs and primates are snatched from the wild for laboratory use. Zoos, aquariums, and trade in "exotic pets" also effect ecosystems, and damage and destroy native populations when animals such as primates and cetaceans are taken from the wild. While spay/neuter is considered an animal activist's cause, unaltered dogs and cats contribute to domestic overpopulations that destroy wildlife and damage local ecosystems. Animal agriculture causes more environmental damage than any other industry—it makes absolutely no sense for environmentalists to hate bovines with hamburger in hand.

Similarly, animal advocates need to rethink how they choose to live—especially consumer choices. Plastic bags and large cars use petroleum and pollute, packaging requires water and destroys trees, plastics clog oceans, palm oil takes out rainforests, and incandescent light bulbs require more energy. With regard to energy use, animal advocates ought to strive to reduce consumption to prevent the damming of even one more river where herons and kingfishers hunt, the building of windmills on even one more prairie where bats and birds fly, the leveling of even one more forest where deer or orangutans dwell. All creatures depend on their particular environment—undammed rivers, intact forests, uncluttered oceans, and stable temperatures—in order to survive, and therefore animal advocates ought to be environmentally conscious on behalf of nonhuman animals. Additionally, those who are concerned about the lives or suffering of rabbits, deer, song birds, mice, moles, and small lizards and snakes, will keep dogs and cats indoors and/or on leashes to make sure they do not harass or harm wildlife.

Despite the vital importance of both earth and animal advocacy, philosophers (particularly in environmental ethics) have actively bolstered a divide between the two. A great deal of philosophical energy has been directed toward differences, emphasizing questions such as:

- What is the proper focus of moral attention, species or the individual?
- What is rightly viewed as "natural"?
- Which aspects or elements of the natural environment are morally considerable, and why? (Are streams and soils morally considerable? What about plants, feral cats, invasive species, and farmed animals?)
- Where—and why—might we draw a line between morally considerable and non-morally considerable individuals, species, and/or aspects of the natural world? (Is sentience essential for moral considerability? Which beings are sentient? Is sentience a matter of degree, or an all-or-nothing phenomenon?)

As the first example aptly demonstrates, these questions build an armory of difference around two presumably mutually exclusive possibilities. A more reasonable approach would assume a middle ground and strive for balance, a measure of synthesis and integration. Each of these philosophical concerns seeks to exclude, inquiring as to what might be placed outside a particular construct, such as "natural" or "sentient." Environmentalists, who tend to focus moral attention specifically on the natural world, thereby exclude certain species (such as farmed animals) and geographical areas (such as cities) from their moral sphere. If domestic chickens are deemed unnatural, for example, they are of no concern to environmentalists—except inasmuch as they harm the "natural" environment. Meanwhile, animal advocates often focus on sentience, neglecting vital habitat and sometimes bogging down on discussions of ants and microbes that do nothing to improve the plight of cattle or elephants, or protect vital habitats such as the ocean or rainforest.

## Case Studies

Different foci, goals, and methods are perhaps most evident in real-life scenarios. The following two cases (ongoing in Texas and Oregon) exemplify philosophical differences expressed in divergent concerns and conflicting approaches.

### *Big Bend Ranch State Park, Texas*

Animal and earth advocates are butting heads over burros on 316,000 acres near the Rio Grande. Abandoned by Texas ranchers long ago, donkeys have become part of the landscape, wandering in and out of Big Bend Ranch State Park at will. Those focused on ecosystems accuse feral burros of fouling the park's precious springs and creeks with manure and urine, disrupting local native food chains, and competing with endemic species—some of which are endangered—for scarce water and fodder (Texas Parks and Wildlife, n.d.).

Texas Parks and Wildlife tried unsuccessfully to trap and relocate the burros, so they drew their guns in 2007, killing 71 burros before the public caught wind of the slaughter (Humane Society of the United States, 2012). Texas Parks and Wildlife then turned to a California NGO, Peaceful Valley Donkey Rescue, but not even one burro was relocated two years later. In 2010 Texas Parks and Wildlife again resorted to high-powered weapons, killing 59 burros and igniting a public relations wildfire.

In the eyes of some, including The Wild Burro Protection League, wild burros ought to be protected because they are a "heritage species," a beast of burden that played a critical role in settling the area (Scharrer, 2012). Feral burros are protected in nearby Big Bend National Park "by a 40-year-old federal ban" preventing anyone from harming "living symbols and pioneer spirit of the West" (Blaney, 2012). Those concerned about the burros note that they are peaceful, largely defenseless herbivores, that they fertilize the landscape and keep the area free of dead-wood, reducing fire risks. The controversy was yet more heated when the burro contingent learned that Texas Parks and Wildlife only seeks to eliminate burros in order to bring in bighorn sheep, a big-money, big-game species, but the "Desert Bighorn Council won't release the bighorn while burros are present" (Jonsson, 2012).

In 2011, the Humane Society of the United States (HSUS) offered to "devise a nonlethal plan" to extract the remaining 300 burros (Blaney, 2012). It remains to be seen if/how HSUS will remove and rehome 300 burros … and somehow prevent other burros from wandering into Big Bend Ranch State Park. Until such a time, those concerned about Big Bend ecosystems and those concerned about local burros remain decidedly at odds.

### *Fishing Interests in Oregon*

The Columbia River was once rich with salmon (Chinook, Coho, sockeye, chum, pink) and steelhead trout (U.S. Fish and Wildlife Services, n.d.). When Europeans first arrived in Oregon, 10 to 16 million "salmon and steelhead returned to the river each year to spawn" (Northwest Power and Conservation Council, 2008). Settlers in the late nineteenth century noted that fishing interests in the region were "beyond calculation," offering an "almost inexhaustible" supply of fish (Northwest Power and Conservation Council, 2008).

A century later, overfishing and habitat destruction (dams, pollution, etc.) have reduced Columbia River fish runs by as much as 14 million fish per year (Northwest Power and Conservation Council, 2008; U.S. Fish and Wildlife Services, n.d.). In the Northwest and California, 214 (of about 400) subpopulations of salmon, steelhead, and cutthroat trout "are at risk of extinction," while 106 subpopulations "are already extinct" (U.S. Fish and Wildlife Services, n.d.; Northwest Power and Conservation Council, 2008). Pink salmon have been extirpated from the Columbia River, and each remaining salmon species is listed as Threatened—except sockeye, listed as Endangered (Northwest Power and Conservation Council, 2008; Specie, n.d.). We have become so used to the terms "Threatened" and "Endangered" that we perhaps fail to grasp their meaning: Endangered means "likely to become extinct"; Threatened means "likely to become Endangered in the near future" (U.S. Fish and Wildlife Services, n.d.).

In the proper season, beleaguered fish populations battle their way up the Columbia to spawn, but 150 miles up the Columbia, the Bonneville Dam turns fish into easy prey for sea lions (Northwest Power and Conservation Council, 2008). Not to be out-maneuvered when it comes to fishing interests, the Army Corps of Engineers stationed:

> observers with spotting scopes along the deck of Bonneville Dam to record the number of salmon and steelhead consumed by sea lions between January and May when ESA [Endangered Species Act]-listed runs are present. In 2002, they observed 31 sea lions consume 448 salmon and steelhead. In 2008, observers reported that 103 sea lions ate more than 4,243 salmon and steelhead in the same limited area below the dam. Another estimate, based on California sea lions' metabolic needs, suggests that 100 animals feeding in that area consume at least 13,000 salmon each spring.
>
> (Washington Department of Fish and Wildlife, n.d.).

Funding was quickly provided to protect Threatened/Endangered fish from sharp-toothed sea lions. Washington and Oregon Fish and Wildlife Services (FWS):

> each received an annual grant of $150,000 from the Pacific States Marine Fisheries Commission to support hazing and sea lion removal below Bonneville Dam. Each state agency … also contributed approximately $15,000 for early-season hazing efforts designed to protect sturgeon in the same area. The Army Corps of Engineers provides approximately $150,000 per year to document predation, haze sea lions and conduct fieldwork related to sea lion predation. The Corps also invested more than $3 million to install heavy bars and sonic devices to keep sea lions out of fishways and ladders at Bonneville Dam.
>
> (Northwest Power and Conservation Council, 2008)

Washington FWS promptly shot 40 California sea lions, but these were quickly replaced by other sea lions coming to feed—40 dead sea lions and no fewer fish-eaters below the Bonneville Dam. There was, however, one noticeable difference—an increased density of protected Steller sea lions. Like salmon below the Bonneville Dam, Steller sea lions are listed as Threatened (and are thereby protected) and must be left to gobble up fish, even Threatened/Endangered salmon beneath the Bonneville Dam.

Those concerned about sea lions requested relocation, but relocated sea lions quickly returned to the rich source of sustenance below the Bonneville Dam. To date, there has been "[n]o halt to sea lion killing" (Daily World, 2012). In fact, there has been no halt to *any* killing: People, sea lions, and salmon continue to congregate carnivorously on the Columbia River.

## Connecting the Dots: Common Ground, Joint Action

Burros at Big Bend and sea lions at Bonneville Dam exemplify a different focus and distinct goals, addressed with contrasting methods, between environmentalists and animal activists: Environmentalists seek to protect ecosystems and endangered species, and are generally willing to kill non-endangered individuals in the process—especially domesticated animals, feral domestics, non-native, or "invasive" species. Animal advocates seek to protect every individual—even those who consume Threatened/Endangered species. While environmentalists are eager to include streams and valleys and endangered species in our moral landscape, animal activists are more interested in protecting cattle and cats and burros and seals.

Despite core differences, which have received more than their share of attention, environmentalists and animal advocates have much in common. Most critically, both groups share an interest in expanding our moral circle (Noske, 2007). Animal activists and environmentalists share a fundamental, core interest that shapes both camps, and which ought to draw them together toward shared goals. The desire to eradicate anthropomorphism/speciesism/humano-centrism constitutes critical common ground: "The environmental movement and the animal protection movement are at the very least cousins and, even more important, natural allies" (Waldau, 2011: 122). Habitat is another plot of prime realty shared by both groups, though frequently obfuscated by our dualistic urge to separate "animal" from "environment." In truth, all beings are utterly dependent on their environment such that it makes no sense to view individuals as divorced from their essential surroundings. "Environmentalists have obvious allies in advocates of … animal rights" (Sagoff, 2012: 308).

In light of these vital points of alliance, joint action makes sense, and offers many advantages, most notably people-power. More activists bring more ideas, more skills, more financial backing, and more boots on the ground for outreach, education, and hands-on projects. A larger group is also more apt to influence legislators—when advocating for change in a democracy, numbers count.

Working together also provides the advantage of cross-fertilization, likely improving both camps. Earth and animal activists will be encouraged to realign behaviors (especially consumption patterns) with core commitments: environmentalists will be encouraged to shift to a plant based diet; animal advocates will find incentive to shop organic and seek products with less packaging, cloth bags in hand.

## Common Ground Practical Application: Bonneville Dam

With regard to fish and sea lions below the Bonneville Dam, what shape might joint action take? Would environmentalists and animal advocates seek nonlethal means of protecting salmon from the sea lions? Might we find ways to reduce energy consumption so that the dam could be dismantled? Is there some more fundamental cause of the problem that we have overlooked? With regard to the Bonneville Dam, environmentalists are concerned about Threatened/Endangered fish populations; the root concern for animal advocates is the protection of sea lions—what constitutes the most likely shared plot of ground?

Root causes are usually a good place to look for common ground. The root cause of both concerns (Threatened/Endangered salmon and the shooting of sea lions) is human consumption of fish from the Columbia River. Sea lions are innocent—they can't choose veggie stir fry; they *must* eat fish to survive, and they have no way of knowing that (thanks to humans) certain fish are Threatened/Endangered. While sea lion fish consumption in the region is natural and essential for sea lion survival, such consumption is entirely unnecessary for human health and welfare. Human fish consumption and resultant commercial fishing operations are the root cause of salmon and steelhead depletion in the Columbia River—and around the world. On what reasonable grounds would sincere, informed environmentalists or animal advocates refuse to join forces in a campaign to protect Threatened fish from the snapping teeth of humanity?

The most viable solution to the problem of Threatened/Endangered fish and sea lions at the Bonneville Dam is for human beings to:

- stop eating Threatened/Endangered fish—and stop consuming fish more generally inasmuch as our voracious consumption of fish damages ecosystems, depletes sea life, and destroys individual fish (see Chapter 10);
- encourage others to protect fish through alternative food choices via education and outreach;
- lobby against fishing industries and practices that harm Threatened/Endangered fish (and other sea life).

Changing our diet is less costly than shooting sea lions, and a much more effective way to solve the Bonneville Dam problem. It is also the responsible and honest approach—sea lions are not the problem, we are.

Is there a similar core problem that undergirds the conflict over burros and ecosystems at Big Bend Ranch State Park?

## Animals and Environment: Common Ground

In a world with so many desperate problems facing nonhuman animals and the environment, and also because systems of oppression link animal exploitation and environmental degradation, it is ineffective to ask, "Who is right?" and "Where can we draw lines?" We would be much more successful if we would ask, "How can we work together?" Earth and animal activists have much to gain by cooperating, and much to learn from one another. Moreover, fragmentation and division between social justice activists only serves to help mutual adversaries—there *are* enemies in this battle (such as unscrupulous large corporations), but earth and animal activists are not enemies because they share core common concerns and very fundamental goals.

Bearing this in mind, essays in this anthology highlight common ground between earth and animal advocates. Authors in Malaysia, Greece, Canada, Norway, Uganda, and The United States—including Native American, Chicana Indigenae, and Indian-American voices—unveil the interconnections between earth and animal activism, and demonstrate how activists from both camps might work together to protect earth and nonhuman animals—and why it makes perfect sense to do so. Meanwhile, each section begins with a poem that invites readers to ponder our effect on and place in the universe.

## Part I: Establishing and Exploring Common Ground

### *Section I. Foundations: Theoretical Connections*

Foundations and frameworks are critical for activists, most notably because we are often called to provide *reasons* for our commitments and our actions. Reasons generally require activists to articulate motivations and larger purposes. Part I, Section I investigates and elucidates foundations and frameworks that support and validate earth and animal activism as a single cause.

In the first essay of Section I, Carol Glasser makes a case for total liberation, exploring systems of oppression that affect the natural world, nonhuman animals, and disempowered human beings. Carrie Freeman places animal advocacy front and center not only for environmental activism, but for all social justice movements. Josephine Donovan explores ecofeminism and an ethic of care rooted in respect, responsibility and attentiveness. The final essay of the first section turns attention to religion—the oldest, most pervasive foundation and framework connecting earth and animal activism.

### *Section II. Common Ground: Wildlife and Wilderness*

Sections II and III focus on the most obvious contemporary issues common to earth and animal advocates: ecosystems/habitat and diet. Section II explores vital connections between wildlife and wilderness. Most fundamentally, all creatures depend on habitat such that the two cannot reasonably be considered independently. In the first essay of Section II, I debunk a handful of misconceptions and myths surrounding sport hunting, highlight the damaging effects that hunting and hunting policies have on wildlife and ecosystems. Valarie Chalcraft explores the cycle of misery and violence affecting both elephants and humans, paying particular attention to forces that threaten Asian elephants. Anja Heister and I expose the ecologically disruptive effects of Montana's government-run Furbearer Trapping Program, designed by and for trappers. Melanie Martin investigates how patriarchal nature films affect our understanding of and encounters with wilderness and wildlife, including feminization of land and masculinization of animals and scientific "facts" that diminish and objectify nonhuman animals.

### *Section III. Common Ground: Dietary Choice*

Essays in Section III focus on dietary choice and environmental degradation exposing the intimate connections between the two. In the first essay, John Halley uses a mathematical model to calculate the ecological footprint of several dietary options: omnivore, vegetarian, and vegan. Next, Bethany Dopp and I investigate the effects of industrial fishing on sealife, ocean habitat, and ocean ecosystems. The final two essays examine animal agriculture: Chris Hunt describes both the inherent cruelty of factory farms and how animal agriculture pollutes air and water—causing sexless fish, acid rain, and climate change; I expose connections between animal agriculture and freshwater depletion, deforestation, soil degradation, wildlife, and land use.

### *IV. Common Ground: Raising Questions, Pondering Connections*

This group of essays explores conflicts and connections between earth and animal activists, offering more questions than answers (because there *are* more questions than answers). Wilderness advocate Randall Gloege presents animal advocate criticisms of environmentalists, and poses three fundamental questions regarding the nature of humanity, each central to our relationship with animals and the natural world. Jon Swenson, head of the Scandinavian Brown Bear Research Project, describes his scientific, utilitarian approach, and how this places him at odds with animal activists. In the process, he ponders a handful of questions that lie at the heart of the historic divide between animal and earth activists. Environmentalist Bernard Quetchenbach takes readers to a spectacular Montana wilderness riddled with dead and dying evergreens—destroyed by climate change and bark beetles—which heightens his awareness of "imbalance" and pushes him to ponder the path less taken. Cara Chamberlain explores and celebrates bison as "scapegoats, martyrs, icons," and also as "wild bovines." In the process, she suggests ways that we might re-envision our relationship with nature.

## Part II: Politics, Organized Activism, and Personal Encounters

### *Section I. Foundations: Community and Politics*

After describing a factory farm and introducing readers to the indigenous Malaysian Dayak people, displaced (along with rainforests and wildlife) by the voracious palm oil industry, Debra Erenberg suggests that we look to the International Declaration of Human Rights as one possible avenue for bringing much-needed change. Charlotte Laws points to the giant pink elephant in the middle of Congress—democracy is a government by and for *people*—blatantly humanocentric—and outlines a utilitarian "omniocracy," essential to the cause of both earth and animal activists. Daniel Kirjner, Jennifer Gross, Nathan Baillet, and I explore connections between environmental degradation, consumption, and reproduction.

### *Section II. Bringing Change: Activists and NGOs*

This section features activists who simultaneously work to liberate earth and animals, demonstrating both the common sense and the effectiveness of an integrated approach. Four members of the Raincoast Conservation Foundation discover—and successfully capitalize on—the power of ethics, economics, and human compassion to protect bears and ecosystems. Haida Bolton, who founded a youth camp in Uganda, explains how the camp helps local youth to foster environmental values and respect for wildlife. Employed by a Malaysian environmental organization, Phaik Kee Lim describes her work educating the public and pressuring businesses and governments around the world on behalf of penguins, coral reefs, snakes, freshwater reserves, and orangutans.

### *Section III. Bringing Change: Personal Encounters and Reflections*

The final section explores personal journeys that exemplify the interrelated nature of earth and animal activism. Bethany Dopp, who took a zoo internship at ZooMontana, walks readers through zoo facilities, exposing what goes on behind closed doors, and explaining how this experience led her to question zoo ethics, particularly caging carnivores. Scientist and social justice advocate, environmentalist and animal activist, Xylem Galadhon clarifies how seemingly contradictory causes and interests support a complete, harmonious, and meaningful life. Deric Shannon describes his winding path from Marx and anarchy through feminism to holistic revolution, which ultimately forced him to engage with both earth and animal liberation—however reluctantly.

## Note on Word Choice

Environmentalists tend to use the words anthropocentric, human-centered, humanocentric, or human chauvinism to reference attitudes that hold humans above other aspects of the natural world. Animal advocates tend to use the term "speciesism" to reference attitudes that elevate human beings over other species. These terms, used interchangeably, are particularly important in this anthology because the rejection of anthropocentric/humanocentric/speciesist attitudes are foundational to both earth and animal liberation. Moreover, no other social justice cause shares this core concern.

This anthology attempts to make visible and also to reverse speciesist/humanocentric linguistic habits. For example, the English language most often refers to nonhumans as if they were things rather than individuals. For example, we might say, "the dog that chased the ball was black." I have tried to note these conventions, and change "that" to "who": "The dog *who* chased the ball was black."

People also tend to refer to nonhuman animals as "it" or sometimes as "he"—regardless of an individual's sex. Inasmuch as we do not refer to human animals as "it," we ought not to refer to other animals as "it." Almost all species include males and females (among other karyotypes). A one-sex-fits-all vocabulary objectifies cattle and dogs and pigmy lemurs, denying their individuality. Where animal agriculture is concerned, this tendency obscures the truth—obviously the egg and dairy industry exploits only female animals to collect their nursing milk and reproductive eggs.

Moreover, "animal" includes humans: We are animals, mammals, primates. Therefore, I have encouraged authors to use the more cumbersome but correct "nonhuman animal" or "other animals"—except when "animal" is used in conjunction with a second word, such as "animal liberation," "animal welfare," "wild animal," "animal advocate," "animal testing," and so on. And there is no such creature as a "farm animal"—except human beings, who have spent considerable time farming down through history. Other species, such as turkeys and pigs, are *exploited* on farms, by humans. As such, they are "farmed" animals. Similarly, there is no such thing as a "veal calf" or a "lab animal," though there are millions of calves and mice who are systematically exploited by ranchers, researchers, and consumers. "Seafood" also disguises the truth—sea creatures are individuals who happen to be exploited by some humans for profit and the pleasure of the palate, and should not be defined by their eatability any more than should any other creature: We are all eatable, it is just a matter of who might eat us—large mammals such as bears and sharks, for example, or microbes and worms.

Finally, many animal activists and environmentalists use the term "breeding" to refer to human reproduction. This is an appropriate application, inasmuch as we use "breeding" to refer to reproduction in animals—given that we are animals.

But among such activists, "breeding" is too often applied only to women, thereby using the term in a sexist manner, without mention of men. As a matter of clarity, justice, and appropriate application, "breeding" ought to be applied to all who breed—bovines and fishes, women and men.

## Note

Many authors in this anthology are busy activists; some come to English as a second (or third) language. I often worked extensively with authors to create essays for anthologies, offering their voices to readers while keeping their hands as free as possible to continue their work on behalf of earth and animals.

## References


Blaney, B., 2012. *Big Bend Ranch State Park Burro-Killing Policy Suspended in Texas.* [Online] Available at: http://www.huffingtonpost.com/2012/03/20/big-bend-ranch-state-park-burros_n_1366863.html [Accessed 19 December 2012].

Daily World, 2012. *No Halt to Sea Lion Killing.* [Online] Available at: http://thedailyworld.com/sections/newswire/northwest/federal-judge-says-no-halt-sea-lion-killing.html [Accessed 23 January 2013].

Humane Society of the United States, 2012. *Texas Suspends Killing of Ferral Burros in Park.* [Online] Available at: http://hsus.typepad.com/wayne/2012/03/texas-burros.html [Accessed 06 June 2012].

Jonsson, K., 2012. *Straight From the Horse's Heart: Wild Burro Protection League-Texans Protest Wild Burro Endangerment.* [Online] Available at: http://rtfitchauthor.com/tag/wild-burro-protection-league/ [Accessed 20 May 2013].

Northwest Power and Conservation Council, 2008. *Columbia River History: Salmon and Steelhead.* [Online] Available at: http://www.nwcouncil.org/history/salmonandsteelhead.asp [Accessed 23 January 2013].

Noske, B., 2007. Ethics and Animal Protection: Human-Animal Kinship for Animal Rightists and Deep Greens. In: M. Bekoff, ed. *Encyclopedia of Human-Animal Relationships: A Global Exploration of our Connections with Animals.* Westport: Greenwood Press, pp. 762–769.

Sagoff, M., 2012. Do We Consume Too Much?. In: D. Schmitdz and E. Willott, eds *Environmental Ethics: What Really Matters, What Really Works.* New York: Oxford University Press, pp. 302–323.

Scharrer, G., 2012. *Houston Chronicle: Fans Protest Wild Donkeys Being Killed at Big Bend Park.* [Online] Available at: http://www.chron.com/news/houston-texas/article/Protesters-deliver-petitions-on-donkeys-2616000.php [Accessed 20 May 2013].

Specie, E., n.d. *Endangered Species in Oregon.* [Online] Available at: http://ecos.fws.gov/tess_public/pub/stateListingAndOccurrenceIndividual.jsp?state=OR&s8fid=112761032792&s8fid=112762573902 [Accessed 23 January 2013].

Texas Parks and Wildlife, n.d. *Facts About Feral Burros at Big Bend Ranch State Park.* [Online] Available at: http://www.tpwd.state.tx.us/burros/ [Accessed 06 June 2012].

U.S. Fish and Wildlife Services, n.d. *Species: Pacific Salmon (Oncorhynchus Spp).* [Online] Available at: http://www.fws.gov/species/species_accounts/bio_salm.html[Accessed 23 January 2013].

Waldau, P., 2011. *Animal Rights: What Everyone Needs to Know.* New York: Oxford University Press.

Washington Department of Fish and Wildlife, n.d. *Columbia River Sea Lion Management.* [Online] Available at: http://wdfw.wa.gov/conservation/sealions/questions.html [Accessed 6 June 2012].

# PART I

# Establishing and Exploring Common Ground

# 1

# CONFLICT AND ACCORD

## A Critical Review of Theory and Methods for Earth and Animal Advocacy

*Lisa Kemmerer and Daniel Kirjner*

In March of 2014, The Humane Society of the United States hosted a conference in Washington DC titled "The Science of Animal Thinking and Emotion," in which a psychologist commented skeptically that, despite many improvements brought on by animal advocacy, there was no significant change in meat consumption. Human cravings for meat remain, driving animal exploitation. During Q&A an activist asked: "If you don't believe people will stop eating meat for moral reasons, do you have hope for artificial laboratory meat?" The scholar responded "Lab meats are too expensive and I hear they don't taste very good. I think the real solution is creating animals that can't feel pain."

Indeed we hated him. Such a response could not come from someone who cared deeply about animals. Nor could it come from someone concerned with the environment. This response could only come from a man focused on his own privilege: He envisioned changing cattle and pigs and turkeys to fit his needs rather than changing his diet. It is harder to inspire, invite, and lead change than it is to find excuses for conformity. His cynical pragmatism underscored his anthropocentric outlook—a vision of the universe in which human beings commanded front and center over all else.

Anthropocentrism leads people to believe that nonhuman animals and the natural world are merely here for our purposes—"livestock" (living stocks of consumable goods) warehoused and slaughtered, "resources" (forests and seas and soils) waiting for people to claim and exploit them for human purposes. This attitude is as harmful to chickens and dogs as it is to streams and trees, and is pervasive among peoples of industrial, capitalistic nations. Why, then, earth and animal advocates failed to unite against anthropocentrism?

This chapter explores the interwoven history of animal and earth advocacy, surveying key philosophical underpinnings, historic divides, and shared tactics that stem from a mutual concern for things generally viewed as outside the "western" human world.

## Conflict and Accord

Despite years of separation and noncooperation, animal and earth advocates want the same essential thing—an end to anthropocentrism, an end to human domination that leads to exploitation and devastation (Callicott, 1989). If those fighting for the earth and nonhuman animals share such a fundamental connection, what prevents union?

The simple answer seems to be anthropocentrism *within* both movements (as exemplified by the speaker at the Washington DC conference). For example, indifference to the environment is common among animal advocates, who tend to feel entitled to consume at will, without regard for energy use or what they have taken from forests, oceans, or under the earth's crust. Many animal advocates fail to consider packaging when they purchase, are unwilling to pay extra for organic, and prefer processed to simple foods. One of our vegan friends used to say proudly that he cared about neither his health nor the Amazon—only the animals mattered to him. While he had come to see that nonhuman animals had value apart from human exploitation, he had failed to recognize the same value in waterways and forests. Such animal activists fail to recognize that they treat trees and marshes just as most consumers treat pigs and chickens. Similarly, environmentalists cling to the anthropocentric vision that allows them to decide—only with regard to nonhuman species—who lives and who dies. They feel empowered to kill human-introduced burros to protect a fragile desert ecosystem, that seals and sea lions who eat endangered fish—endangered because humans have fished out the earth's waters—must be killed (see "Big Bend Ranch State Park, Texas" and "Fishing Interests in Oregon" in the Introduction, page 4). In fact, such environmentalists are likely to still be eating fish (and cows and turkeys) even though salmon would not be at risk were it not for our taste for fish flesh.

Despite underlying anthropocentrism within these movements, both stand united in their distaste for anthropocentrism, and this is not the only core ideal shared between earth and animal advocates. In fact, both movements have more similarities than differences. Moreover, academics have often worked in both fields. For example, philosopher Peter Singer, among the most famous of animal ethicists, focuses frequently on environmental issues (Singer, 1975, 1993a, 1993b, 2004). Tom Regan, also among the most renowned of animal ethicists, has also offered much to environmental ethics (*All that Dwell Therein* and *Essays on Animal Rights and Environmental Ethics*, 1982; see also Regan, 1983, 2004). Regan's writings on inherent value are particularly pertinent to environmental ethics. Many other well-known scholars work in both animal studies and environmental studies, including Mark Bekoff, Greta Gaard, Mary Midgley, James Baird Callicott, Stephen Clark, and Marti Kheel. It is now common to meet people who are working both in environmental and animal advocacy organizations, and it is increasingly possible to find organizations that work both on behalf of nonhuman animals and on behalf of the environment, such as the Food Empowerment Project (http://www.foodispower.org).

## Historical Roots: Philosophical Cross-Fertilization and Activists' Discord

Historian Lynn White Jr, in "The Historical Roots of our Ecological Crisis"(1967), criticized Christianity and technology for environmental degradation. White called attention to a Bible-based instrumental approach to the natural world, as well as to the capitalist marriage between science and technology, which provided material means to realize Christian man's anthropocentric, utilitarian rule over the rest of "creation" (White, 1967).

In light of more recent work, White's assessment of the Bible was narrow—there is much in the Judeo-Christian scriptures calling for an earth-friendly approach to all of "creation" (Kaufman and Braun, 2004; Kemmerer, 2012; Linzey, 1992; Schwartz 2001). Yet despite teachings of care-taking and compassion that are critical to Judeo-Christian ethics, anthropocentrism has long been central to the Christian outlook. For example, in *Conquest of America: The Question of the Other*, Tzvetan Todorov analyzes the writings of Christopher Columbus, in which this famous explorer describes his first impressions of the Americas. Columbus viewed the "New World" as Eden, a land of plenty, created for man (wealthy European men, to be exact) by the hand of God. Religious delusions shaped Columbus' worldview such that he imagined every landscape, and all that stood on the land, as his for the taking. This view placed both nonhuman animals and the natural environment at risk (along with women, non-Christians, non-Europeans, and the poor) (Todorov, 1996).

Previous to White, sociologist Max Weber explored how Protestant Christianity in Europe fostered the perfect moral medium for capitalism (Weber, [1905] 2002). His most famous book, *The Protestant Ethic and the Spirit of Capitalism*, notes the Protestant tendency to view self-achievement and materialism as aspects of spiritual ascension, standing in stark contrast to earlier Catholic ethics, which encouraged poverty and self-denial. The new Protestant ethic created an ideal atmosphere for human beings to exploit everything around them, from ore and timber to bovines and burros.

In light of dangerous and damaging religious views highlighted by White and Weber, it is not surprising that earth and animal advocates were organizing by the end of the nineteenth century (White, 1967; DeMello, 2012). For example, the American Society for the Prevention of Cruelty to Animals (ASPCA) was founded in 1866, the National Anti-Vivisection Society was founded in 1875, The Audubon Society was founded in 1883, and the Sierra Club was founded in

1892. But as with many social justice movements, they came into their own in the 1960s, revealing revolutionary tendencies.

In 1962 Rachel Carson published an inflammatory book called *Silent Spring* ([1962] 2002) in which she attacked both the pesticide and food industries for poisoning the natural world and animals (Carson, [1962] 2002). Though rooted in a scientist's knowledge of the natural world, her writings stirred the emotions of those who love nature and animals. Discredited among more conventional, male scientists, Carson faced massive opposition—but she ignited a spark that became part of a flaming environmental revolution (Lear, 2009).

In the aftermath of the upheaval brought on by Carson's work, another courageous woman, Frances Moore Lappé, challenged a giant among giants—the meat industry (Lappé, [1971] 2011). In *Diet for a Small Planet,* Lappé documented the many ways that a flesh-based diet is a recipe for environmental disaster. She noted the sheer volume of vegetables required to feed farmed animals—vegetables that could more efficiently feed hungry humans. Lappé also noted that the meat industry consumes excessive amounts of fossil fuels and water, and causes soil degradation (especially because of heavy dependence on monocultures like corn and soy). Lappé noted that our choice to consume meat is both environmentally imprudent and damaging to "long term food security" (Lappé, [1971] 2011: 51). She even debunked the "Protein Myth"—the common and erroneous association of flesh with protein required by the human body.

The power of *Silent Spring* and *Diet for a Small Planet* both stemmed from and bolstered a growing awareness of the effects of human power and indifference on nonhumans and nature. Despite the brilliant work of Carson and Lappé (books that ought to be standard reading for those working to change the American meat culture, still dependent on deadly monocultures and pesticides) the meat industry is still one of the most powerful industries on the planet, and a startling portion of the English-reading population continues to consume flesh and to associate animal products with (male) strength and power, portraying animal protein as essential to human health (Adams, 1990, 2004; Luke, 2007). Meanwhile, the earth is in shambles (literally and figuratively—historically, "shambles" referred to "a place where meat is butchered and sold," a situation of "bloodshed or carnage" (Anon, n.d).

Peter Singer might best be credited with doing for animal advocacy what Carson accomplished for the environmental movement. In 1975, Singer published *Animal Liberation*, likely the most influential book for both animal advocacy and animal studies. Revisiting Jeremy Bentham's utilitarian moral theory, Singer presented an ethical framework that included nonhuman animals (Singer, 1975). Bentham made clear that neither reason nor language are necessary for moral considerability. Underscoring the application of utilitarian philosophy to nonhumans, Singer quoted Bentham—"The question is not, Can they reason? nor, Can they talk? but, Can they suffer?" (Singer, 1975: 23). Utilitarianism, rooted in the task of achieving the greatest good for the greatest number (achieving the greatest happiness or the greatest satisfaction of interests, for example) must include all who can be made happy, all who have interests. As with toddlers and other human beings who cannot reason or communicate, sentience—the ability to feel and therefore to suffer—is the core requirement. Singer demonstrated that utilitarian moral theory protects non-humans from such things as vivisection or exploitation in food industries.

*Animal Liberation*, which inspired many people to take up the torch for nonhuman animals, carried the term speciesism (coined by Richard Ryder in 1970) into the mainstream of animal advocacy (Ryder, 2005). Singer defines speciesism as "prejudice or an attitude of bias in favor of the interests of members of one's own species and against those members of other species" (Singer, 1975: 6). Speciesism is the tendency among humans to "make a distinction with regard to how individuals ought to be treated *based solely on species*, without regard to morally relevant similarities and distinctions" (Kemmerer, 2006a). Analogous with racism and sexism, speciesism is a way of thinking that allows one group of individuals to assert self-assumed privilege over another group of individuals despite the fact that the two groups do not differ in morally relevant ways (Pluhar, 1995).

Tom Regan, also foundational to the shaping of the contemporary animal advocacy movement, used rights theory to build a parallel case for including animals in our moral sphere. In *The Case for Animal Rights* (1983), Regan argues that most mammals aged one year or more are "subjects-of-a-life," who prefer not to be harmed, and who therefore ought to be protected from harm by basic rights. To be a "subject-of-a-life," an individual must be "characterized by":

> beliefs and desires; perception, memory, and a sense of the future, including their own future; and emotional life together with feelings of pleasure and pain; preference and welfare interests; the ability to initiate action in pursuit of their desires and goals; a psychophysical identity over time and an individual welfare in the sense that their experiential life fares well or ill for them, logically independent of their utility for others and logically independent of their being the object of anyone else's interests.
>
> (Regan, 1983: 243)

Regan argues that the lives of such individuals can be made better or worse depending on what we do to or with them, and that they therefore hold inherent value, and are entitled to have their basic rights protected, preventing us from exploiting or otherwise harming them as we do in our animal labs and food industries (Regan, 1983).

Like the works of Carson and Lappé, the works of Singer and Regan are relevant both to animal studies and advocacy and environmental studies and advocacy. For example, environmental ethics has been deeply influenced by the work of those in the field of animal ethics (Callicott, 1989; Goodpaster, 1978). Moral theories that take nonhuman animals into consideration have expanded the scope and vision of many ethicists—especially those in search of compelling philosophical arguments for protecting mountains and mushrooms and all things from the world of nature (Callicott, 2005). Environmentalists such as Paul Taylor and Kenneth Goodpaster, for example, have responded to Singer. Biocentrists—those who would extend moral standing to all living things, including shrubs and clams—accuse Singer of sentientism, of arbitrarily excluding all things lacking sentience from moral consideration (Schmidtz and Willott, 2012). Meanwhile, Kenneth Goodpaster rejects Regan's "subject-of-a-life," arguing that moral standing should be extended to all living things with "interests that could be represented," including grasses, trees, and flowers, all of which need water and sunlight (Goodpaster, 1978).

Environmental philosopher Paul Taylor, author of *Respect for Nature* (1986), was also deeply influenced by animal ethicists, especially the work of Tom Regan. Like most environmental philosophers, his interests extend beyond the animal kingdom to include other aspects of the natural world. While Taylor questions Regan's assertion that animals can hold moral rights, he agrees that nonhuman animals have a "good of their own," that they can be made better or worse off, and that they therefore have "inherent worth" (Taylor, 1986). In *Respect for Nature*, Taylor builds on Regan's work, asserting that every living thing seeks a "good of its own," and thereby qualifies as a teleological entity endowed with inherent worth, and is therefore worthy of moral consideration (Taylor, 1986).

Lively discussions between philosophers working on behalf of the environment and philosophers working on behalf of nonhuman animals have fostered cross-fertilization, and have stimulated, enhanced, and advanced both fields. Unfortunately, this healthy debate has not touched most activists working in the streets and offices outside of academia. Many anthropocentric animal advocates are perhaps too satisfied with their comfortable, urban life to make even small sacrifices on behalf of the environment, such as consuming less, eliminating earth-wrecking products such as bottled water and processed foods, and tending toward organic. Some animal advocates neglect not only the earth, but the overwhelming majority of animals, preferring specific mammals whom they consider appealing, such as horses, whales, baby seals, tigers, dogs, and/or cats. For environmentalists, the situation is similar. Many prefer to employ cloth shopping bags rather than rethink what they put into their bags, notably dairy, eggs, and flesh. Others view environmental advocacy as a matter of trade-offs, where all but humanity are expendable on behalf of certain aspects of natural the world.

In 1973 Arne Naess coined the term deep ecology, asserting that care for the earth requires a much more fundamental, sweeping change. Rather than patchwork repairs to address mounting environmental crises, he indicated that we must change how we understand our place in the world. For Naess, deep ecology is not just about solving a growing series of crises, but about establishing a sustainable way of life grounded in a worldview that entails respect for the diversity of life (Naess, 1973). He envisioned a world-wide revolution.

The core principles of deep ecology are compatible with the ideals of animal advocacy (Somma, 2006).

Among academics, particularly in the field of philosophy, there has been much collaboration and respectful discussion between those who primarily align with animal advocacy and those more inclined toward environmental advocacy—as well as a measure of animosity. Outside scholarly circles there has been little acknowledgement of shared roots and little inclination toward respectful discussion of core philosophical principles. Activists have tended to focus on prickly differences, such as the environmentalist's tendency to support hunting—even labeling hunting as environmentally friendly (Kheel, 1996). In the United States, the environmental movement and hunting have been linked since the beginnings of U.S. government conservation (Kemmerer, 2014). Theodore Roosevelt and Aldo Leopold, critical agents in the formation of environmental policy and philosophy in the United States, were both avid hunters. The first U.S. environmental laws were enacted by Roosevelt in reaction to hunter concerns about the decline and possible extinction of "game" animals (DeMello, 2012; Kemmerer, 2014). The purpose of these initial U.S. conservation policies was utilitarian, designed to satisfy human interests (males, more specifically): Roosevelt and Leopold were primarily concerned that target species thrive to satisfy the pleasures of hunters. Leopold's love of wilderness and his delight in killing wildlife shed much light on the historic divide between animal and earth advocates. This schism between earth and animal advocates, and the tendency for environmentalists and environmental organizations to align with hunters, is immortalized in a photo of Aldo Leopold, the supposed father of the environmental movement, proudly dangling a dead bird by his or her legs. Leopold viewed meat eating among humans as natural, as expressed in his "land pyramid," in which human beings share "an intermediate layer with the bears, raccoons, and squirrels which eat both meat and vegetables" (Leopold, 1966). Leopold's writings, which call for love and respect for nature and recall heart-rending lessons learned from the natural world, also speak much of his love for hunting—he was quite happy, at least early in life, to eliminate any species that hindered his utilitarian interest in wildlife (Leopold, 1966).

Industrialized humanity has too often exploited that which is "loved" and "respected," just as did Leopold the hunter. In his human-centered vision, he imagined hunting to be a perfect way to experience "communion" with nature, a way of better understanding animals and coming to know wilderness (Lorbiecki, 1999). But just as males will never understand women by stalking and assaulting them, humans can never come to know other animals by hunting and killing them. Be that as it may, hunters today continue to talk of their respect for nature and wildlife—even asserting their love for the animals that they kill (Kheel, 2008). Respect and love that permit exploitation and gratuitous killing are a bastardization of both concepts. In short, respect and love that permit exploitation and gratuitous killing are not respect and love.

Fortunately, Leopold's ideas evolved somewhat as he aged, and he came to recognize nature as worthy of preserving in and of itself; he even came to recognize the importance of keeping a few nonhuman predators around, even though they kill at least some hunter target species. And though Leopold never recognized hunting for what it is—unnecessary domination and destruction of others—his vision of the "biotic community" enlarged moral boundaries to include "soils, waters, plants *and animals*," which Leopold referred to collectively as "the land" (Leopold, 1966, italics added). Despite inviting nonhuman animals into the moral circle, Leopold's work has done much to hinder the united front that animal and earth advocates might otherwise have created: The prominent place of hunters in the environmental movement has most likely kept many animal advocates from joining or working with environmental organizations.

Despite this historic divide, animal advocates and environmentalists would do well to focus on and embrace their shared history, and take an example from scholars, especially philosophers, whose understandings, progress, and productivity are richer due to cross-fertilization that has pushed the boundaries of both fields. Between environmentalists and animal advocates, the most fundamental ideals are far more similar than different. Like philosophers, activists would do well to stand back and explore the interwoven ideas that support both movements, including a strong distaste for anthropocentrism, with a willingness to reexamine personal habits and choices, in order to inspire and revitalize both movements, as well as to search for new and more effective directions and methods.

## Exploring Connections

A joint effort on behalf of the earth and nonhuman animals is both feasible and reasonable. In fact, grassroots organizations, scholars, and masked liberationists are already thinking, speaking up, and acting out on behalf of both. The purpose of this anthology is, therefore, not to build bridges between two distinct causes, but rather to point out that what we currently tend to envision as two distinct causes are, in many key ways, one cause.

### *Science and Technology*

Technology in the hands of a greedy humanity has had unpredicted outcomes: climate change, mass extinctions, deforestation, and overpopulations (human and nonhuman, purposefully created and unintentional). Scientists appear to be on the verge of naming the first human-driven geological era—the Anthropocene (Crutzen and Stoermer, 2000). Apparently humans are a phenomenon beyond our control. Both those in the humanities and those in the sciences sometimes turn to science in hopes of solving problems caused by the earth- and animal-ravaging ways of humanity. Indeed, science offers insights that are critical if we are to extend moral protections to earth and nonhuman animals, but sciences are also part of the problem.

Ursula Heise, who works both for the department of English and at the Institute of the Environment and Sustainability at UCLA, leans toward science, protesting "diversity" (as celebrated in the humanities) as a delusional misappropriation of a biological concept. She encourages people to move beyond local and cultural particularities to a transnational vision of ecology (Heise, 2008). Heise envisions science as central, but her view is dismissive of societies that do not see the world through the lens of science—societies that many earth and animal activists have looked to as harbingers of hope for re-envisioning and reclaiming ways of living that are healthier for both nonhuman animals and the earth. Maybe a "delusion" of diversity is necessary to help industrialized nations move beyond the science bias that has privileged the ideas and actions of an elite few. In a nation whose lifestyle tends to be fundamentally exploitative, there is much to be gained from celebrating diversity—especially different cultures that offer the hope of more functional and sustainable visions of our place in the universe (Gaard, 2010).

Ecologist and evolutionary biologist Mark Bekoff, working from within the "hard" sciences, uses his status as a scientist to debunk myths that have elevated the hard sciences above other academic fields. For instance, he writes,

> Of course, science is not value-free—we all come to our lives with a point of view. But it took some time for me to come to this realization because of the heavy indoctrination and arrogance of my training. Indeed, if science were value-free, my critics would leave me alone.
>
> (Bekoff, 2007: 3)

Working both on behalf of animals and the earth, Bekoff seems to recognize that protecting and saving animals is the best way to preserve the earth—and that earth and animal advocates are in the same camp. Working to inject emotion and humility into the world of science, Bekoff is helping to build a bridge that is essential to earth and animal advocacy. Bekoff is a breath of fresh air for earth and animal advocates alike. He is prolific, and immensely popular. He highlights the importance of an emotional connection and respect for nature and animals (Goodall and Bekoff, 2002).

Sociologist Richard Twine is also working to push science in different directions. He questions our vision of health as internal to the human body, noting that bodies are part of the environment. Twine encourages bioethics to adopt a more ecological approach—health is not merely physical; health is also environmental and social. According to Twine, the indifference that bioethics demonstrates towards nature is directly linked to our patriarchal, dualistic vision of the world:

> Inserted into a Western historical dualistic narrative, biomedicine represents the cultural domination of human physicality, which is assumed to symbolize nature. But the traditional biomedical view that the body simply represents

> nature uncritically reproduces a dualistic view that not only separates mind from body, but also discounts the social, economic and cultural construction of the body.
>
> (Twine, 2005: 290)

Twine helps readers to see how an inadequate vision of bioethics is connected with inadequate moral theories, both of which focus narrowly on human beings, shutting out the larger interconnected world.

Traditionally, science has been more closely connected with the environmental movement than with animal advocacy—with environmental studies and ecology than with animal studies and animal ethics. Science is no less important to understanding nonhuman animals as explored in animal studies than it is to understanding ecosystems as explored in environmental ethics. The importance of science (biology in particular) constitutes yet another unbreakable bond between earth and animal advocacy.

### Biopolitics and the Capability Approach

In lectures given in 1977–78 that were eventually published in *Security, Territory, Population: Lectures at the Collège de France*, French philosopher Michel Foucault pondered justice and politics, calling attention to a transformation in the dynamic of power in contemporary Western societies (Foucault, 2007). In the past, Foucault observed, state power was rooted in coercive control justified by assumed God-given rights and abstract ideals of justice. In contrast, modern nation states manage the lives of entire populations with mechanisms like public security and healthcare. While government violence and coercion continue, Foucault noted a trend toward political strategies that target human biological needs, which he called biopower (Foucault, 2007).

Foucault's notion of biopolitics has intrigued many scholars interested in pondering the state's responsibility to nature and animals (Bazzicalupo and Clò, 2006; Dubreuil and Eagle, 2006; Wadiwel, 2002; Stanescu, 2013). Unfortunately, many of these discussions remain anthropocentric—"for all their attention to biology and life, and indeed the politicization of such, they give little consideration to the subjection of animals within the regimes of biopower they critique" (Chrulew, 2012: 53). Nonetheless, Foucault's conception of biopolitics opened the door for political responsibility that reaches beyond humanity; "it will be important to ask whether and how their analyses of biopolitics might be elaborated in eco- and zoo-political terms (Chrulew, 2012: 53).

Philosophical explorations of justice have also extended a hand to nonhuman animals. In *Frontiers of Justice*, philosopher Martha Nussbaum includes a chapter on "justice for nonhuman animals" in which she offers a critique of major moral theories. She notes that many philosophical theories were developed long before our current understandings of animal cognition and emotional bonding. In light of such current information, Nussbaum fosters a more inclusive alternative, the "capability approach," which focuses on potentialities. If animals have potentialities that can be fulfilled, Nussbaum argues, justice should provide them with the necessary protections and autonomy to do so (Nussbaum, 2006a). Nussbaum's capability approach

> suggests that it is appropriate for each nation to include in its constitution or other founding statement of principle a commitment to regarding nonhuman animals as subjects of political justice and to treating them in accordance with their dignity.
>
> (Nussbaum, 2006b: 6)

Foucault and Nussbaum expanded the political horizons with regard to earth and animals.

## *Direct Action*

Environment and animal welfare laws, along with organizations lobbying on behalf of earth or animals, have proliferated since the sixties, especially in the United States (Tarlock, 2009). Lobbying for the environment or nonhuman animals is more common than ever before (Tischler, 2012), but there are many activists who have moved beyond lobbying to engage in direct action. For decades activists in both camps have risked liberty (sometimes even their lives) to protect and defend animals or the environment. Many activists believe that their struggle on behalf of earth or animals is a revolution, a war against institutionalized power (Schnurer, 2004; Atwood, 2004; Watson, 1995):

> Animal liberation is … a war. A long, hard, bloody war in which all the countless millions of its victims have, so far been on one side only, have been defenseless and innocent, whose tragedy was being born nonhuman. The oceans, the land and the sky should be free to all rather than be the domain of whoever is most powerful in the human world. The methods to achieve a just world are many and varied, but all tactics are important.
>
> (Webb, 2004: 80)

The climate for direct action in the United States changed radically after 9/11, when fear engulfed the nation and citizens favored, or at least were willing to tolerate, a conservative approach to dissent, even to the extent of limiting and curtailing civil liberties. The number one casualties of the nation's paranoia were, certainly not by coincidence, the biggest enemies of the biggest corporate powers—earth and animal activists. Demonized by the press, politicians, and powerful corporations, anonymous activists and well-known militants speak their mind in two fat anthologies that give voice to direct action in both movements. The first one, called *Freedom Fighters* (2004) is directed more toward animal advocacy, while the second, *Igniting a Revolution* (2006), turns more toward environmental activism. That said, these books stand as perfect testimony to the melding of earth and animal activism—each book includes and speaks on behalf of both movements.

Repressive legislation targeting such activists stems from "alliances between corporations and professional lobbying groups," with intent to prevent any challenge that might threaten production and profits (Best, 2004: 313). Consequently, The Animal Liberation Front (ALF) and The Earth Liberation Front (ELF)—social justice activists—are now labeled *top* domestic terrorists in the United States (Best, 2007). Terrorists target noncombatants (Walzer, 2004: 51): neither earth nor animal activists are terrorists—they do not target noncombatants. In fact, they don't target anyone—they go to great lengths to avoid harming life. Animal and earth activists do not even harm those who are guilty of gross atrocities against nonhuman animals and the earth. They do, however, target property, and in a nation of corporate capitalism, there is nothing quite so alarming to those in power.

Liberation Front practices are sometimes intended to discourage oppressors, but considering how many animals are killed at any given moment in slaughterhouses and labs, and how many animals are in excruciating pain and discomfort in animal labs, on fur farms, on factory farms, and in roadside zoos, for example, these activities on behalf of the defenseless are comparatively mild (Bernstein, 2004). Steven Best, who views animal activism as a war between a terrorist state and social justice activists, writes:

> In this sick and violent society, property is more sacred than life, and thus only those who destroy property are branded as criminals while the real terrorists perpetuate the "banality of evil" (Hannah Arendt) through the daily affairs of torture and killing.
>
> (Best, 2004: 334)

Reacting to the harsh labels of a conservative society, Rod Coronado, known for sinking whaling ships and sabotaging hunts, rejects the suggestion that direct action on behalf of earth or animals constitutes a "radical movement." He views such direct action as a pragmatic measure against an unjust and cruel multi-billion dollar industry that is backed by the U.S. government: labeling the ALF and ELF as "extreme" suggests that animals and the natural world "are not valuable enough to warrant extra-legal means" (Coronado, 2004: 180).

Offering a dissenting voice, Tom Regan discourages property damage as a form of violence (see Regan, 2004). Regan argues that illegal tactics turn the public against activists, who otherwise hold the high moral ground. But those engaged in direct action note that history tells a different story—activists only turn the tide in their favor by ruffling feathers. Even Jesus is reported to have used direct action that involved property damage, overturning the tables of the money changers (Kemmerer, 2006b). Many liberationists are committed to both legal and illegal actions with intent to build a cause that is not easily sidelined—even for great powers such as animal industries and their government allies. ELF and ALF activists cross legal lines side by side, refusing to compromise their integrity, determined to bring change for earth and animals (Walton and Widay, 2006).

Liberationists in both camps share much in common, as evidenced by the writings of scholars as diverse as Mark Somma and Are Naess. Mark Somma comments on the importance of self-realization to both earth and animal activism, often manifest when an activist deliberately changes his or her ways of living:

> The potent combination of changing beliefs and behaviors, such as the decision to switch to a vegetarian or vegan diet, encourages personal growth and steers individuals away from anthropocentric orientations and toward biocentricity. The changes in belief and behavior meet deep ecology's admonition to "self-realization."
>
> (Somma, 2006: 41)

When Arne Naess introduced the term "deep ecology," he indicated that necessary changes on behalf of the environment must be both profound and revolutionary (Naess, 1973). He argued that industrialized nations have to change the way we understand and relate to the world. Western culture is built on rigid dichotomies that prevent an appreciation of diversity (Shiva, 1988). Viewing the world in terms of mutually exclusive opposites radically reduces our ability to recognize and appreciate interconnections and mutuality. Anthropocentrism is a reflection of this problem; environmentalists and animal activists alike are fighting the weight of anthropocentrism by means of direct action.

## *Ecofeminism*

Stemming from the understanding that the oppression of women and the oppression of the natural world suffer under the same yoke, ecofeminism has much to offer the alliance between earth and animal activism. Some ecofeminists—inherently onboard with earth liberationists—have extended the borders of their already inclusive philosophical framework across species boundaries (Gaard, 2011). Feminists accuse ecofeminists of essentialism because some writers and thinkers in this field have associated women with nature, carrying the "fear of contamination-by-association" which has proven "just too strong" for an alliance between these two groups (Gaard, 2011: 27). As a result, ecofeminist articles have generally been rejected by feminist journals and academic conferences—ecofeminists have been marginalized by the feminist community, a group that ought to have been their closest allies, and as a result, have struggled for years to be taken seriously. As a consequence, some ecofeminists chose to place their work under a different label:

> [M]ost feminists working on the intersection of feminism and the environment thought it better to rename their approach to distinguish it from essentialist feminism and thereby gained wider audience, hence the proliferation of terms like "ecological feminism" (Warren 1991, 1994), "feminist environmentalism" (Argwal, 1992; Seager, 1993), "social ecofeminism" (Heller, 1999; King, 1989), "critical feminist eco-socialism" (Plumwood, 2002), or simply "gender and the environment."
>
> (Gaard, 2011: 27)

Unfortunately, ecofeminism continues to be marginalized. For example, a student at a Brazilian university, who was drawn in by the writings of Marti Kheel, wanted to include some ecofeminist analysis in her dissertation, which focused on animal trafficking in Brazil. Her advisor responded: "Ecofeminism is essentialist and shallow—not academic … and now they are comparing women to animals! You will never get published!" The enthusiastic student had to abandon her interest in applying ecofeminism to the topic of animal trafficking.

In a sea of humans building theories, feminism is a deconstruction site, and this ought to help feminists to revisit ecofeminism. Historian Mary Poovey once said:

> Deconstruction is a critical component of the political work I am outlining here, but unless it is deployed up on itself, it will trap us in a practice that once more glorifies the "feminine" instead of giving us the means to explode binary logic and make the social construction of (sexed) identities a project of pressing political concern. If deconstruction took feminism seriously, it wouldn't look like deconstruction anymore. If feminism took deconstruction at its word, we could begin to dismantle the system that assigns to all women a single identity and a marginal place.
>
> (Poovey, 1988: 65)

Poovey defines feminism as meta-critical and critical at the same time: It deconstructs itself even as it deconstructs everything around. What better philosophy for breaking barriers, whether between men and women, humans and nonhuman animals, animal activists and environmentalism, or feminism and ecofeminism? This feminist tendency has carried over to ecofeminism: "Exploding" dichotomies has been one of the greatest triumphs of ecofeminism (Poovey, 1988).

Carol Adams is a critical catalyst for deconstruction in books such as *The Sexual Politics of Meat* (1990) and *The Pornography of Meat* (2003) and her anthology with Josephine Donovan, *Animals and Women* (1995), all of which give practical expression to ecofeminism (Adams, 1990, 2004). Adams questions prejudice against nonhumans among feminists (and also within mainstream culture), and overtly connects sexism and speciesism. For example, Adams notes that animals are often tortured by men in situations of domestic abuse (Adams, 1995). The bold connections that she draws between speciesism and sexism have paved the way for other ecofeminist authors, perhaps most notably Greta Gaard, who pulled topics such as queer theory, ecocriticsm, and pedagogy into the ecofeminist mix (Gaard, 1997, 2009, 2010).

Despite the historic divide with feminists, ecofeminism is critical to feminist deconstruction, and can help feminists to become aware of their own abuse of power in relation to the natural world—especially with regard to nonhuman animals. Ecofeminism, perhaps better than any other social justice theory, explains why earth and animal activists ought to work together.

Ecofeminism is not just a revolutionary theory, but a philosophy that is simultaneously a call to action, especially self-transformation. Ecofeminism is a way of seeing and acting, of living and interacting. Ecofeminists are struggling to extricate privileges and barriers through and within many different social justice movements, working on behalf of animals and the earth more generally, while fighting sexism, classism, racism, heterosexism, and ageism.

*Sister Species* (Kemmerer, 2011) offers a collection of writings from female activists who reveal the living breath—and muscle—of ecofeminism. Authors in this collection offer narratives in which they discover themselves and come into their own as activists. *Sister Species* effectively demonstrates "that there is a reward for courage and determination in the face of helplessness and suffering: walking into pain in the hope of bringing change moves a person from helplessness and despair to empowerment and activism" (Kemmerer, 2011: 27). For example, activist Allison Lance explains her fearless fury when confronted by hunters and fishers who ravage the forests and seas; scholar and animal activist Elizabeth Farians describes her unflappable patience in the interest of offering a course in animal studies at a Christian university that clearly devalued both Elizabeth—as a woman—and nonhuman animals; the founder of United Poultry Concerns, Karen Davis, relates her experience as an extremely sensitive person growing up in a world of ignorance, violence, and gross insensitivity, explaining how she emerged from this dark time to become one of the nation's most effective animal activists (Lance, 2011; Farians, 2011; Davis, 2011). *Sister Species* also offers a more-than-human experience, understandings that require more than one species. For example, Hope Ferdowsian describes how her father wept over Charlie, an abused and neglected dog, whose past sufferings were as clearly visible in his actions as were his mother's long-term suffering as a persecuted minority (Ferdowsian, 2011). Breeze Harper describes how racism that she experienced helped her to empathize with nonhuman animals. She encourages animal activists to notice and eradicate white privilege as part of our struggle for justice (Harper, 2011). *Sister Species* presents nonhumans as protagonists, changing the world around them even as determined activists work to change the lives of other animals. This is perhaps most visible in Sangamithra Iyer's essay, in which she writes of a female chimpanzee, Washoe, exploited by researchers who wanted to know, among other things, if she would pass sign language to her offspring. In the process, Washoe gave birth to two stillborns and lost a baby when she was in labor. Iyer writes of Washoe's pain and empathy—on learning that a pregnant undergrad had suffered a miscarriage, Washoe signed, "CRY," followed later by "PLEASE PERSON HUG" (Iyer, 2011: 91). Washoe eventually did pass sign language to her offspring, but it seems unlikely that she will be able to pass her wealth of empathy to an exploitive humanity.

## *Androcentrism*

The ecofeminist understanding of patriarchal systems of oppression—a scaffold supporting both sexism and environmental degradation—was revolutionary. In the process, ecofeminists called attention to the role of masculinist cultural norms, especially aggression.

The late André Collard's book (finished by Joyce Contrucci) *Rape of the Wild*, argues that male violence is rooted in antagonism and domination, and that everything that falls under the male conqueror's gaze comes to be viewed as feminine—not feminine in any positive sense, but feminine in the sense of being subjected to man's power and control. Collard also noted that gender-related exploitation of women and nature spills over to smite animals (Collard and Contrucci, 1989).

Philosopher and ecofeminist Marti Kheel made a similar point in her article for the classic anthology *Reweaving the World* (1990), where she noted that deep ecology and ecofeminism are markedly similar at their core: deep ecology criticizes anthropocentrism (a human-centered focus) while ecofeminism questions androcentrism (a male-centered focus). Like Collard, Kheel calls attention to the male fixation on dominating those who are viewed as "other," whether women or elk or hogs, approaching everything that exists in the world as a sexual conquest. In her book *Nature Ethics* (2008), Kheel explores the hunter mentality as a male problem, and criticizes ecofeminists who support "ecological hunters," denouncing the holism of deep ecology as a masculinist fantasy (Kheel, 1990, 1996, 2008). Kheel encourages (as does Jim Mason in the same anthology) a social deconstruction of patriarchy with intent to build more inclusive, stronger advocacy groups, and a more just platform for direct action (Kheel, 2006; Mason, 2006).

Kheel also questions the archetype of the male eco-hero as the paladin of nature and animal advocacy (Kheel, 2006). Among activists, Kheel notes that the "predatory ethic that fuels the masculine conquest of the beast is transformed into the ethic of heroic protection. Again, women and nonhuman animals are seen as devoid of independent identity, passive objects which reinforce the masculine self" (Kheel, 2006: 308). Whether eco-hero or animal liberationist, the vision of the savior is invariably male. Better to transform our understandings of male and female than to transfer patriarchal images to the earth and animal liberation movements. What nonhuman animals and nature need most is not a male savior, but for men to quit trying to dominate everything they encounter.

As an ecofeminist, Marti Kheel worked for decades at the borders of environmental ethics, feminism, and animal studies, denouncing androcentrism. She had a strong influence over many other scholars, including Brian Luke, who exposes ugly masculinist ideals that support and perpetuate sport hunting and the myth of the eco-hunter (Luke, 2007).

## *The Ethic of Care*

Starting in the fifties, psychologist Lawrence Kohlberg posed specific scenarios to boys of various ages in order to explore stages of moral development. Kohlberg assumed that his resultant scale of moral development, rooted in moral theories developed by males (most notably Kant), using only boys from the Chicago area, applied to all of humanity. The moral reasoning of girls—when assessed—tended to fall below that of boys in Kohlberg's stages of moral development. Carol Gilligan, who worked with Kohlberg, noticed "the imagery of violence in the boys' responses, depicting a world of dangerous confrontation and explosive connection," compared with the girls' responses, which present a "world of care and protection, a life lived with others whom 'you may love as much or even more than you love yourself'" (Gilligan, 1982: 34). In response to her own reflections, studies, and findings, Gilligan wrote *In a Different Voice* (1982), offering the foundations for what has come to be known as the feminist ethic of care.

The feminist ethic of care (developed over time by scholars such as Gilligan, Nel Noddings, and Josephine Donovan) emerged as a reaction to male-dominated moral theories, rooted in such hypothetical notions as absolute rights, autonomous individuals, and social contracts, which fail to reflect the family-centered, home environment that tends to be most familiar to women. In contrast with male moral theories, the feminist ethic of care considers relations, context, emotion, care, narrative, and attentiveness—aspects of humanity that are more effective (and therefore more essential) within families, among close friends, and in tight-knit communities (Donovan and Adams, 2007). Interestingly, likely the most famous portion of Aldo Leopold's writing—a turning point for him in his understanding of animals and the natural world—is a narrative where he describes his emotive response to watching a wolf die (Lorbiecki, 1999). Even men who appear to be trapped in narrow, masculinist roles are sometimes able to access the powerful moral tools of the feminist ethic of care.

The development of the feminist ethic of care provides additional insights into our general moral indifference to nonhuman animals and the natural world—why our standard moral theories, rooted in utilitarianism and rights, established by men and stemming from the male point of view, are inadequate. Relations, context, emotion, care, and attentiveness are no less critical to our interactions with other animals and the natural world than they are to our interactions with one another. *The Feminist Care Tradition in Animal Ethics* (2007), edited by Josephine Donovan and Carol Adams, gives voice to some of the most important ecofeminists, including Adams and Donovan alongside Marti Kheel, Deane Curtain, Lori Gruen, and Brian Luke.

## Conclusion

The happiness of individual animals (including humans) is intimately connected with the health of a well-protected environment; stable ecosystems necessarily include healthy, well-protected nonhuman animals (Kheel, 1995). Both the earth in its entirety and nonhuman animals as individuals need human representation if they are to be protected against human stupidity, greed, and indifference. Yet anthropocentrism and human privilege—the very problems that threaten earth and animals—prevent earth and animal activists from joining forces. Environmentalists who refuse to make significant dietary changes are a dime a dozen, as are animal rights activists who cannot be bothered to reduce, reuse, and recycle.

Perhaps, in the spirit of ecofeminism and the feminist ethic of care, we ought to ponder what Washoe's narrative might look like if she were explaining her needs to earth and animal activists. It is likely that she would want expansive, intact forests for her beloved friends and relatives in West Africa, lost to her so long ago, when her mother was killed and she was kidnapped and sold into science in North America. It also seems reasonable to suppose that she would desperately want laws put in place to prevent nonhuman animals from being taken from the wild and protecting those who have already been kidnapped and sold into exploitation. Washoe's likely narrative exposes critical overlap between earth and animal advocacy: Chimpanzees need laws protecting them from exploitation for science *and* laws protecting their forest homes. Similarly, both farmed animals and wild ecosystems—if we listen to what they would likely say—cry out for an end to factory farming and hunting, trade in exotic species and industrial fishing.

Working together does not require a seamless harmony between advocates and organizations any more than does friendship or marriage. Animal activists and environmentalists are encouraged to disagree—to work out differences in a way that fosters growth and change on both sides. But moving forward as separate (and antagonistic) movements is foolhardy. It's time to stop focusing on differences and start bending to the needs of the earth and animals.

Returning to the psychologist who spoke at the Humane Society's conference, consider the Washington speaker featured in the initial paragraphs of this essay, who suggested that animal biology be manipulated to create non-sentient animals so that we might go on eating meat—presumably without guilt—into perpetuity. Following the logic of the speaker, our best choice of action against child labor would be to find a way to manipulate the minds of children so that they like working from dawn until dusk at menial tasks. Similarly, our best choice of action in response to sexual assault is to alter women so that they do not mind being assaulted. Clearly these are not reasonable solutions to violence and exploitation. Though the speaker indicated that he cared both about animals and the environment, his proposed solution helped neither: cattle without sensation will be a hazard to themselves and will be no less harmful to the environment. The *moral* response to our exploitation and destruction of earth and animals is not to attempt to change the exploited so that they are insentient and indifferent to our cruelty and greed, but rather to stop our cruel exploitation and greed-driven destruction of other animals and the earth. Returning to the psychologist who spoke at the Humane Society's spring conference on "Science of Animal Thinking and Emotion," who suggested perhaps what most closely unites animals and earth advocacy is not the intent to change everything around us, but a commitment to the painful journey to change ourselves.

## Bibliography


Adams, C. J. (1990) *The Sexual Politics of Meat: A Feminist-Vegetarian Critical Theory*. London: Continuum International Public Group.

Adams, C. J. (2004) *The Pornography of Meat*. London: Bloomsbury Academic.

Adams, C. J. (1995) "Woman-Battering and Harm to Animals," in Carol J. Adams and Josephine Donovan (eds) *Animals and Woman: Feminist Theoretical Explorations*. Durham: Duke University Press. pp. 55–84.

Adams, C. J. and Donovan, J. (eds) (1995) *Animals and Women: Feminist Theoretical Explorations*. Durham: Duke University Press.

Agarwal, B. (1992) "The Gender and Environment Debate: Lessons from India." *Feminist Studies* 18 (1), 119–59.

Anon (n.d.) "Shambles." *The Free Dictionary* [online]. Available from: http://www.thefreedictionary.com/shambles (Accessed 15 June 2014).

Atwood, N. (2004) "Revolutionary Process and the ALF," in Steven Best and Anthony Nocella II (eds) *Terrorists or Freedom Fighters?: Reflections on the Liberation of Animals*. New York: Lantern Books. pp. 272–276.

Bazzicalupo, L. and Clò, C. (2006) "The Ambivalences of Biopolitics." *diacritics*. 36 (2), 109–116.

Bekoff, M. (2007) *Animals Matter: A Biologist Explains Why We Should Treat Animals with Compassion and Respect*. Boston: Shambhala Publications.

Bernstein, M. (2004) "Legitimizing Liberation," in Steven Best and Anthony Nocella II (eds) *Terrorists or Freedom Fighters?: Reflections on the Liberation of Animals*. New York: Lantern Books. pp. 93–105.

Best, S. (2004) "It's War! The Escalating Battle Between Activists and the Corporate-State Complex," in Steven Best and Anthony Nocella II (eds) *Terrorists or Freedom Fighters?: Reflections on the Liberation of Animals*. New York: Lantern Books. pp. 300–340.

Best, S. (2007) "The Animal Enterprise Terrorism Act: New, Improved, and ACLU Approved." *The International Journal of Inclusive Democracy*. 3 (3) [online]. Available from: http://www.inclusivedemocracy.org/journal/vol3/vol3_no3_best.htm (Accessed 22 June 2014).

Best, S. and Nocella II, A. J. N. (eds) (2004) *Terrorists or Freedom Fighters*. New York: Lantern Books.

Best, S. and Nocella II, A. J. N. (eds) (2006) *Igniting a Revolution: Voices in Defense of the Earth*. Oakland: AK Press.

Callicott, J. B. (1989) "Animal Rights and The Environmental Ethics: Back Together Again," in *In Defense of the Land Ethic: Essays in Environmental Philosophy*. New York: State University of New York Press. pp. 163–169.

Callicott, J. B. (2005) "Introduction," in Michael E. Zimmerman, J. B. Callicott, G. Sessions, K. J Warren and J. Clark (eds) *Environmental Philosophy: From Animal Rights to Radical Ecology*. Upper Saddle River: Prentice Hall. pp. 5–15.

Callicott, J. B. (2009) "What Can the Humanities Contribute to a New Consciousness in Harmony with Nature?" in Stephen R. Kellert and James Gustav Speth (eds) *The Coming Transformation: Values to Sustain Humans and Natural Communities*. New Haven: Yale School of Forestry and Environmental Studies. pp. 278–298.

Carson, R. ([1962] 2002) *Silent Spring*. Boston: Mariner Books.

Chrulew, M. (2012) "Animals in Biopolitical Theory: Between Agamben and Negri." *New Formations*. 76 (1), 53–67.

Collard, A. and Contrucci, J. (1989) *Rape of the Wild: Man's Violence against Animals and the Earth*. Bloomington: Indiana University Press.

Coronado, R. (2004) "Direct Actions Speak Louder than Words," in Steven Best and Anthony Nocella II (eds) *Terrorists or Freedom Fighters?: Reflections on the Liberation of Animals*. New York: Lantern Books. pp. 178–184.

Crutzen, P. J. and Stoermer, E. F. (2000) "The Anthropocene." *Global Change Newsletter*. May.

Davis, K. (2011) "From Hunting Grounds to Chicken Rights: My Story in an Eggshell," in Lisa Kemmerer (ed.) *Sister Species: Woman, Animals, and Social Justice*. Champaign: University of Illinois Press. pp. 127–140.

DeMello, M. (2012) *Animals and Society: An Introduction to Human-Animal Studies*. New York: Columbia University Press.

Donovan, J. and Adams, C. J. (eds) (2007) *The Feminist Care Tradition in Animal Ethics*. New York: Columbia University Press.

Dubreuil, L. and Eagle, C. C. (2006) "Leaving Politics: Bios, Zōē, Life." *Diacritics*. 36 (2), 83–98.

Farians, E. J. (2011) "Theology and Animals," in Lisa Kemmerer (ed.) *Sister Species: Woman, Animals and Social Justice*. Champaign: University of Illinois Press. pp. 102–109.

Ferdowsian, H. (2011) "Compassion without borders," in Lisa Kemmerer (ed.) *Sister Species: Woman, Animals and Social Justice*. Champaign: University of Illinois Press. pp. 97–101.

Foucault, M. (2007) *Security, Territory, Population: Lectures at the Collège de France 1977–1978*. New York: Palgrave Macmillan.

François, T. (2011) "From Roots to Angels' Wings," in Lisa Kemmerer (ed.) *Sister Species: Woman, Animals and Social Justice*. Champaign: University of Illinois Press. pp. 57–64.

Gaard, G. (ed.) (1993) *Ecofeminism: Woman, Animals, Nature*. Philadelphia: Temple University Press.

Gaard, G. (1997) "Toward a Queer Ecofeminism." *Hypatia*. 12 (1), 114–137.

Gaard, G. (2009) "Children's Environmental Literature: From Ecocriticism to Ecopedagogy." *Neohelicon*. 36 (2), 321–334.

Gaard, G. (2010) "New Directions for Ecofeminism: Toward a More Feminist Ecocriticism." *Interdisciplinary Studies in Literature and Environment*. 17 (4), 643–665.

Gaard, G. (2011) 'Ecofeminism Revisited: Rejecting Essentialism and Re-placing Species in a Material Feminist Environmentalism." *Feminist Formations*. 23 (2), 26–53.

Gilligan, C. (1982) *In a Different Voice: Psychological Theory and Women's Development*. Boston: Harvard University Press.

Goodall, J. and Bekoff, M. (2002) *The Ten Trusts: What We Must Do to Care for The Animals We Love*. New York: HarperCollins.

Goodpaster, K. E. (1978) "On Being Morally Considerable." *The Journal of Philosophy*. 75 (6), 308–325.

Harper, A. B. (2011) "Connections: Speciesism, Racism, and Whiteness as the Norm," in Lisa Kemmerer (ed.) *Sister Species: Woman, Animals and Social Justice*. Champaign: University of Illinois Press. pp. 72–78.

Heise, U. K. (2008) "Ecocriticism and the Transnational Turn in American Studies." *American Literary History*. 20 (1–2), 381–404.

Heller, C. (1999) *Ecology of Everyday Life: Rethinking the Desire for Nature*. Montreal: Black Rose Books.

Iyer, S. (2011) "Small Small Redemption," in Lisa Kemmerer (ed.) *Sister Species: Woman, Animals and Social Justice*. Champaign: University of Illinois Press. pp. 87–96.

Kaufman, S. R. and Braun, N. (2004) *Good News for All Creation: Vegetarianism as Christian Stewardship*. Cleveland: Vegetarian Advocated Press.

Kemmerer, L. (2006a) *In Search of Consistency: Ethics and Animals*. Leiden: Brill Academic.

Kemmerer, L. (2006b) "In the Beginning God Created the Earth and 'Ecoterrorism'," in Steven Best and Anthony Nocella II (eds) *Igniting Revolution: Voices in Defense of the Earth*. Oakland: AK Press. pp. 156–169.

Kemmerer, L. (2008) "Just War and Warrior Activists." *Green Theory and Praxis: A Journal of Ecological Politics*. 4 (2), 25–49.

Kemmerer, L. (ed.) (2011) *Sister Species: Women, Animals, and Social Justice*. Champaign: University of Illinois Press.

Kemmerer, L. (2012) *Animals and World Religions*. Oxford: Oxford University Press.

Kemmerer, L. (2014) *Eating Earth*. Oxford: Oxford University Press.

Kheel, M. (1990) "Ecofeminism and Deep Ecology: Reflections on Identity and Difference," in Irene Diamond and Gloria Feman Orenstein (eds) *Reweaving the World: The Emergence of Ecofeminism*. San Francisco: Sierra Club Books. pp. 128–137.

Kheel, M. (1995) "License to Kill: An Ecofeminist Critique of Hunters' Discourse," in Carol J. Adams and Josephine Donovan (eds) *Animals and Woman: Feminist Theoretical Explorations*. Durham: Duke University Press. pp. 85–125.

Kheel, M. (1996) "The Killing Game: An Ecofeminist Critique of Hunting." *Journal of the Philosophy of Sport*. 23 (1), 30–44.

Kheel, M. (2006) "Direct Action and the Heroic Ideal: An Ecofeminist Critique," in Steven Best and Anthony Nocella II (eds) *Igniting Revolution: Voices in Defense of the Earth*. Oakland: AK Press. pp. 306–318.

Kheel, M. (2008) *Nature Ethics: An Ecofeminist Perspective*. Lanham: Rowman and Littlefield.

King, Y. (1989) "The Ecology of Feminism and the Feminism of Ecology," in Judith Plant (ed.) *Healing the Wounds: The Promise of Ecofeminism*. Philadelphia, PA: New Society Press. pp. 18–28.

Lance, A. (2011) "A Magical Talisman," in Lisa Kemmerer (ed.) *Sister Species: Woman, Animals and Social Justice*. Champaign: University of Illinois Press. pp. 161–172.

Lappé, F. M. ([1971] 2011) *Diet for a Small Planet*. New York: Ballantine Books.

Lear, L. (2009) *Rachel Carson: Witness for Nature*. Boston: Houghton Mifflin Harcourt.

Leopold, A. (1915) "The Varmint Question." *Pine Cone*. 12 (1), 47–49.

Leopold, A. (1966) *A Sand County Almanac*. Oxford: Oxford University Press.

Linzey, A. (1992) "The Arrogance of Humanism," in Richard Ryder (ed.) *Animal Welfare and the Environment*. Melksham: Duckworth. pp. 68–72.

Lorbiecki, M. (1999) *Aldo Leopold: A Fierce Green Fire*. New York: Oxford University Press.

Luke, B. (2007) *Brutal: Manhood and the Exploitation of Animals*. Urbana: University of Illinois Press.

Mason, J. (2006) "The Animal Question: Uncovering the Roots of Our Domination of Nature and Each Other," in Steven Best and Anthony Nocella II (eds.) *Igniting Revolution: Voices in Defense of the Earth*. Oakland: AK Press. pp. 178–187.

Midgley, M. (2001) *Gaia: The Next Big Idea*. London: Demos.
Naess, A. (1973) "The Shallow and the Deep, Long-Range Ecology Movement: A Summary." *Inquiry*. (16), 95–100.
Nussbaum, M. C. (2006a) *Frontiers of Justice: Disability, Nationality, Species Membership*. Cambridge: Belknap Press.
Nussbaum, M. C. (2006b) "The Moral Status of Animals." *Chronicle of Higher Education*. 3 February. [online]. Available from: http://chronicle.com/article/The-Moral-Status-of-Animals/25792 (Accessed 10 June 2014).
Pluhar, E. B. (1995) *The Moral Significance of Human and Nonhuman Animals*. Durham: Duke University Press.
Plumwood, V. (2002) *Environmental Culture: The Ecological Crisis of Reason*. New York: Routledge.
Poovey, M. (1988) "Feminism and Deconstruction." *Feminist Studies*. 14 (1), 51–65.
Regan, T. (1982) *All that Dwell Therein: Animal Rights and Environmental Ethics*. Berkeley: University of California Press.
Regan, T. (1983) *The Case for Animal Rights*. Berkeley: University of California Press.
Regan, T. (2004) *Empty Cages: Facing the Challenge of Animal Rights*. Lanham: Rowman and Littlefield.
Ryder, R. (2005) "All Beings that Feel Pain Deserve Human Rights." *The Guardian*. 6 August. [online]. Available from: http://www.theguardian.com/uk/2005/aug/06/animalwelfare (Accessed 21 June 2014).
Seager, J. (1993) *Earth Follies: Coming to Feminist Terms with the Global Environmental Crisis*. New York: Routledge.
Schmidtz, D. and Willott, E. (2012) "Respect for Nature. Introduction: The Last Man and the Search for Objective Value," in *Environmental Ethics: What Really Matters, What Really Works*. New York: Oxford University Press. pp. 42–47.
Schnurer, M. (2004) "At the Gates of Hell: The ALF and the Legacy of Holocaust Resistance," in Steven Best and Anthony Nocella II (eds.) *Terrorists or Freedom Fighters?: Reflections on the Liberation of Animals*. New York: Lantern Books. pp. 106–127.
Schwartz, R. H. (2001) *Judaism and Vegetarianism*. New York: Lantern Books.
Shiva, V. (1988) *Staying Alive: Women, Ecology and Development*. London: Zed Books.
Singer, P. (1975) *Animal Liberation: A New Ethics for Our Treatment of Animals*. New York: Random House.
Singer, P. (1993a) *How Are We to Live?: Ethics in an Age of Self-Interest*. Amherst: Prometheus Books.
Singer, P. (1993b) *Practical Ethics*. Cambridge: Cambridge University Press.
Singer, P. (2004) *One World: The Ethics of Globalization*. New Haven: Yale University Press.
Somma, M. (2006) "Revolutionary Environmentalism: An Introduction," in Steven Best and Anthony Nocella II (eds) *Igniting Revolution: Voices in Defense of the Earth*. Oakland: AK Press. pp. 37–46.
Stanescu, J. (2013) "Beyond Biopolitics: Animal Studies, Factory Farms, and the Advent of Deading Life." *PhaenEx*. 8 (2), 135–160.
Tarlock, D. (2009) "History of Environmental Law," in Dan Tarlock (ed.) *Environmental Laws and Their Enforcement*. Oxford: EOLSS Publishers. pp. 29–36.
Taylor, P. (1981) "The Ethics of Respect for Nature." *Environmental Ethics*. 3 (3), 197–218.
Taylor, P. (1986) *Respect for Nature: A Theory of Environmental Ethics*. Princeton: Princeton University Press.
Tischler, J. (2012) "A Brief History of Animal Law, Part II (1985–2011)." *Stanford Journal of Animal Law and Policy (SJALP)*. 5 (1), 27–77.
Todorov, T. (1996) *The Conquest of America*. New York: HarperCollins.

Twine, R. (2005) "Constructing Critical Bioethics by Deconstructing Culture/Nature Dualism. "*Medicine, Health Care, and Philosophy*. 8 (3), 285–295.

Wadiwel, D. J. (2002) "Cows and Sovereignty: Biopower and Animal Life." *Borderlands*. 1 (2). [online]. Available from: http://www.borderlands.net.au/vol1no2_2002/wadiwel_cows.html (Accessed 22 June 2014).

Walton, M. and Widay, J. (2006) "Shades of Green: Examining Cooperation Between Radical and Mainstream Environmentalists," in Steven Best and Anthony Nocella II (eds) *Igniting A Revolution: Voices in Defense of the Earth*. Oakland, CA: AK Press.

Walzer, M. (2004) *Arguing about War*. New Haven: Yale University Press.

Warren K. J. (ed.) (1991) "Ecological Feminism," special issue of *Hypatia: A Journal of Feminist Philosophy* 6 (1).

Warren K. J. (ed.) (1994) *Ecological Feminism*. New York: Routledge.

Watson, P. (1995) "Tora! Tora! Tora!," in J Sterba (ed.) *Earth Ethics*. New Jersey: Prentice Hall. pp. 341–346.

Webb, R. (2004) "Animal Liberation—By 'Whatever Means Necessary'," in Steven Best and Anthony Nocella II (eds) *Terrorists or Freedom Fighters?: Reflections on the Liberation of Animals*. New York, NY: Lantern Books. pp. 75–80.

Weber, M. ([1905] 2002) *The Protestant Ethic and the Spirit of Capitalism: and Other Writings*. 3rd edition. London: Penguin Group US.

White, L. (1967) "The Historical Roots of Our Ecological Crisis." *Science*. 155 (3767), 1203–1207.

# SECTION I

# Foundations

## Theoretical Connections

**Poem:** Randall Gloege

### The Pika and the Desert Boot

Where day burned cold and bright
near a rough cliff edge,
not far from a foot-hardened path,
I met a furry, diminutive beast
who took my right tan suede desert boot
for another little furry. I kept quiet
while the pika approached my size 12,
then perched attentively on hind paws.
For me, it was affection at first sight.

Understand, this happened long ago
when there were still glaciers in the Park,
and not just boulders, but snow, ice, water and cold
in abundance. Inviting clumps of grass
poked up everywhere.

Now no more ready access to tasty succulents
in flower-strewn meadows; far fewer the hay caches
piled in chinks in the talus; less common those babies
kept snug from harm. The bewildering heat climbs skyward,
and my curious pika friends, imprisoned in their rock islands,
must choose between shriveling up
or boiling into capricious oblivion.

# 2

# BEYOND INTERSECTIONALITY TO TOTAL LIBERATION

*Carol L. Glasser*

In October of 2012 I attended what was advertised as the "World's largest slaughterhouse demonstration." An estimated 80 activists showed up at Farmer John, a slaughterhouse near Los Angeles, California where over 6,000 pigs are killed each day. We chanted, held signs and encouraged passersby to stop eating meat.

It was a hot day and one well-meaning activist brought cases of bottled water to give out to the activists. Ann Bradley, an animal activist at the demonstration, asked that these single-use water bottles not be handed out and be returned to the store, citing the harm they cause to nonhuman animals and the environment. But the animal activists rallied around the bottles of water, responding that the water was a nice gesture and those who had forgotten to bring their own refillable water bottles needed it to stay hydrated. The debate spilled onto Facebook, where the punch line of almost every comment was that the day was "for the pigs" and not other issues.

This story is emblematic of the problematic schism between the animal and Earth liberation movements. When I asked Ann Bradley about her actions that day, she explained (personal communication, 27 April 2014): "Environment is habitat. If we open every cage, shutter all slaughterhouses and end animal exploitation while destroying our fragile planet we achieve nothing. Environment is not intersection, it is animal rights."

## Connected Oppressions

The exploitation of the Earth, nonhuman animals and human animals are all connected through a single overarching system of oppression, what Patricia Hill Collins calls the *matrix of domination* (Collins 2000). The matrix of domination highlights that all inequalities are organized by the same "structural, disciplinary, hegemonic, and interpersonal domains of power" (Collins 2000, p 18). Though oppressed groups may appear unrelated to one another, all oppressions are enmeshed:

> [T]he oppression of various devalued groups in human societies is not independent and unrelated; rather, the arrangements that lead to various forms of oppression are integrated in such a way that the exploitation of one group frequently augments and compounds the mistreatment of others.
>
> (Nibert 2002, p 13)

The interconnection of oppressions is especially relevant in the case of animal and Earth exploitation. Not only do the same structures oppress both, but the exploitation of one directly injures the other in obvious ways. The privatization of water, the overuse of disposable plastics, and animal slaughter all have severe negative consequences for both the Earth and nonhuman animals.

Efforts to control and privatize water have resulted in dams that have emptied water basins, dried up agricultural communities, and changed surrounding ecosystems, displacing nonhuman and human animals (Shiva 2002). In 2012, the United States generated 32 million tons of plastic waste, 14 million tons of which was from plastic containers and packaging such as single-use water bottles, and only 9 percent of this plastic was recycled (EPA 2014). The sale of bottled water in the United States has grown eightfold since the 1970s and the average American now drinks about 30 gallons of bottled water each year, most of it bought in single-use bottles (Gleick 2010, pp 5-6). There is now so much plastic debris in the ocean that in the Great Pacific Garbage Patch, a vortex in the Pacific Ocean with an accumulation of debris and chemical waste, plastic outweighs plankton by a ratio of six to one (Grant 2009).

This marine debris composed of plastic waste has negatively impacted animals of at least 267 species, including "86% of all sea turtle species, 44% of all seabird species, and 43% of all marine mammal species" (Denuncio et al. 2011, p 1836). On the Midway Islands in the Pacific Ocean, home to a number of endangered species, one-third of all Albatross chicks die, many because they are fed plastic by their parents who have mistaken it for food (O'Neill 2011). In 2013 a sperm whale died from a stomach blockage due to ingesting plastic—s/he had over 17 kg (37 lbs) of waste in the stomach (Tremlett 2013).

The slaughter of pigs for the consumption of their flesh is also a problem for both nonhuman animals and the environment. As Americans have developed an insatiable hunger for animals' flesh, animal agriculture has grown tremendously and developed into confined animal feeding operations (CAFOs), or factory farms, where thousands of animals are bred and raised in extreme confinement. These methods are inhumane and also harm the environment.

In 2013 over 112 million pigs were killed for food in the U.S. alone (USDA 2014), most of them living in cramped deplorable conditions. Shortly after birth, factory farmed pigs are taken from their mothers, their tails are cut off, their teeth are cut short and males are castrated—all without anesthesia. These pigs will spend their entire shortened lives in cramped environments that do not allow them to engage in natural behaviors like rooting and wallowing. After this they are transported and slaughtered in harsh and painful conditions.

This intensive "production" of animals also comes with a heavy environmental toll. A pig can produce up to 1,700 pounds of solid waste per year. In 2006 there was 9.6 million tons of solid waste from hog farms produced in the eastern part of North Carolina alone (Edwards and Driscoll 2009). This waste releases ammonia into the air and contaminates ground water (EPA 2012). Hog waste is often stored in open-air lagoons prone to spills that have devastated communities. In 1995 one such spill occurred in North Carolina, polluting the New River and killing about 10 million fish (NRDC 2013).

Like the Earth and nonhuman animals, humans are harmed by factory farming, particularly poorer people. In North Carolina large pig farms are seven times more likely to be in low-income areas than high-income areas (Wing et al. 2000). Living in proximity to CAFOs is associated with having higher rates of asthma (Sigurdarson and Kline 2006), reproductive problems, and cancers. Further, the privatization of water has led to water shortages in India; in the maquiladoras in Mexico water is so scarce that children drink soda rather than water because it is more affordable (Shiva 2002). Poor people and people of color bear the brunt of the environmental damage caused by animal exploitation because CAFOs and slaughterhouses are more likely to be in their communities. Slaughterhouse work in the U.S., done mostly by poor immigrant and black workers, was identified by Human Rights Watch as one of the most dangerous jobs in the country (HRW 2005). The violent nature of the industry and the abuses that surround the treatment of workers have negative impacts on surrounding communities; communities where slaughterhouses are a major employer have increased arrest rates for violent crimes, rape, and other sex offenses (Fitzgerald et al. 2009). Plastic waste, animal slaughter, and privatization of water provide just a few examples of the many ways that the Earth and animals (human and nonhuman) mutually suffer from injustice.

## The Structure of Oppression

Ecofeminists have long identified the interconnected nature of the oppression of disadvantaged humans, the Earth, and nonhuman animals. Ecofeminists emphasize that inequality is maintained through the use of *value hierarchies*, *value dualisms*, and a *logic of domination* (Adams 1996). Value dualisms create either–or thinking that establishes only two opposite ways of being, in which only one is valued. All that are not on the valued end of this dichotomy are defined in relation to what they are *not* as opposed to what they *are*. Value hierarchies create up–down thinking that allows for certain positions, ideas, and groups to be valued over others. This dichotomous and rank-ordered way of thinking also erases complexities, nuances, and variability. For example, whiteness is the privileged racial category (value hierarchy) and while there are many culturally established races and ethnicities, people who are not white get easily lumped into the "people of color" category regardless of whether that is how they personally choose to identify (value dualism). There is only one valued category—"white"—and anything that deviates from it is devalued as "not-white." This is why Barack

Obama was not touted as the United States' first biracial President, but rather as the first black president.

This logic of domination justifies and institutionalizes up–down and either–or ways of thinking. The logic of domination is: "a value-hierarchical way of thinking which explains, justifies, and maintains the subordination of an 'inferior' group by a 'superior' group on the grounds of the (alleged) inferiority or superiority of the respective group" (Karen J. Warren, quoted in Adams 1996, p 79–80).

Man–nature and human–animal dichotomies form the basis of nonhuman animal and Earth exploitation. These false value dualisms separate hu(man)s from nature and nonhuman animals, privilege humans over the rest of the Earth and other animals, and accept that humans can dominate the Earth and nonhuman animals. These ideologies of human supremacy over the Earth and nonhuman animals are called anthropocentricism and speciesism, respectively. Anthropocentricism and speciesism reinforce one another and rationalize human supremacy.

## Movement Schisms

Value dualisms and value hierarchies also create divisiveness and a lack of coalition building between animal and Earth liberation movements, which further serves those in power at the expense of mobilizing (for) the oppressed. Just as human supremacy relies on false value dualisms and hierarchies between humans and nature/animals, the divide between the animal and Earth liberation movements exists because of an acceptance of false value dualisms and hierarchies that separate nature from animals. But humans are animals and both human and nonhuman animals rely on the Earth for survival. There is no single human trait used to justify speciesism that cannot be found in some other animal species. For example, there is ample evidence that other animals experience culture, joy, and pain, have intelligence and friendships, and even play games (e.g. see Bekoff 2007, n.d.). As we try to manipulate and control food production we are killing off animal and plant diversity and potentially endangering our own long term access to food (GRID n.d.).

Anthropocentricism and speciesism stem from a logic of domination that justifies human supremacy, benefitting a few of the most privileged people while devaluing and oppressing everyone else. This is how oppression maintains itself—it allows various oppressed groups to see their struggles as independent and to feel that there is a limit on the amount of passion and compassion that are available. This encourages a false perception that some forms of exploitation are more important than others—a view that the animal and Earth liberation movements have largely accepted. But everyone loses in these "oppression Olympics" (Martinez 1993)—the exploited fight among each other, leaving the structures of domination intact.

## Total Liberation

When it comes to oppression, a logic of domination is "the glue that holds it all together" (Warren, quoted in Adams 1996). A logic of *total liberation* has the potential to unbind this oppression. Total liberation is: "[t]he process of understanding human, animal, and earth liberation movements in relation to one another and forming alliances around interrelated issues such as democracy and ecology, sustainability and veganism, and social justice and animal rights" (Best 2014, Intro.).

As a first step toward total liberation, we must adopt an *ethic of intersectionality* as the basis of all social justice struggles. Kimberlé Crenshaw (1989, 1991) introduced the term intersectionality to describe the way in which different forms of oppression simultaneously impact disadvantaged groups. Importantly, the effects of oppression are not additive. For example, a person is not separately and predictably disadvantaged by being poor, being a person of color, and being a woman. All of these disadvantaged positions interact and the effects of oppression multiply. Consequently, a poor white woman, a poor white man, and a poor Hispanic woman all have different experiences of poverty.

To embrace an ethic of intersectionality, the concept of intersectionality must be understood not only on an individual level, but also on a systemic level via an understanding of the matrix of domination: "Intersectional paradigms remind us that oppression cannot be reduced to one fundamental type and that oppressions work together in producing injustices. In contrast, the matrix of domination refers to how these intersecting oppressions are actually organized" (Collins 2000, p 18).

An ethic of intersectionality pays attention to both, addressing individual problems while acknowledging that these struggles stem from systemic issues. An *ethic of intersectionality* acknowledges that:

- Individuals may be affected by multiple oppressions at once.
- Experiences of oppression differ within any oppressed group.
- All social justice causes are fighting the same oppressive structures.
- Any organization that supports any part of this oppressive structure will ultimately strengthen forces of oppression and delay liberation.

One way to practice an ethic of intersectionality is through *solidarity*, rejecting the notion that "one form of oppression is central and others merely peripheral." Rather, solidarity highlights "that all resistance to domination is essential and necessary" (Colling et al. 2014, p 64).

Activists and social justice organizations must sometimes choose a single campaign or issue at which to direct their resources. It is possible to have single-issue campaigns that are pragmatically focused on one type of exploitation, rooted within a specific sociocultural context, but that still need to ideologically embrace an ethic of intersectionality, embracing the mutually reinforcing nature of oppression, thereby avoiding the exploitation of others.

Without solidarity, social movements might embrace the oppression of others in the process of promoting their own cause. Solidarity avoids a strategy that privileges some oppressions over others or embraces one form of oppression while trying to combat another. This type of solidarity prevents serving meat at a climate justice meeting or handing out single-use water bottles at a slaughterhouse protest.

More importantly, solidarity leads movements to support one another. For example, it encourages animal activists to show up to a May Day protest and to carry a sign that reads "Animal Activists in Support of Worker Rights" (rather than one that reads "Animals are the Most Exploited Workers").

*Coalition building* between social justice organizations is necessary for total liberation. It moves beyond an ideological understanding of other oppressions and seeks out others affected by oppression to work together and understand each other's histories of oppression and strategies for liberation. Coalition building between Earth and animal liberation movements is imperative if either movement is to succeed, particularly given the mutual dependence between animals and their ecosystems.

When movements are willing to work together, they are also more likely to make gains. Coalition building expands and sustains the momentum that is necessary to combat the institutionalized economic and political structures that are destroying both animals and the Earth. Instead of viewing other campaigns and movements as opponents competing for scarce resources, coalition building allows activists to recognize other social justice movements as allies. This enables advocacy groups to build a wide base of support and connect multiple minority groups, creating formidable opposition to dominating forces.

I am not suggesting that this will be easy. Intra-movement disagreement occurs regularly in the social justice sector and is important for thinking through ideas and possibilities. There will always be debates and schisms, but the goal of total liberation encourages solidarity and coalitions; it is possible to work together despite these differences. Total liberation acknowledges that as long as the structures of oppression are intact no one can be free. Total liberation requires inclusive approaches to social justice that recognize the commonalities of the oppressed rather than the differences.

Speciesism and anthropocentricism are the ideological foundations that support the exploitation of animals and the Earth. These ideologies are maintained by value hierarchies and value dualisms; constructs that support all oppressions. Ideologies that support oppression pit exploited groups and social justice advocates against one another, distracting from the structures of oppression that support an overarching matrix of domination. This system of up–down and either–or thinking allows organizations working for the liberation of animals and the Earth to remain disparate and at odds. However, embracing an ethic of intersectionality, and a commitment to coalition building and total liberation, provides an alternative vision and a more productive way forward.

We can't just free the pigs. We also have to clean up and stop the run-off from the factory farms and slaughterhouses, help the people physically suffering from the environmental degradation it causes, change the conditions that propel slaughterhouse workers to take jobs in which they are exploited and abused, end global trends of water privatization, make reparations to the communities whose access to clean water has been taken away, reduce plastic production and consumption, clean the oceans, and rehabilitate animals injured by plastic debris.

Ann Bradley was right. Unless each of these problems is resolved the problem of the pigs and the plastic water bottles will remain unsolved. Though activists are likely to work on only one of these issues at a time, they can still support other activists and their campaigns. We must all be allies in the struggle for total liberation.

Ann Bradley took a stand for animals at Farmer John slaughterhouse that the other activists did not—she demanded that the pigs be freed and also have a place to call home. Her actions changed me that day. She made me an activist for total liberation, rather than just for animal rights.

## Discussion Questions

1 What links are expressed in this essay between animal and Earth advocacy?

2 What other social justice causes are brought into play, and how are these relevant to Earth and animal activism?

3 What choices do you make in your daily life that reflect speciesism or anthropocentricism? Are you willing to change these behaviors? Why or why not?

4 In what three specific ways might the animal and Earth liberation movements take actions to help you to change such speciesist, anthropocentric behaviors?

5 Do you know anyone who is an environmentalist or animal activist who consumes animal products or bottled water? How might you approach them to encourage them to change these behaviors? What other common, damaging behaviors might we encourage others to change in similar ways?

## Essay Questions

1 How does the poem at the beginning of this section, "The Pika and the Desert Boot," demonstrate the vital connections between Earth and animal activism?

2 Write a letter to a local animal or environmental organization explaining some basic links between Earth and animal advocacy, asking them what actions they are taking or might take to reach across the imaginary divide that separates these two social justice causes.

3 Write a letter to a local animal or environmental organization explaining some basic links between various social justice causes, asking them what actions they are taking or might take in solidarity with other social justice causes.

4 Can you think of anyone in your past who has helped you to understand something vital about your life with regard to animals or the environment? Write a letter explaining how their words/actions have changed your way of thinking and thereby altered your way of living.

## Suggested Further Reading

### *Books*

Nibert D. 2002. *Animal rights/human rights*, Rowman and Littlefield Publishers, New York.
Schlosser E. 2002, *Fast food nation*, Houghton Mifflin Harcourt, New York.
Shiva V. 2002, *Water wars: Privatization, pollution, and profit*, South End Press, Cambridge.

## Blogs

*Because we must.* http://www.becausewemust.org/
*Appetite for justice* by The Food Empowerment Project. http://appetiteforjustice.blogspot.com/

## References

Adams, C.J. 1996, "Bringing peace home: A feminist philosophical perspective on the abuse of women, children, and pet animals," in K.J. Warren and D.L. Cady (eds) *Bringing Peace Home: Feminism, Violence, and Nature*, Hypatia, Villanova, PA.
Bekoff, M. n.d., "Animal emotions blog," *Psychology Today*. [online] Viewed 28 April 2014. http://www.psychologytoday.com/blog/animal-emotions
Bekoff, M. 2007, *The emotional lives of animals*, New World Library, Novato.
Best, S. 2014, "Introduction: Crisis and the crossroads of history," *Total Liberation: Revolution for the 21st Century*, [online] Viewed 28 April 2014. http://drstevebest.wordpress.com/2014/01/28/new-book-announcement/
Colling, S., S. Parson and A. Arrigoni 2014, "Until all are free: Total liberation through revolutionary decolonization, groundless solidarity, and a relationship framework," in A.J. Nocella, J. Sorenson, K. Socha and A. Matsuoka (eds) *Defining Critical Animal Studies: An Intersectional Social Justice Approach for Liberation*, Peter Lang, New York.
Collins, P.H. 2000, *Black Feminist Thought*, Routledge, New York.
Crenshaw, K. 1989, "Demarginalizing the intersection of race and sex: A black feminist critique of antidiscrimination doctrine, feminist theory, and antiracist politics," *University of Chicago Legal Forum*, pp 139–167.
Crenshaw, K. 1991, "Mapping the margins: Intersectionality, identity politics, and violence against women of color", *Stanford Law Review*, vol. 43 no. 6, pp 1241–1299.
Denuncio, P., R. Bastida, M. Dassis, G. Giardino, M. Gerpe and D. Rodriguez 2011, "Plastic ingestion in Franciscana dolphins, *Pontoporia blainvillei* (Gervais and d'Orbigny, 1844), from Argentina," *Marine Pollution Bulletin*, vol. 62, pp 1836–1841.
Edwards, B. and A. Driscoll 2009, "From farms to factories: The environmental consequences of swine industrialization in North Carolina," in K.A. Gould and T.L. Lewis (eds) *Twenty Lessons in Environmental Sociology*, Oxford University Press.
EPA 2012, "Pork Production," *Environmental Protection Agency website*. Viewed 26 April 2012. http://www.epa.gov/oecaagct/ag101/printpork.html#impact
EPA 2014, "Plastics," *Environmental Protection Agency website*. Viewed 26 April 2014 http://www.epa.gov/osw/conserve/materials/plastics.htm#facts
Fitzgerald, A.J., L. Kalof and T. Dietz, 2009, "Slaughterhouses and increased crime rates: An empirical analysis of the spillover from 'The Jungle' into the surrounding community," *Organization and Environment*, vol. 22 no. 2, pp 1–27.
Gleick, P.H. 2010, *Bottled and Sold: The Story Behind our Obsession with Bottled Water*, Island Press, Washington, DC.

Grant, R. 2009, "Drowning in plastic: The Great Pacific Garbage Patch is twice the size of France," *The Telegraph.* 24 April, [online] viewed 27 April 2014, http://www.telegraph.co.uk/earth/environment/5208645/Drowning-in-plastic-The -Great-Pacific-Garbage-Patch-is-twice-the-size-of-France.html

GRID n.d., "The environmental food crisis," *GRID Arendal website* Viewed 28 April 2014. http://www.grida.no/publications/rr/food-crisis/page/3569.aspx

HRW 2005, "Blood, sweat, and fear," *Human Rights Watch website.* Viewed 26 April 2014. http://www.hrw.org/en/reports/2005/01/24/blood-sweat-and-fear

Martínez, E. 1993, "Beyond Black/White: The racisms of our times," *Social Justice* vol. 20 no. 1/2, pp 22–34.

Nibert, D. 2002, *Animal Rights/Human Rights: Entanglements of Oppression and Liberation,* Rowman and Littlefield Publishers, Boulder

NRDC 2013, "Facts about pollution from livestock farms," *National Resources Defense Council website.* Viewed 26 April 2014. http://www.nrdc.org/water/pollution/ffarms.asp

O'Neill C. 2011, "How soda caps are killing birds," *National Public Radio website,* 1 November, viewed 28 April 2014. http://www.npr.org/blogs/pictureshow/2011/10/31/141879837/how-soda-caps-are-killing-birds

Shiva, V. 2002, *Water Wars: Privatization, Pollution, and Profit,* South End Press, Cambridge.

Sigurdarson, S.T. and J.N. Kline 2006, "School proximity to concentrated animal feeding operations and prevalence of asthma in students," *Chest,* vol. 129 no. 6, pp.1486–1491.

Tremlett, G. 2013, "Spanish sperm whale death linked to UK supermarket supplier's plastic, *The Guardian,* 8 March, viewed on 28 April 2014. http://www.theguardian.com/world/2013/mar/08/spain-sperm-whale-death-swallowed-plastic

USDA 2014, "Livestock slaughter: Summary 2013," *United States Department of Agriculture website.* Viewed 26 April 2014. http://usda.mannlib.cornell.edu/usda/current/LiveSlauSu/LiveSlauSu-04-21-2014.pdf

Wing, S., D. Cole and S. Grant 2000, "Environmental injustice in North Carolina's hog industry," *Environmental Health Perspectives* vol. 108, no. 3, pp 225–231.

# 3

# EARTHLINGS SEEKING JUSTICE

## Integrity, Consistency, and Collaboration

*Carrie P. Freeman*

Animal rights advocacy is necessary to save Earthlings—every last one of us. This perhaps provocative premise is intended to privilege the importance of this social movement at the outset—one that is often marginalized in comparison to human rights and environmentalism, yet encompasses both by default. Environmental advocacy needs to embrace animal advocacy in order to protect human beings in this age of environmental crisis, an age that demands problem-solving and sacrifice. This essay situates animal advocacy[1] as the vital bridge connecting the struggle to protect the rights of human beings with the struggle to protect all living beings.

First, this essay theorizes why animal advocacy is marginalized, and explains why the movement should be considered central to a sustainable society that maintains justice for all sentient beings. Focusing on common ground between animal advocacy, human rights, and environmental advocacy an ideological basis is proposed on which these movements can coalesce to resist the ever-increasing corporate exploitation of life. The essay ends by utilizing exploitation of farmed animals as an example of how a justice ethic—a shift away from raising and eating animals—would support the work of environmentalists, animal advocates, and other social justice advocates.

### Why is Animal Advocacy Marginalized?

In the classes I teach, I rarely need to ask students why human rights are important or why we care about other human beings. At least in education, the moral basis of respect for humanity is now taken as a given. But when one broaches the topic of environmental or animal protection, it is often necessary to establish why nature or other species matter. Certainly, there are always anthropocentric (human-centered) reasons why other species matter. Humans need clean air, clean water, and food, and hence we need wetlands, bees, earthworms, trees, etc.—we are all ecologically inter-dependent. We could just leave it at that when establishing a foundation for the importance of the natural environment to humanity, or we could assess the value of other species through an economic lens, considering how domesticated and wild/free species support industries and sustain human beings. But these self-interested perspectives don't value other animals for who they are (their inherent value)[2] as opposed to what they can do for us (their instrumental value).

Without embracing animal ethics, environmentalists can often stand within this anthropocentric (what's-in-it-for-me) model, primarily asserting environmental goals in terms of human benefits. This appeal to human self-interest seems to hold much appeal to humanity, and of course we are the producers and consumers of all campaigns. Environmental advocacy also holds a measure of respect for social justice causes focusing on human aspirations and needs (as long as they are moving toward sustainable practices). But neither environmentalists nor humanitarians seem to pay much attention to justice issues involving nonhuman animals.

In fact, it is often viewed as a liability for either camp to be associated with animal advocacy. Perhaps such a focus seems to belittle human issues (discrimination, starvation, genocide) or to fundamentally threaten humanity's sense of entitlement over the more-than-human world (especially when it comes to such things as animal agriculture and hunting, or competing with wild animals for trees, fresh water, land, and so on). While social movements are, by definition, charitable and altruistic, they must compete for dollars and other scarce resources, casting a wide net for support (McAdam et al., 1996). Thus, they are reticent to risk (over)extending their concerns to issues that may be controversial or offensive to donors/supporters (Snow et al. 1986). But why is animal advocacy *in particular* controversial and, thus, marginalized?

On a pragmatic level, animal advocacy challenges and threatens to change (and in many cases end) how we exploit other animals—essentially, our assumed right to use, own, domesticate, breed, eat, and kill other animals.[3] In particular, consider how animal advocacy threatens industries of agribusiness, fishing, captive animal entertainment, scientific experimentation, and fashion—industries where nonhuman animals are "products" from which corporations and individuals profit. Additionally, animal advocacy seeks to end the standard human practice of overlooking the interests of nonhumans, in favor of human interests, in most decision-making processes, especially with regard to habitat.

On a socio-psychological level, the idea of respecting nonhuman animals threatens our own self-perception as an exalted species that, either biologically or by divine right, is "superior" to other animals, and is thus entitled to a dominant status (Schmidtz 2002; Scully 2002; Taylor 1993). Western culture, in particular, has cultivated a notion of the human (a rational, civilized person) in opposition to the animal (an uncivilized, less developed creature within nature), even to the point of strategically referring to certain humans as "animal" to deny them privileges unjustly reserved only for "humanity" (Adams 1990; Spiegel 1996).

These hierarchical dichotomies (human/animal and even culture/nature) require deconstruction before humans can ideologically appreciate and embrace animal and earth liberation (Freeman 2010a).

## Why is Animal Advocacy Central to Creating a Just Notion of Sustainability?

While the life-preserving goals of environmental advocacy may make it evident why other social movements ought to be concerned about the environment, it is perhaps less clear why environmental advocacy needs other social movements. The mainstream environmental movement seeks to protect humans and many specific animal species. But environmentalists do not provide a sound justification for limiting privileges to certain species (such as humans, whales, elephants, etc.), while at the same time using lethal or violent methods against many nonhuman animals. Environmental advocates therefore need to include animal ethics in their platform as a matter of consistency. Once environmentalists respect all sentient subjects, their protection of the human animal and other favored species will be morally consistent.

Similarly, social justice advocates working on behalf of humans also need to embrace animal advocacy as a matter of moral consistency. Humans may face severe disadvantages if the powers that be decide to "manage" our species as we do other "invasive" or dangerously destructive species (Freeman 2010b; Taylor 1993), which would equate to what Tom Regan (2002) referred to as environmental "fascism" (107). Granted, this is an egoistic rationale, and there are more altruistic reasons for humans to extend rights or freedoms to fellow animals, such as empathy and acknowledging our kinship (Kheel 2008; Regan 1983; Singer 1990; Steiner 2008). In any event, we ought to live up to our own self-image as a compassionate and just species rather than cope with the shame, albeit perhaps subconscious shame, that comes with our (ab)use and exploitation of fellow animals and nature.

Animal advocacy also leans on the power of environmental ethics and human rights. The animal protection movement seeks to protect both free and domesticated nonhumans. Animal advocates therefore rely on environmental advocates to protect habitat and to extend moral standing to the more-than-human world, and they rely on human rights inasmuch as animal activists extend certain extant, well-established human rights to nonhuman animals.

Furthermore, each of these three social movements target similar opponents, especially corporations (and their government enablers), that exploit living beings and resources in a competitive global market. These powerful entities are the common adversaries of many non-profit organizations that work to protect the vulnerable—humans, other animals, and the environment—from corporate abuse. One way for social movements to garner more power to face these mighty foes is to join forces. As an allied force, movements on behalf of humans, other animals, and the natural world are more likely to democratically influence government and corporate practices.

Despite a traditional/historic separation between these three movements, the good news for animals (including humans) is that the ethical basis and ideals of the animal advocacy movement overlap with advocacy on behalf of the earth and humanity, lighting a common path for all living beings. Most fundamentally, they all share an inherent respect for life that precludes objectification and exploitation. Each seeks to promote fair play and responsible and respectful interactions with the diversity of life.

While each cause has its particular focus—human rights directs concern primarily at humanity, animal advocacy looks to the wellbeing of sentient individuals, and earth activists focus on certain species and ecosystems (Sagoff 1993)—if the concerns of all three movements were taken into consideration, joint action would call for humanity to:

- respect sentient individuals (members of the animal kingdom, including human beings), allowing them to live free from suffering caused by exploitation and oppression, and
- respect life-supporting ecological systems, sustaining the natural world and all living beings (including human beings).[4]

Advocacy for humans, nonhuman animals, and the environment are all part of a wider social *justice* movement. Each form of advocacy holds that humans should meet survival needs in ways that are fair, responsible, and moderate—calling us to share the planet and, in so doing, to avoid oppression and exploitation.

## Applying this Justice Ethic to the Issue of Farming and Eating Animals

Animal advocates lament the environmental movement's reluctance to fully embrace veganism. Animal activists have increasingly called for coalition-building between animal advocates and environmental advocates to fight the proliferation of factory farms, but environmentalists tend to shy away from such coalitions, fearing that they might alienate members of their organizations (some of whom hunt or are engaged in animal agriculture) (Holt, 2008). *E Magazine* editor Jim Motavalli (2002) investigated this "great divide," describing environmental goals as broader and more flexible, and animal advocacy goals as narrower and absolutist. Whereas most animal advocates call for a vegan diet, the focus of environmental activists and organizations does not generally include concern for farmed animals, thus many environmental advocates simply encourage people to reduce meat consumption (Motavalli 2002).

When it comes to promoting a vegan diet, many animal advocacy organizations take an abolitionist approach because their priority is protecting the lives of individual animals. Earth activists, on the other hand, often promote reform ("sustainable" use of animals for food), consistent with the movement's tendency to focus on reducing consumptive practices, mitigating health risks to humans, and showing a modicum of decency for the welfare of farmed animals. Environmental organizations tend to protect individual animals from being killed only when those animals are human, endangered, or charismatic (particularly mega-fauna, marine mammals, and apex predators)—none of whom are used in animal agriculture (Freeman 2010b).

Environmental advocates contradict themselves when they claim to be holistic, prioritizing the health and wellbeing of a particular whole species, while expressing concern for the suffering and death of privileged *individuals* only within certain species, such as humans, dolphins, turtles, seals, sharks, and wolves. In order to coherently assert that these particular individuals should not be objectified or exploited, environmental advocates must justify this glaring inconsistency by explaining why only individuals from specific species ought to be prioritized (in instances where survival does not dictate killing to survive). Furthermore, environmental advocates who exclude domesticated species yet respect the lives of *human* animals (who are certainly not wild or endangered) must account for this equally glaring inconsistency.

It is evident that environmentalists offer privilege to the human animal in many ways. For example, even though humans are responsible for the lion's share of environmental destruction, and even though humans are an over-populated species that ought, by definition, to be considered "invasive" and "non-native" across most of the planet, humans are (rightly) never targeted for culling or outright destruction (nor should any other animals be targeted for culling). Environmental advocates lean on human/animal and culture/nature dualisms in accepting this traditional exclusion of humanity, thereby adopting the environmentally damaging mindset that lies at the core of abuse and exploitation of the environment (Freeman 2010b).

This anthropocentric privilege was demonstrated in a 2010 study that I conducted, in which I analyzed web messages from 15 environmental organizations to see what they had to say about animal-based vs plant-based diets (Freeman 2010b). My research demonstrated that environmental organizations often acknowledge the unsustainable nature of an animal-based diet, yet yield to consumer preference for animal products. Consequently, such organizations merely encourage those consuming animal products to do so in a more environmentally sensitive manner, reducing meat consumption or simply choosing more "sustainable" options, including farmed fish, organic, or grass-fed beef—none of which are fully sustainable—without calling for more substantial dietary change. Approached apprehensively, if at all, veganism was often referenced as an extreme and unrealistic ideal. But solutions posed ought to be relevant to the actual causes of severe environmental degradation, therefore environmental organizations ought to encourage shifting to a plant-based diet (Freeman 2010b).

In doing so, environmental advocates should prioritize the following (acknowledging that these may vary regionally): plant-based, organic, non-genetically modified, local, and fair-trade. While in some regions or instances, certain animal products might be ecologically sustainable, they are not going to be morally justifiable, most notably, when such options involve the needless farming (and killing) of sentient individuals. Based on the inherent value of each farmed animal—as someone not something—earth activists should not cater to conventional dietary preferences, or show flexibility and tolerance with regard to farming, hunting, and eating other living beings. Determining *necessity* is essential: Whenever killing is unnecessary for human survival, such violence is less likely to be morally justifiable, even if deemed more sustainable than other animal agriculture options (Freeman 2010b).

Environmental activists should also question the naturalness, necessity, and fairness of animal agriculture. Most seek to protect that which is natural, but environmental discourse has not yet explained how farming other species is "natural." And while all types of agriculture affect the natural world and living beings, *plant-based* agriculture is less destructive and, in any event, is necessary to feed our vast human population (Singer and Mason 2006). Advocating for plant-based agriculture and against animal agriculture would be morally consistent with environmental advocates' anti-exploitative stance, extending this core ideal beyond protecting nature and individual humans to include individual domesticated nonhuman animals. Environmental advocates ought to take a stand, for the sake of consistency, against *domesticating anyone* (Hall 2010; Kheel 2008; Nibert 2013). Incorporating animal ethics into an environmental platform increases the logical integrity of environmental advocates' respect for nature and for human rights.

If environmental activists (especially ones who identify with deep ecology or ecofeminism and seek to deconstruct dualisms that privilege human domination over nature) are to establish and maintain ideological consistency, they must do more than merely suggest that we all try to adopt Meatless Mondays, purchase grass-fed dairy products, or switch to non-GMO farmed fish. Environmental organizations ought to strongly promote a plant-based diet as part of their core platform for an ethic of justice, asking humanity to share the earth and not to exploit other animals (Freeman 2010b).

## Conclusion

Advocacy groups for the environment, nonhuman animals, and humans are all life-affirming at their core, honoring intrinsic value in living beings and nature. By embracing *both* animal ethics and environmental ethics (such as through veganism), social justice activists demonstrate increased logical consistency, promoting fairness to all life forms (Hall 2010; Kheel 2008). The future of all Earthlings depends on humanity's willingness and ability to choose responsible coexistence—sharing our planet and respecting the diversity of life.

## Discussion Questions

1 Are animal rights logically necessary for those engaged in human rights? Why or why not?

2 Is animal advocacy a necessary component of other life-affirming social justice causes, such as organizations working for earth liberation, on behalf of those physically challenged, or those striving for equality between races or sexes? If not, why not? If so, why and in what ways?

3 What do you think are the three primary advantages of working with other social justice advocates? Do you see any equally significant drawbacks—drawbacks that outweigh advantages?

4 The slogan for People for the Ethical Treatment of Animals is, "Animals are not ours to eat, wear, experiment on, use for entertainment, or abuse in any other way." The slogan for Earth First is, "No Compromise in Defense of Mother Earth." The slogan for the Food Empowerment Project is, "Because your food choices can change the world." If you were forming a social justice campaign on behalf of all earthlings, human and nonhuman, what would your slogan be?

## Essay Questions

1 Propose three campaigns on which activists for humans, nonhuman animals, and the environment might join forces. For one of these campaigns, outline strategies and goals, and primary tasks to be accomplished by earth, animal, or human advocates respectively, according to the expertise of each.

2 What arguments would you pose to defend privileging the rights of the human animal over the rights of other animals? What arguments would you pose to defend privileging the rights of other animals over the rights of the human animal?

3 You are founding an organization on behalf of earth, animals, and humanity. Write a statement of purpose for your new organization, outlining your motivations, specific goals, and methods. Include an explanation as to why your group is intersectional—working across concerns that most people envision as distinct and separate.

## Suggested Further Reading

Hall, Lee. 2010. *On their own terms: Bringing animal-rights philosophy down to Earth*. Darien: Nectar Bat Press.

Kheel, Marti. 2008. *Nature ethics: An ecofeminist perspective*. Lanham: Rowman and Littlefield.

Mason, Jim. 1993. *An unnatural order: Why we are destroying the planet and each other*. New York: Continuum.

## Notes

1 By "animal advocacy" I mean the animal protection movement broadly, but especially what is commonly referred to as the "animal rights movement" to end the use of nonhuman animals as resources for food, entertainment, clothing, and research, which is based on a moral philosophy that the interests of all sentient beings matter.
2 I like Lee Hall's notion we should respect animals "on their own terms" (the name of her 2010 book).
3 Animal advocacy is largely about liberation—letting nonhuman animals live free of human control (Hall 2010; Kheel 2008). Partly for this reason I choose to describe "wild" animals as "free."
4 As an example of this blended ethic, Dale Jamieson (2002) contends we can respect nature as a home for all animal beings.

## References


Adams, Carol J. 1990. *The sexual politics of meat: A feminist-vegetarian critical theory*. New York: Continuum.

Freeman, Carrie P. 2010a. Embracing humanimality: Deconstructing the human/animal dichotomy. In Greg Goodale and Jason Edward Black, eds, *Arguments about animal ethics,* 11–30. Lanham: Lexington Books.

Freeman, Carrie P. 2010b. Meat's place on the campaign menu: How U.S. environmental discourse negotiates vegetarianism. *Environmental Communication* 4.3: 255–276.

Hall, Lee. 2010. *On their own terms: Bringing animal-rights philosophy down to Earth*. Darien: Nectar Bat Press.

Holt, David M. 2008. Unlikely allies against factory farms: Animal rights advocates and environmentalists. *Agriculture and Human Values* 25.2: 169–171.

Jamieson, Dale. 2002. *Morality's progress: Essays on humans, other animals, and the rest of nature*. New York: Oxford University Press.

Kheel, Marti. 2008. *Nature ethics: An ecofeminist perspective*. Lanham: Rowman and Littlefield.

McAdam, Doug, John D. McCarthy, and Mayer Zald. 1996. *Comparative perspectives on social movements: Political opportunities, mobilizing structures, and cultural framings*. Cambridge: Cambridge University Press.

Motavalli, Jim. 2002. Across the great divide. *E Magazine,* 34–39.

Nibert, David Alan. 2013. *Animal oppression and human violence: Domesecration, capitalism, and global conflict*. New York: Columbia University Press.

Regan, Tom. 1983. *The case for animal rights*. Berkeley: University of California Press.

Regan, Tom. 2002. How to worry about endangered species. In David Schmidtz and Elizabeth Willott, eds, *Environmental ethics: What really matters, what really works,* 105–108. New York: Oxford University Press.

Sagoff, Mark. 1993. Animal liberation and environmental ethics: Bad marriage, quick divorce. In Michael E. Zimmerman, ed., *Environmental philosophy: From animal rights to radical ecology*, 84–94. Englewood Cliffs: Prentice-Hall.

Schmidtz, David. 2002. Are all species equal? In David Schmidtz and Elizabeth Willott, eds, *Environmental ethics: What really matters, what really works,* 96–103. New York: Oxford University Press.

Scully, Matthew. 2002. *Dominion: The power of man, the suffering of animals, and the call to mercy*. New York: St. Martin's Press.

Singer, Peter. 1990. *Animal liberation*. 2nd ed. New York: Random House.

Singer, Peter, and Jim Mason. 2006. *The ethics of what we eat: Why our food choices matter*. New York: Rodale.

Snow, David, Burke Rochford, Steven Worden, and Robert Benford. 1986. Frame alignment processes, micromobilization, and movement participation. *American Sociological Review* 51.4: 464–481.

Spiegel, Marjorie. 1996. *The dreaded comparison: Human and animal slavery.* New York: Mirror Books.

Steiner, Gary. 2008. *Animals and the moral community: Mental life, moral status, and kinship.* New York: Columbia University Press.

Taylor, Paul. 1993. The ethics of respect for nature. In Michael E Zimmerman, ed., *Environmental philosophy: From animal rights to radical ecology*, 67–81. Englewood Cliffs: Prentice-Hall.

# 4

# CARING FOR EARTH AND HER CREATURES

*Josephine Donovan*

> When I was small, my mother often told me that animals, insects, and plants are to be treated with the kind of respect one customarily accords to high-status adults.
>
> Paula Gunn Allen

> We must respond to a "thou-ness" in all beings … not just as "I to it," but as "I to thou," to the spirit, the life energy that lies in every being in its own form of existence.
>
> Rosemary Radford Ruether

Most ethical theories about how humans should treat the earth and her creatures focus on one or the other: either on the natural world or on animals. Ecofeminism focuses on both, recognizing them as linked, inherently, and within a political web cast by patriarchal colonization. Ecofeminists seek to liberate nature and animals from this web of oppression; first, by challenging the powers who create and sustain the web, and second, by calling for a new way of relating to animals and the natural world, in which we hear their voices and pay attention to what they are telling us.

Ecofeminism emerged in the latter twentieth century as an outgrowth of second-wave feminism and in reaction against the catastrophic destruction of the natural world and all living creatures by the depredations of modernity, in particular, by industrialism and corporate capitalism. The term *ecofeminism* was coined in 1974 by French feminist Françoise d'Eaubonne, who called for an "ecological revolution" in which "the planet in the feminine gender would become green again for all" (d'Eaubonne 1981, 236). D'Eaubonne, like many French feminists and American radical cultural feminists, believed that women historically have had a separate, more humane, less violent, and more nurturing value-system than men and that an ethic derived from this value-system provides fresh ways of thinking about and treating the natural world, offering alternatives to the dominative instrumentalism of patriarchal capitalism. The ecofeminist approach would replace destructive, polluting practices with a nurturing, attentive, and caring attitude, which would enable the earth to regenerate, to "become green again."

Following d'Eaubonne, several American feminists pioneered ecofeminist theory. They recognized that a crucial factor in the destruction of nature was an arrogant assumption on the part of privileged white Western males that they had the right to control, develop, tame, dominate, and exploit—that is, to colonize—the natural world and the living creatures therein. It was, feminists realized, the same presumptuous notion that accorded these men the right to dominate and subordinate women and nonwhite racial groups. Ecofeminists thus agreed, as Karen Warren noted in 1991, "that the wrongful and interconnected domination of women and nature … must be eliminated" (Warren 1991, 1).

However, as a vibrant, effective political movement, ecofeminism began not in the Western academic world but in "Third World" regions through local, grass-roots protest movements led by indigenous women. The first and most famous of these was India's celebrated Chipko movement wherein organized groups of women blocked development projects that entailed destruction of woodlands; they did this by using Gandhian tactics of nonviolent resistance, hugging trees to protect them against loggers' saws. In a compelling ecofeminist analysis of this movement, Indian ecofeminist Vandana Shiva noted, "violation of nature, which seems intrinsic to the dominant development model, is also associated with violence to women who depend on nature for drawing sustenance for themselves, their families, their societies." The Chipko women "challenged the western concept of nature as an object of exploitation and have protected her as Prakriti, the living force that supports life." By their actions the Chipko women called into question the capitalist profit-oriented model of development, replacing it, Shiva notes, with a concept of sustainable economics that responds to peoples' needs (Shiva 1988, xvi–xvii).

The Chipko movement thus dramatized how women's lives are directly affected by development projects that destroy the environment. This is especially the case with women who depend on the surrounding environment for food, water, and wood-fuel. As Rosemary Radford Ruether notes,

> Deforestation means women walk twice as far each day to gather wood. Drought means women walk twice as far each day seeking water. Pollution means a struggle for clean water …; it means children in shantytowns dying of dehydration from unclean water.
>
> (Ruether 1996, 6)

Other examples of grass-roots ecofeminist activism include the antinuclear Greenham Common movement in England in the 1980s; the Clayoquot Sound Peace Camp on Vancouver Island, Canada; the Green Belt movement in Africa led by Kenyan Wangari Muta Maathai (who won the 2004 Nobel Peace Prize); and the anti-corporate actions of Diane Wilson, a Texan who took on Union Carbide and Dow Chemical and with Medea Benjamin helped found the women's peace action and social justice organization Code Pink in 2004 (see Godfrey 2005).

In 2009 Kenyan ecofeminist Wahu Kaara rearticulated the ecofeminist vision in her address to the peoples' assembly at the Copenhagen "Global Warming" summit. "The 21st century," she proclaimed, "belongs to ecosocialist ecofeminism." She continued:

> Women, as the custodians of life, are the first to be conquered and to be occupied. And to be exploited … [by] the patriarchal market economy [which] is a system that has continued to kill life, by conquering, occupying, and commodifying any manifestation of life for profit.

Kaara, like d'Eaubonne, calls for "'a Mother World,' … because 'a Father World' has failed."

> A Father world is guided by greed, hatred, wars. A Father world has put in place an intricate global web of military hardware to perpetuate violence in conquering and occupation for profit. In this web of capitalists, military hardware propels their agri-business, which connects with the insurance industry, which connects with the pharmaceuticals, and which connects with the banking system, which has already collapsed before our eyes. And to revive it … they have taken money from the poor people to resuscitate the banks. … If the climate crisis was the banks, they would have a solution.
>
> (Kaara 2010, 109)

Kaara thus identifies the links that undergird the global corporate economy, which Ruether terms "simply the latest stage of Western colonialist imperialism" (Ruether 2005, 1). It is also the latest manifestation of a destructive dominative and violent patriarchal value-system, what Kaara calls the "Father World." Thus a central feature of contemporary ecofeminism is a political critique of and activist protest against this "Father World" of global corporate capitalism.

At the same time many ecofeminists, myself included, are calling for a reorientation of our attitude toward the natural world and animal life, one that involves caring respect for nonhuman animals as beings whose right to life is no less valid than our own, who are living subjects capable of expressing their own feelings and wishes, which we are obliged to respect, and to whom we owe attentive, empathetic concern. Therefore, instead of viewing the natural world and the animals therein as commodities (objects to be sold and transformed into consumable products for profit—for example, trees into logs, cows into meat, foxes into fur), these ecofeminists advocate a life-affirming ethic of care and respect for all life-forms as individuals whose value lies not in their market price but in their own desire to live.

A *sûtra* from the ancient Jain religion of India notes that "*All beings are fond of life; they like pleasure and hate pain, shun destruction and like to live, they long to live. To all, life is dear*" (Chapple 1986, 217). This is also a basic tenet of an ecofeminist ethic of care (see Aristarkhova 2012).

Feminist ethic-of-care theory originated from Carol Gilligan's *In A Different Voice* (1982), wherein Gilligan identified a women's ethic as "concerned with the activity of care … responsibility and relationships," in contrast to a man's "conception of morality as fairness," more concerned with "rights and rules" (19). Gilligan called the feminine conception "a morality of responsibility," as opposed to the masculine "morality of rights," which emphasizes "separation rather than connection." She noted that while the masculine ethical concern with rights, rules, and an abstract ideal of justice seemed like "a math problem with humans" (28), the feminine approach offered a more flexible, situational, and particularized ethic.

As with feminism in general, ethic-of-care theory resists hierarchical dominative dualisms that establish the powerful (humans, males, whites) over the subordinate (animals, women, people of color). Dualistic thinking took an especially deadly turn in the early modern era when French philosopher Descartes proposed that the universe was formed of two basic components: matter and spirit (or mind). As matter was assumed to be life-less, devoid of energizing spirit (unlike in much premodern thought, which is animist), matter was held to be of lesser value than mind, spirit, or reason. In this viewpoint, which continues to inform modern scientific thought, animals are reduced to mere things, machine-like automatons lacking inner spirit, sensitivity, or feelings. As Thomas Kelch (2007, 229–49) points out, it is this view that supports the current common-law conception of animals as property (see also Schrengohst 2011).

By contrast, the feminist ethic of care sees animals as individuals with feelings, who can communicate their feelings, and to whom humans therefore have moral obligations. An ethic of care also recognizes the diversity of individual animals; each has a particular situational history—one size doesn't fit all. Insofar as possible, in our ethical decisions about animals we must pay attention to these particularities—which we learn from the animals themselves in dialogical communication with them.

Of course, there may be times when we have to override the animal's wishes—where there are compelling ethical reasons to do so—in the best interests of the animal. There are other times when we are best not to intervene, as is often the case with wild animals (see Clement 2007). In any event, we are obliged, ecofeminists maintain, to respect the needs and wishes of all life-forms insofar as we can, and to prevent human harm to them wherever possible.

Some will argue that this is an impractical ethic—especially with respect to plant forms. Trees have to be cut down to make houses; plants have to be harvested for food. Ecofeminists reply that there are ways to mitigate the damage. Houses can be recycled. McMansions are not necessary. Food can be grown organically to protect the vegetation and preserve the environment. In short, ecofeminists favor alternative models that are less destructive and more life-affirming.

Ecofeminist care theorists have differentiated their work from two other approaches to eco- and animal ethics: on the one hand, "animal rights" and utilitarian theories; on the other, "deep ecology" theory. Animal rights theory was developed largely by Tom Regan (1983) and the utilitarian approach by Peter Singer (1975). While embracing the core idea of both Regan and Singer, that nonhuman animals are subjects worthy of ethical respect, feminist care theorists have questioned some of their premises (see especially Donovan and Adams, 2007), particularly their rigorous rationalism and their untenable emphasis on similarities between humans and animals. Regan, for example, argues that because animals—especially primates—display intellectual capabilities similar to humans', they are entitled to certain basic rights. But this requirement of intellectual similarity leaves out the vast array of animals who do not reason or communicate like humans but who demonstrably experience emotions—such as fear, excitement, even in some cases, love—that are similar to humans' and to which humans can relate. Are not these creatures entitled to be free from human harm? Singer, on the other hand, argues that rights or moral consideration are owed to creatures who can suffer and feel pain. However, as an adherent of utilitarianism—a philosophy that makes ethical decisions on the basis of mathematical formulas ("the greatest good for the greatest number"), Singer necessarily subsumes the interests of the individual suffering creature to a mass calculation. If a greater number of creatures will be helped by an action than harmed, the smaller number can be harmed or killed, according to utilitarian theory. Ecofeminist ethic-of-care theory rejects this kind of calculation, insisting that individual suffering is ethically relevant and cannot be overridden by mathematical formulas.

In its emphasis on numbers over individuals, utilitarianism resembles "deep ecology" theory. That theory, which has heavily influenced many current environmentalists, ensued largely from Aldo Leopold's conception of the "land ethic" in *A Sand County Almanac* (1949). Leopold argues that nature, conceived as a "biotic community," has the right to exist apart from human interests. Ethic-of-care theorists would generally agree with this assertion; however, Leopold's theory, like utilitarianism, concludes by subordinating an individual's rights and interests (whether those of a human, a pig, a snake, or a tree) to the needs of the totality. Tom Regan criticized this view as "environmental fascism" (Regan 1983, 362), a characterization most care theorists would affirm. Ecofeminists have also criticized deep ecology theory for its sexist assumptions, its covert hierarchy (wild animals are privileged over domestic), and its ethical neglect of individual suffering (see especially Kheel 1990 and 2007). Ecofeminist care theorists, unlike many environmentalists, include domestic animals in their circle of concern (see especially Davis 1995).

Ecofeminist care theory calls for a special kind of attentiveness and empathy, an exercise of the moral imagination, in order to understand what nonhumans communicate, and so as to be able to include that information in our ethical decision making. This vein of care theory derives from Simone Weil's concept of "attentive love," which she articulated in 1942.

> The love of our neighbor in all its fullness simply means being able to say to him [or her]: "What are you going through?" It is a recognition that the sufferer exists, not only as a unit in a collection, or a specimen from the social

> category labeled "unfortunate," but as [an individual], exactly like us, who was one day stamped with a special mark by affliction. For this reason it is enough, but it is indispensable, to know how to look at him [or her] in a certain way. This way of looking is first of all attentive.
>
> (Weil 1977, 51)

"Attentive love" is a moral reorientation that requires developing one's powers of attention; it is a discipline similar to that exercised by great artists or scholars. (Weil used the idea originally in an essay on the discipline of scholarship.) This reorientation forces our attention toward others, to what *they* are experiencing. It requires us to see another creature as a subject who has needs separate from our own.

Jewish theologian Martin Buber exemplifies this kind of loving attention in his meditation on a tree. Buber writes:

> I contemplate a tree.
> I can accept it as a picture. …
> I can feel it as a movement. …
> I can assign it to a species. …
> But it can also happen, if grace and will are joined, that as I contemplate the tree I am drawn into a relation, and the tree ceases to be an It. …
> Does the tree have consciousness, similar to our own? I have no experience of that. … What I encounter is neither the soul of a tree nor a dryad, but the tree itself.
>
> (Buber 1970, 57–59)

In his theory of environmental ethics Paul Taylor calls for a similar meditative attentiveness to the particular reality of individual organisms as a basis for our relationship to nonhuman life-forms.

> As one becomes more and more familiar with the organism being observed, one acquires a sharpened awareness of the particular way it is living its life. One may become fascinated by it and even get to be involved in its good and bad fortunes. The organism comes to mean something to one as a unique, irreplaceable individual. … This progressive development from objective, detached knowledge to the recognition of individuality … to a full awareness of an organism's standpoint, is a process of heightening our consciousness of what it means to be an individual living thing.
>
> (Taylor 1986, 120–21)

Taylor further maintains that we must be:

> "open" to the full existence and nature of the organism … let the individuality of the organism come before us, undistorted by our likes and dislikes, our

hopes and fears, our interests, wants, and needs. As far as it is humanly possible to do so, we comprehend the organism as it is in itself, not as we want it to be.

(120)

The ecofeminist ethic of care similarly requires this kind of attentive listening to environmental voices who do not speak in human languages, but who nevertheless speak in languages that we can understand. It is incumbent upon us humans to learn those languages. Much ethological work is being done in the new field of interspecies communication; for example, G. A. Bradshaw's work on elephant communicative signs (2009). While much more needs to be done, a growing body of theory already endorses what Patrick Murphy calls an "ecofeminist dialogics" (1991, 50), that is, understanding the natural world through dialogue with it. Earlier, German phenomenologist Max Scheler (*The Nature of Sympathy*) spoke similarly about the necessity for learning the "*universal grammar*" of nonhuman lifeforms (1970, 11). Indeed, over a century ago American writer Sarah Orne Jewett speculated about the possibility of learning the language of nonhumans, asking,

> Who is going to be the linguist who learns the first word of an old crow's warning to his mate … ? [H]ow long we shall have to go to school when people are expected to talk to the trees, and birds, and beasts in their own language! … It is not necessary to tame [creatures] before they can be familiar and responsive, we can meet them on their own ground.
>
> (1881, 4–5)

Our relationship with animals, in short, should not be the "conquest of an alien object," Ruether notes, but "the conversation of two subjects." We must recognize "that the 'other' has a 'nature' of her own that needs to be respected and with which one must enter into conversation" (1975, 195–96).

Ecofeminist care theory thus holds that we should not hunt, destroy, or butcher living creatures; nor market them as commodities, nor experiment on them, or in any way treat them as feelingless objects, because we know they do not want to be treated this way. They run away, they hide, they cry out in pain, they shiver in fear. These are communicative signs that are unmistakable.

Paying dialogical attention to individual living creatures also requires attentiveness to the political, social, and historical context in which an individual exists. Beyond emotional awareness and attentive communication—though these are important aspects of the feminist ethic of care—one must also be aware of the chain of cultural and political connections that doom animals to death or harm. And one must act to prevent them.

Some might think, for example, that a package of ground beef wrapped in plastic and sitting on a supermarket counter is not a relevant subject for environmentalists. But if we know the chain of connections that link a living cow to a supermarket package of her flesh, we understand how bad cattle-raising is for the environment in general, causing the destruction of rain forests, pollution of water run-off, and contributing to greenhouse gases and global warming. At the same time, we also see, if we care to look, that a living creature was forced to live and die in cruel and inhumane conditions, a living creature who did not want to be so treated. This animal's misery should be of as much concern to environmentalists as it is to animal activists.

Others may say that road kill, to take another example, is not a relevant subject for environmentalists. Road kill is viewed as "collateral damage," which most people simply ignore as they speed along the highway. But road kill is part of the destruction wrought by the environmentally devastating automobile. In response to the recent fad of road-kill toys and candy, Dennis Soron (in "Road Kill: Commodity Fetishism and Structural Violence") observes,

> By transforming the animal's desecrated body into a spectacle and offering it up as a consumable thing abstracted from the violent encounter that caused its death, such commodities both bear witness to and dissolve responsibility for one of the most apparent consequences of our collective attachment to another commodity: the automobile. Automobile-oriented land use has become a primary threat to the integrity of ecosystems and animal habitat.
> (2011, 56)

An ecofeminist ethic of caring attention forces us to deconstruct the mass term "road kill" and see these dead animals as individuals who were but recently following their daily pursuits and who undoubtedly did not wish to meet their violent deaths. The ethic of care also considers the wider environmental issue raised by Soron, the destructive impact of the automobile, and advocates alternatives, such as fewer cars, fewer highways, attention to habitats and ecosystems, protective structures such as fences and crossings (tunnels or bridges), and/or warning sounds, which deflect animals from roads, to prevent these unnecessary deaths.

Another example of the importance of political context in what seems an innocuous practice—buying supermarket strawberries—is offered by Ruether in *Woman Healing Earth*: "A wealthy minority can enjoy strawberries in winter, winged to their glittering supermarkets by a global food procurement system, while those who pick and pack strawberries lack the money for bread and are dying from pesticide poisoning" (1996, 5).

Strawberries are thus linked to agribusiness with its heavy use of pesticides, cheap labor, and energy-wasteful, polluting transportation systems. At the same time, one sees in these neatly packaged strawberries, the oppressed farm worker, forced to endure unhealthy, impoverished conditions. When aware of the political context, conscientious activists will work to change or abolish the (capitalist) system that requires such destructive production practices and will buy local, organic produce whenever possible. An ecofeminist ethic of care is therefore attentive to the suffering individual, the environment, *and* to the political and economic context of everyday normalized activities and products, such as supermarket items and road kill. This ethic requires us to take personal responsibility for our own participation, if any, in these practices.

Care theory thus depends on a dialogical communication with nonhuman life-forms, and a practice of listening to voices we've been trained not to hear, and seeing realities we've been conditioned not to see. It means seeing, like Buber, the individual tree *and also* the logging industry, global development, and deforestation. Care theory means seeing the downed cow in the slaughterhouse pen *and* the destructive environmental impact of the farming and dairy industry. In this way care theory highlights links among animal exploitation, human oppression, and environmental degradation, recognizing each as inseparable from the other. An ecofeminist ethic of care provides an alternative nurturant vision that, by paying attention to their individual voices, allows living beings to flourish on their own terms. If we really listen to what animals—indeed, all life-forms—are telling us, we will end factory farms and agribusiness, capitalist exploitation and pollution of the natural world, animal experimentation, hunting, habitat destruction, and meat-eating—in short, all practices that cause animal or environmental harm, destruction, or death. Thus will the planet "become green again for all."

## Discussion Questions

1 Donovan quotes Wahu Kaara stating that women are always the first to be conquered, occupied, and exploited. Is this true, or do nonhuman animals best qualify for this unfortunate position?

2 How does ecofeminist care theory differ from other theories of animal ethics and environmental ethics?

3 How is "road kill" an ecofeminist environmental issue? Why is a cow in a slaughterhouse an ecofeminist issue?

4 Donovan notes that suffering is always of concern, and cannot be justified by "mathematical formulas." Do you agree, or do you think there are instances in which utilitarianism might be applied—instances in which the suffering of some individuals might be overlooked on behalf of the many? Is suffering a legitimate concern for environmental ethics?

## Essay Questions

1 Why is veganism/vegetarianism the preferred ecofeminist diet?

2 Offer an ecofeminist critique of industrial agriculture.

3 Describe the ways in which global warming and other environmental degradations stem from a "Father World."

4 What is the importance of an ethic of care with regard to nonhuman animals (especially wild animals), ecosystems, and the environment? What are the problems you see with this theory?

## Suggested Further Reading

Adams, Carol J. and Lori Gruen, eds 2014. *Ecofeminism: Feminist Intersections with Other Animals and the Earth*. New York: Bloomsbury.

Kheel, Marti. 2008. *Nature Ethics: An Ecofeminist Perspective*. Lanham, MD: Rowman and Littlefield.

Kowalski, Gary. 1991. *The Souls of Animals*. Walpole, NH: Stillpoint.

Luke, Brian. 2007. *Brutal: Manhood and the Exploitation of Animals*. Urbana, IL: University of Illinois Press.

Thomas, Elizabeth Marshall. 2009. *The Hidden Life of Deer*. New York: HarperCollins.

## Note

The epigraphs are from Paula Gunn Allen, *The Sacred Hoop* (1986, 1) and Rosemary Radford Ruether, *Sexism and God-Talk* (1983, 87).



## References

Allen, Paula Gunn. 1986. *The Sacred Hoop: Recovering the Feminine in American Indian Traditions*. Boston: Beacon.

Aristarkhova, Irina. 2012. "Thou Shalt Not Harm All Living Beings: Feminism, Jainism, and Animals." *Hypatia* 27 (3): 636–50.

Bradshaw, G. A. 2009. *Elephants on the Edge: What Animals Teach Us about Humanity*. New Haven: Yale University Press.

Buber, Martin. 1970. *I and Thou*. New York: Scribner.

Chapple, Christopher. 1986. "Noninjury to Animals: Jaina and Buddhist Perspectives." In Tom Regan, ed. *Animal Sacrifices: Religious Perspectives on the Use of Animals in Science*, 213–35. Philadelphia: Temple University Press.

Clement, Grace. 2007. "The Ethic of Care and the Problem of Wild Animals." In Josephine Donovan and Carol J. Adams, eds. *The Feminist Care Tradition in Animal Ethics*, 301–15. New York: Columbia University Press.

Davis, Karen. 1995. "Thinking Like a Chicken: Farm Animals and the Feminine Connection." In Carol J. Adams and Josephine Donovan, eds. *Animals and Women: Feminist Theoretical Explorations*, 192–212. Durham, NC: Duke University Press.

D'Eaubonne, Françoise. 1981. "Feminism or Death." In Elaine Marks and Isabelle de Courtivron, eds. *New French Feminisms*, 64–7. New York: Schocken.

Donovan, Josephine and Carol J. Adams, eds. 2007. *The Feminist Care Tradition in Animal Ethics*. New York: Columbia University Press.

Gilligan, Carol. 1982. *In a Different Voice: Psychological Theory and Women's Development*. Cambridge: Harvard University Press.

Godfrey, Phoebe. 2005. "Diane Wilson vs. Union Carbide: Ecofeminism and the Elitist Charge of 'Essentialism.'" *Capitalism Nature Socialism* 16 (4): 37–55.

Jewett, Sarah Orne. 1881. "River Driftwood." In *Country By-Ways*, 1–33. Boston: Houghton Mifflin.

Kaara, Wahu. 2010. "Reclaiming Peoples' Power in Copenhagen 2009: A Victory for Ecosocialist Ecofeminism." *Capitalism Nature Socialism* 21 (2): 108–11.
Kelch, Thomas G. 2007. "Toward a Non-Property Status for Animals." In Josephine Donovan and Carol J. Adams, eds. *The Feminist Care Tradition in Animal Ethics*, 229–49. New York: Columbia University Press.
Kheel, Marti. 1990. "Ecofeminism and Deep Ecology: Reflections on Identity and Difference." In Irene Diamond and Gloria Feman Orenstein, eds. *Reweaving the World: The Emergence of Ecofeminism*, 128–37. San Francisco: Sierra Club.
Kheel, Marti. 2007. "The Liberation of Nature: A Circular Affair." In Josephine Donovan and Carol J. Adams, eds. *The Feminist Care Tradition in Animal Ethics*, 39–57. New York: Columbia University Press
Leopold, Aldo. 1949. *A Sand County Almanac*. Oxford: Oxford University Press.
Murphy, Patrick. 1991. "Prolegomenon for an Ecofeminist Dialogics." In Dale M. Bauer and Susan Jaret McKinstry, eds. *Feminism, Bakhtin and the Dialogic*, 39–56. Albany: State University of New York Press.
Regan, Tom. 1983. *The Case for Animal Rights*. Berkeley: University of California Press.
Ruether, Rosemary Radford. 1975. *New Woman/New Earth: Sexism and Human Liberation*. New York: Seabury.
Ruether, Rosemary Radford. 1983. *Sexism and God-Talk: Toward a Feminist Theology*. Boston: Beacon.
Ruether, Rosemary Radford, ed. 1996. *Woman Healing Earth: Third World Women on Ecology, Feminism, and Religion*. Maryknoll, NY: Orbis.
Ruether, Rosemary Radford. 2005. *Integrating Ecofeminism, Globalization, and World Religions*. Lanham, MD: Rowman and Littlefield.
Scheler, Max. 1970. *The Nature of Sympathy*. Trans. Peter Heath. Hamden, CT: Archon.
Schrengohst, Karina L. 2011. "Animal Law–Cultivating Compassionate Law." *Western New England Law Review* 33 (3): 854–900.
Shiva, Vandana. 1988. *Staying Alive: Women, Ecology and Development*. London: Zed.
Singer, Peter. 1975. *Animal Liberation*. New York: Avon.
Soron, Dennis. 2011. "Road Kill: Commodity Fetishism and Structural Violence." In John Sanbonmatsu, ed. *Critical Theory and Animal Liberation*, 55–69. Lanham, MD: Rowman and Littlefield.
Taylor, Paul W. 1986. *Respect for Nature: A Theory of Environmental Ethics*. Princeton, NJ: Princeton University Press.
Warren, Karen J. 1991. Introduction, "Ecological Feminism." *Hypatia* 6 (1): 1–2.
Weil, Simone. 1977. "Reflections on the Right Use of School Studies with a View to the Love of God." In George E. Panichas, ed. *The Simone Weil Reader*, 44–52. New York: McKay.

# 5

# ACTIVISM AND ASIAN WISDOM

## Oneness, Interdependence, and Harmony

*Lisa Kemmerer*

The Earth's crust is composed of bedrock that is generally hidden under a blanket of snow and ice, soils and vegetation, or water. Hidden by these more malleable, transient layers, bedrock shapes the surface of the earth.

Similarly, human understandings are shaped by bedrock views, provided by parents, schools, churches, and the larger community, that shape who we are, how we live, and how we understand ourselves, our community, and the world around us. For example, we are likely to view humanity as basically good or fundamentally evil, to understand other living beings as part of our extended community or completely unrelated to us, and to either see the nuclear family as the center of every community or the community as the center of every nuclear family. Despite their importance, we tend to be unaware of these bedrock views.

One of the most critical and decisive determinants of bedrock views is a community's religion. Examining religious teachings and beliefs can provide a window through which to explore a diversity of worldviews—alternative ways of understanding life, our community, and our place in the natural world.

### Diverse Worldviews

Environmentalists and animal advocates have often noted that European peoples and the European diaspora (EPED) tend to have an exploitative relationship toward the larger world—we dominate and destroy other creatures and the natural world with a sense of self-righteous entitlement. Indeed, the EPED worldview places humanity apart from and above the natural world and other living beings. For example, EPED only use "animals" to refer to other species, demonstrating that we do not even recognize ourselves for what we are—animals, mammals, primates. Scientists use the term "lower animals" to refer to nonhuman species, as if humans somehow levitate above alligators and porcupines—as if there really were a food chain or a tree of life in which human beings alone are not hunted and consumed by other animals, and as if the arrival of humanity marked the end of evolution. Similarly, EPED do not tend to view humanity as part of nature; we do not recognize ourselves as part of a biotic community or part of local ecosystems. Perhaps most telling, we usually refer to the wonders of nature as "natural resources," as if waterways and soils and forests were designed to be pantries, purifiers, and sewage systems—as if everything on this green and glorious earth were here *for* us. While misrepresentations of our place in the world may appear self-serving, they are self-destructive.

In contrast, Asian philosophies and religions, which stretch four thousand miles from the steamy southern tip of Sri Lanka to the cold north of China, include a diversity of worldviews rooted in profound visions of deep and abiding unity. Perhaps these alternative worldviews have something to offer us in a world of climate change, mass extinctions, depletion of freshwater, and rampant soil degradation.

## Hindu Oneness

> Thou art the dark-blue bee, thou art the green parrot with red eyes, thou art the thunder-cloud, the seasons, the seas.
>
> (Fourth Adhyana trans. 1884)

The Hindu tradition has long upheld a philosophical vision of oneness rooted in Brahman. Brahman is perhaps best understood as "Ultimate Power," "Great Mystery," or "the One that lies behind all" (Zaehner 1962). Brahman, dwelling in all that exists, the life in all that lives, is a force of unity and oneness (Embree 1988; Nelson 2000, p. 67):

> This Great Being: … dwells in the heart of all creatures as their innermost Self … His hands and feet are everywhere; his eyes and mouths are everywhere. His ears are everywhere. He pervades everything in the universe.
>
> (*Upanishads* trans. 1948)

As the core of every living being, Brahman ensures that "[a]ll reality is ultimately one" (Brodd 2009). In the sacred *Upanishads* (c. 500 BCE), a Hindu teaches his son that a pinch of salt placed in a cup of water cannot be seen or touched, yet pervades every drop of water within the cup. Salt, he notes, is like the subtle essence that runs throughout the universe: That subtle essence comes from Brahman, cannot be perceived or touched, yet pervades the soil, palm trees, human beings, leghorn chickens, elephants, and India's endangered broad-nosed gentle lemurs (*Chandogya* 1962, pp. 104–105). As all rivers flow down to join one great sea, as all drops rise again into the atmosphere and fall once more across the Earth's landscapes, so do all beings take form and yet again dissolve. Each being is ultimately part of a much larger whole (*Chandogya* 1962, p. 102). We readily perceive the singular—a crumbling mushroom, a windblown spruce, the independent creature that we feel ourselves to be. But when we understand the essence beneath the surface—Brahman—we understand that each apparently separate entity is part of something much greater.

All that exists holds the shared essence of Brahman—every being as well as every aspect of the natural world: "By one clod of clay all that is made of clay is known" (*Chandogya* 1962, p. 92). The fundamental unity of a universe undergirded by Brahman allows us to find and know the sacred in a yellow wagtail or mugger crocodile, in a thistle's spiny leaf or a drop of water. The divine voice teaches, "I am not lost to one who sees me in all things and sees all things in me" (trans. Buck 1973).

This bedrock Hindu understanding of Brahman teaches that "all of nature, all of the universe, is sacred" (Kinsley 1995): "The ground is sacred. The rivers are sacred. The sky is sacred. The sun is sacred" (Subramuniyaswami 1993). This Hindu vision of oneness and a sacred universe cultivates humility and respect—Hindus who understand the nature of Brahman will see "God" in all things (Subramuniyaswami 1993). Much-loved Krishna teaches: When a man

> sees me in all and he sees all in me, then I never leave him and he never leaves me. He who in this oneness of love, loves me in whatever he sees, wherever this man may live, in truth this man lives in me. And he is the greatest [devotee], he whose vision is ever one.
>
> (*Bhagavad Gita* trans. 1965)

Oneness undergirds Hindu morality, ideally guiding human interactions with the larger world. Because everything that exists is one in and through Brahman, the natural world is inextricably part of Hindu moral responsibility—those who love Brahman must have "love for all creation" (*Bhagavad Gita* 1965)—men and women, hispid hare and hoolock gibbon, little muddy puddles and vast leafy forests: Hindus are called to respect the entirety of the natural world. An advanced Hindu practitioner "treats a cow, an elephant, a dog, and an outcaste" with the same high regard because *God is all*, and those who are spiritually advanced, those who are *true* devotees of the divine, see "in all creation the presence of God" (Dwivedi 2000, p. 5).

## Buddhism—Co-dependence and Radical Interdependence

> If one strand is disturbed, the whole web is shaken.
>
> (Cook 1977)

Buddhism is fundamentally indebted to Hinduism. Buddhists inherited the Hindu idea of oneness from which a philosophical worldview of radical interdependence emerged as Buddhism grew, evolved, and mixed with Chinese philosophy.

Buddhist philosophy holds that we live in a world of radical inter-identification—nothing and no one is independent. Contemporary Zen monk Thich Nhat Hanh teaches that everything exists in and through every other aspect of the universe: One who looks deeply into a tangerine sees the grocer who stored and sold the fruit, the truck drivers and trucks that transported the fruit to the grocer, the pickers and fields where the fruit ripened and was collected, the sun and raincloud and soil that were essential to the fruit's growth and maturity, and the tiny seed and sapling that grew into a tree that bore tangerines (Hanh 1992). Seed and sapling and tree, soil and sunlight and clouds, harvesters and trucks and a tangerine—all are inter-dependent. Sustenance comes to our hands via the interconnected fingers of the universe.

Buddhist interdependence does not teach that we are all "in this together," but rather that we *are* all this, "rising and falling as one living body" (Cook 1977). Hanh asserts that the tangerine and the cloud "inter-are" (Hanh 1992). Through co-dependence, each entity is identified with all other entities. We do not—and cannot—exist independently: "We have to inter-be with every other thing" (Hanh 1992). If we understand correctly, individuality is a ruse—interbeing is Truth. If we see correctly, we might say:

> I am one with the wonderful pattern of life which radiates out in all directions. … I am the frog swimming in the pond and I am also the snake who needs the body of the frog to nourish [his or her] own body. … I am the forest which is being cut down. I am the rivers and air which are being polluted.
>
> (Allendorf and Byers 1998)

Everything on the planet co-exists; all things are inter-identified. We *are* the toad, we *are* the pond—we and they are one (Hanh 1992). Being is always interbeing; to be is to inter-be.

Buddhist co-dependence and interbeing foster a sense of oneness in which there is no independent action, word, thought, or individual—all things influence and are influenced by all other things. Each being and each act affects every other being and every other act. Buddhist oneness holds that the

> the entire cosmos is a cooperative. The sun, the moon and the stars live together as a cooperative. The same is true for humans and animals, trees and the Earth … [T]he world is a mutual, interdependent, cooperative enterprise.
>
> (Swearer 2004)

This vision of a cooperative, interdependent universe informs Buddhist morality: "Not only do we have to respect the lives of human beings, but we have to respect the lives of animals, vegetables, and minerals" (Hanh 1992). A bedrock worldview of radical interdependence encourages people to think carefully about how they live—about what they buy and throw away, how and where they travel, and how they interact with dogs and mice and chickens and streams and forests and grasses. "'How should I deal with Nature?' We should deal with nature the way we deal with ourselves … ! Harming nature is harming ourselves, and vice versa" (Hanh 1985, p. 41).

Traveling north, Buddhism combined with Chinese philosophy to form an intensely ecologically sensitive branch of Buddhism, Hua-yen, which accentuates

Buddhist oneness and interbeing. The Hua-yen bedrock worldview is perhaps most easily grasped by imagining an infinitely regressing mirror that encompasses the entire universe in "simultaneous *mutual identity* and *mutual intercausality*" (Cook 1977). Because mirrors reflect all, nothing is separate or independent in this "vast web of interdependencies" (Cook 1977).

In a worldview in which each individual is affected by *any* and *every* change, even the slightest change in a far-away place, we cannot live *as we now live* if a fat tick falls from the tail of a coyote on the prairies of Montana. The tick most likely feels that his or her fall from the coyote—a vital source of sustenance—is life-threatening and monumental. In contrast, the tick would likely find tsunamis in Asia to be of absolutely no consequence. Our conception that the tick's tribulations are irrelevant, and the tick's perception that our troubles are irrelevant, both lack grounding in truth. The whole is affected even by what seems a slight and irrelevant event; the ripple effect from each happening is unending and all-encompassing because all things are interdependent.

Buddhist co-dependence and radical interdependence form a bedrock worldview in which humans are part of a single, interrelated process of existence; we inter-are. All things are deeply, inscrutably interconnected. Actions neither begin nor end at the tips of our fingers.

## Daoist Philosophy: Unity, Transformation, Harmony

> The universe and I exist together, and all things and I are one.
>
> (Zhuangzi, in Chan 1963, 186)

Daoist philosophy teaches that every part "of the entire cosmos belongs to one organic whole" interacting as "one self-generating life process" (Tu 1985): "Although the myriad things are many, their order is one" (trans. Chan 1963).

Dao is infinite, eternally changeless, nonbeing (Wu 1991). Dao, "the final source and ground of the universe" runs throughout all that exists, "both the transcendent and the immanent" (Xiaogan 2001, pp. 322–323). Each creature shares Dao; each is shaped and moved by Dao. Dao "abides in all things"—in the ant, in the weeds, in "excrement and urine" (trans. Chan 1963; Jochim 1986). In Daoist philosophy, "a little stalk or a great pillar, … things ribald and shady or things grotesque and strange, the Way [Dao] makes them all into one" (trans. Watson, n.d.).

Transformation is central to Daoism. Individuals are neither isolated nor enduring—everything that exists is in the process of transformation in which the "chain of being is never broken" (Thompson 1996; Tu 1989, p. 70). This eternal process of transformation binds all things—agamid lizard, Chinese elm, euploea, strawberry, human, red-headed vulture, and India's newly discovered chikilidae (legless salamandar)—into a Great Unity of Being (Parkes 1989, p. 91). Transformation binds "all things into one, equalizing all things" (trans. Chan 1963). Pine needles, basalt, noses, and feet become bits and pieces of everything else: "Now a dragon, now a snake, / You transform together with the times, / And never consent to be one thing alone" (Parkes 1989, p. 92). No particular species or individual is favored in this impersonal process of transformation; human beings are of no greater importance than a pebble, mosquito, leaf, or Endangered Bengal tiger (Tu 1989, pp. 71–73; Thompson 1996). In the Daoist worldview, humans are just "one of the myriad kinds of beings" (Wu 1991). Each bar-tailed tree creeper, thistle, and black spined toad shares equally in the Great Unity of Being (Tu 1989, p. 71).

In Daoist philosophy, every link in the web of life is critical to every other link (Ames 1989, p. 120). We form one unity with long-tailed shrikes and crested ibises (one of the most endangered birds in the world), and skinks and Chinese paddlefish (though they have recently gone extinct). The constant flux of this unified universe ties a Manchurian alder to a Chinese sturgeon and a Yoshino cherry tree to a chunk of basalt, all of which are, in turn, tied to everything else.

Harmony is also central to Daoist philosophy. The Daoist universe is ordered—harmonious—so that "alternating forces and phases" shape "rhythms of life" (Kleeman 2001, p. 67). Daoism portrays discord as shallow, like choppy waves that skim across the deep, still ocean on which they travel. Rushing waves break onto the shore and retreat back into the ocean, dispersing into the abiding unity of the sea. A yin-yang symbol portrays the Chinese sense of deeper unity. Though the yin-yang symbol is composed of contrasting black and white, the line that figuratively separates the two colors is curved. There is also a touch of black amid the white, a touch of white amid the black, symbolizing a lack of clear separation. Yin and yang may sometimes be referred to as opposites, but like right and left legs, they belong to one body, they are far more alike than different, and both are an essential part of the larger whole.

Harmony—union, integration, and synthesis—pervades the cosmos. Though we sometimes imagine ourselves to be a wild fleck of foam on the tip of a careening wave, at the end of the day we are all part of the ocean. Daoism encourages both harmony and *bugan wei tianxia xian,* "not daring to be at the forefront of the world." We tend to get caught up in our personal lives, bulldozing along like a herd of rooting hogs, but like the mountain bamboo-partridge and the sweet smelling desert rose, we *only* exist as part of a larger whole. Disharmony only arises from the error of forgetting (or neglecting to notice) the deep unity behind the ruse of individuality. Daoism discourages a shortsighted, self-centered approach to life. It is our duty to live "for the fulfillment of the health and harmony of all living things" (Kirkland 2001, p. 296). Daoism urges us, ever mindful of our place in the Great Unity of Being, to choose a life of harmony, a life in which we blend in seamlessly as part of a larger whole.

## Activism and an Interconnected Universe

Despite an EPED bedrock worldview of hierarchy and dualism, EPED activism is *rooted* in an understanding that small changes lead to much larger changes, that one dedicated voice brings new voices, and that what we buy at the store matters to the larger universe. Environmentalists, for example, understand that neither a species nor a forest can be removed from the planet without far-reaching effects; animal advocates understand that the suffering of any one individual matters to the larger scheme of things. Nonetheless, the EPED activist's sense of oneness and interdependence seems to end here. There is a misguided tendency among EPED activists to believe that "their" cause is *the most important cause*, and many (if not most) activists have worked out a rationale as to why their cause is preeminent. Most activists speak and act as if their social justice cause were independent from and more important than all other causes—or equally problematic, in competition with other social justice causes. This hierarchical, competitive, elitist vision of activism, rooted in distinction and separation, is profoundly unhelpful—and misguided inasmuch as the Eastern worldview is correct—inasmuch as we live in an interconnected universe.

Exploring Eastern philosophy and religions can help to mitigate the EPED tendency to view social justice causes as separate and distinct. Eastern understandings can help us to see that whatever happens to an individual being (farmed animals, for example) affects the natural environment (oceans, wilderness, species, waterways, rainforests, and so on), and whatever happens to the environment effects individual animals (whether eagles, snail darters, polar bears, or humans). Asian philosophies teach that whether we advocate for earth and animals or against corporate capitalism and patriarchy is of little importance *so long as we engage in activism.*

## Conclusion

Bedrock worldviews shape our understanding of life, our community, and our place in the world. Ultimately, earth and animal activism are "based on moral values, not scientific ones, on beauty, ethics, and religion, without which it cannot sustain itself" (Schaller 2011, p. 91).

Eastern philosophies and religions foster a bedrock vision of aesthetics and ethics rooted in oneness, interdependence, transformation, Unity of Being, harmony, and blending in. EPED tend to view humanity as separate and distinct from chickens and lobsters and trees, as above and rightly in control of forests and waterways and flocks of seagulls. This worldview, which appears self-serving, is ultimately self-destructive, and denies the reality of an interconnected universe, an idea that is central to Eastern worldviews.

Bedrock worldviews from Asia can help EPED citizens to re-envision the place of humanity in a universe in which we are wholly intertwined with and dependent on all that exists. When EPED activists grasp the interconnected nature of all things, they will perhaps simultaneously recognize that protecting forests and chickens, children and manatees, gays and indigenous peoples, are all necessary aspects of healing one wounded universe. This vision can reshape our relations with the natural world and connect social justice activists, offering strength in numbers and the power of unity.

## Discussion Questions

1 What do you think are the most important concepts from Asian philosophy and religions with regard to the environment? Why are these important? What about with regard to nonhuman animals? Are they different? If so, why might this be? If not, what is it about these topics that pulls them under a single canopy?

2 What does the author mean when she says that Buddhists believe that their actions neither begin nor end at the tips of our fingers? In what concrete ways do the works of the first authors in this book exemplify and emphasize this point?

3 How are the Eastern concept of *ahimsa* and the Judeo-Christian concept of love the same and different? Which seems preferable as a moral mandate, and why?

4 In what ways does our collective religious background affect your community's attitudes toward and relationship with animals and the environment?

## Essay Questions

1 Compare and contrast two specific religio-philosophical teachings from Asia with two specific Western religio-philosophical ideas.

2 What do you think are the two most important concepts from Asian philosophy and religions with regard to the environment? If you held these views, in what specific ways would they change your behaviors?

3 Offer a few concrete examples of specific religion-based teachings you received early in life (even secular ideas are often rooted in older religious ideals, like dominion), and explain how each is reflected in your daily life. In what ways does your religious background affect your attitude toward and relationship with animals and the environment?

4 List five environmental values that you hold dear. Of these, which is the most important? What concrete aspects of your life might be adjusted to better realize this value in your daily life?

## Suggested Further Reading

Gottlieb, Roger S. (ed.), 2003, *This Sacred Earth: Religion, Nature and Environment*. Routledge, New York and London.

Hanh, Thich Nhat, 1992. *Peace is Every Step: The Path of Mindfulness in Everyday Life*. Bantam, New York.

Kemmerer, Lisa 2012. *Animals and World Religions*. Oxford University Press, Oxford.

Sponsel, Leslie E. 2012, *Spiritual Ecology: A Quiet Revolution*. Praeger, Santa Barbara.

## References

Allendorf, F.S. and Byers, B. 1998, 'Salmon in the Net of Indra: A Buddhist View of Nature and Communities,' in *Worldviews: Environment, Culture, Religion* 2, pp. 37–52.

Ames, R.T. 1989, 'Putting the *Te* Back in Taoism,' in *Nature in Asian Traditions of Thought: Essays in Environmental Philosophy*. eds J.B. Callicott and R.T. Ames, University of New York Press, Albany, pp.113–44.

*Bhagavad Gita* 1965, trans. J. Mascaro, Penguin , Baltimore.

Brodd, J. 2009, *World Religions: A Voyage of Discovery*, St. Mary's Press, Winona MN.

Buck W. (trans) 1973, *Mahabharata*, University of California, Berkeley.

Chan W. (ed. and trans.) 1963, *A Source Book in Chinese Philosophy*, Princeton University Press, Princeton.

*Chandogya Upanishad*, trans. FM Muller 1962, in *The Upanishads* Part I. Dover, New York.

Cook, F.H. 1977, *Hua-yen Buddhism*, The Pennsylvania State University Press, University Park.

Dwivedi, O.P. 2000, 'Dharmic Ecology,' *Hinduism and Ecology: The Intersection of Earth, Sky, and Water*, CK Chapple and ME Tucker (eds), Harvard University Press, Cambridge, pp. 3–22.

Embree, A.T. (ed.) 1988, *Sources of Indian Tradition: From the Beginning to 1800*, Columbia University Press, New York.

*Fourth Adhyana*, trans. M Muller 1884. Available from: http://www.sacred-texts.com/hin/sbe15/sbe15103.htm [Accessed 22 August 2010].

Hanh, T.H. 1985, 'The Individual, Society, and Nature' in *The Path of Compassion: Writings on Socially Engaged Buddhism*, ed. F Eppsteiner. Parallax, Berkeley, pp. 40–46.

Hank, T.H. 1992, *Peace is Every Step: The Path of Mindfulness in Everyday Life*, Bantam, New York.

Jochim, C. 1986, *Chinese Religions*, Prentice-Hall, Englewood Cliffs.

Kinsley, D. 1995, *Ecology and Religion: Ecological Spirituality in Cross-Cultural Perspective*, Prentice-Hall, Englewood Cliffs.

Kirkland, R. 2001, 'Responsible Non-Action' in a Natural World: Perspectives from the Neiye, Zhuangzi, and Daode Jing' in *Daoism and Ecology: Ways within a Cosmic Landscape*, N.J. Girardot, J. Miller and L. Xiaogan (eds), Harvard University Press, Cambridge, pp. 283–304.

Kleeman, T.F. 2001, 'Daoism and the Quest for Order' in *Daoism and Ecology: Ways within a Cosmic Landscape*. N.J. Girardot, J. Miller and L. Xiaogan (eds), Harvard University Press, Cambridge, pp. 61–70.

Nelson, L.E., 2000, 'Reading the Bhagavadgita from an Ecological Perspective' in *Hinduism and Ecology: The Intersection of Earth, Sky, and Water*, C.K. Chapple and M.E. Tucker (eds), Harvard University Press, Cambridge, pp. 127–64.

Parkes, G. 1989, 'Human/Nature in Nietzsche and Taoism' in *Nature in Asian Traditions of Thought: Essays in Environmental Philosophy*, J.B. Callicott and R.T. Ames (eds), University of New York Press, Albany, pp. 79–97

Schaller, G.B. 2011, 'Politics is Killing the Big Cats,' *National Geographic*, December, pp. 89–91.

Subramuniyaswami, S.S. 1993, *Dancing with Siva: Hinduism's Contemporary Catechism*, Himalayan Academy, Concord.

Swearer D.K. 2004, *Becoming the Buddha: The Ritual of Image Consecration in Thailand*, Princeton University Press, Princeton.

Thompson, L.G. 1996, *Chinese Religion: An Introduction*, Wadsworth, Belmont.

Tu, W 1985, *Confucian Thought: Selfhood as Creative Transformation*, State University of New York, Albany.

Tu, W. 1989, 'The Continuity of Being: Chinese Visions of Nature' in *Nature in Asian Traditions of Thought: Essays in Environmental Philosophy*, J.B. Callicott and R.T. Ames (eds), University of New York, Albany, pp. 67–78.

*The Upanishads: Breath of the Eternal* 1948, trans. S. Prabhavananda and F. Manchester, Mentor, New York.

Wu, Y. 1991, *The Taoist Tradition in Chinese Thought*, Ethnographics, Los Angeles.

Xiaogan, L. 2001, 'Non-Action and the Environment Today: A Conceptual and Applied Study of Laozi's Philosophy' in *Daoism and Ecology: Ways within a Cosmic Landscape* Harvard University Press, Cambridge, pp. 315–40.

Watson, B. n.d., *The Complete Works of Chuang Tzu*. Available from: http://www.terebess.hu/english/chuangtzu.html. [Accessed 12 January 2009].

Zaehner R.C. 1962, *Hinduism*, Oxford University Press, Oxford.

# SECTION II

# Common Ground

## Wildlife and Wilderness

> Unless they have the German wall and a moat with a bunch of crocodiles and piranhas, they're not going to contain those woolly tanks.
>
> Montana State Senator John Brenden

**Poem**: Cara Chamberlain

### Bison Sculptures

(Le Tuc d'Audubert Cave c. 15,000 years bp)

Massive, rutting, bearded: two bison—
male after female—lean upright
on supporting rock, sculpted, etched.
Horn curved, back hard, he smells her desire.
She waits, no less impatient. They have been
waiting a long time. Oh happy, happy love!

Of the little that remains is a valley
whose earth sinks, moist, when I leave the road
to find elk thistle and a single boreal frog,
when I forget disdain that flows
impatient over these monsters now—
big and ugly, spreading across
the pastureland, halting traffic.
No bower, no garland for these.

They low, bathe in dust, growl, snort,
and plunge through shallow fords, following
one after another into twilight aspens.
A calf scampers sideways. Their fur
makes a soft wool I'll find in tufts,
sienna and umber, and rich as crumbling willows

past lavender sugar bowls and mauve
checkered lilies, a path to follow
until stopped, tails broken, bodies cracked,
calf tucked away in some museum.
The world has fallen too dark for that raw
grunting, grappling, dying, loving.

# 6

# HUNTING DELUSIONS

*Lisa Kemmerer*

When faced with the truths of buying flesh in the grocery store, some people argue that hunting must be preferable—that hunting is the environmentally friendly way to put "meat" on the table. This chapter explores this assertion.

## Wildlife Services—Ecosystem Manipulation on Behalf of Hunters/Ranchers

In 1913 the U.S. government passed the "Animal Damage Control Act," authorizing the U.S. Secretary of Agriculture to suppress, eradicate, and control any nonhuman animal deemed injurious to human interests, namely the interests of ranchers and hunters (Fox 2009). U.S. Animal Damage Control has since been given a less descriptive title—U.S. Department of Agriculture's Wildlife Services Program. Despite the new name, Wildlife Services continues with the original agenda—killing native wildlife on behalf of ranching and hunting interests.

Wildlife Services kills with/by "poisons, steel-jaw leghold traps, strangulation neck snares, denning (the killing of coyote pups in their dens), hounding, shooting, and aerial gunning," taking out coyotes, foxes, bobcats, otters, wolves, black bears, and mountain lions (Fox 2011). Wildlife Services kills "more than 2.4 million animals each year, including more than 120,000 native carnivores at an annual cost to taxpayers of over $115 million" (Fox 2009, 2011). Such "ecologically short-sighted conduct" has inevitably altered ecosystems, and extirpated species (Robertson 2012).

Because the birthrates of many species vary *directly* in response to external pressures, "culling" is ineffective, guaranteeing greater resurgence:

> [W]hen populations are low with respect to the maximum number of individuals an environment can support. … birth rates (the number of live births per

> female per year) have a tendency to be high. When a population is at or near the maximum number the environment can support, birth rates are low and death rates (the number of animals in the population dying per year) are high.
> (Yarrow 2009)

Killing deer or coyotes ultimately causes each of these species to bear more young than would otherwise be the case. In contrast, immunocontraceptive techniques reduce a variety of populations by 72–86 percent (AWI Quarterly 2012).

It is important to remember that wildlife conservation began *because of* and *for* hunters. Many hunters and U.S. federal and state wildlife agency employees therefore assert that conservation exists—and wildlife remains—thanks to hunters. While there was some truth to this when wildlife management first began, it is no less true that wildlife management was initially necessary *because* of hunters. Without hunters, the U.S. would not have needed conservation in Roosevelt's day. Conservation was initiated to protect hunted animals from hunters—to conserve wildlife for the pleasures of future hunts. In the absence of hunting, conservation would not have been necessary in the early twentieth century, though preservation (preserving wilderness for its own sake, as compared with conservation for human use) would eventually have been necessary due to human population growth and expansion. Preservation under these conditions would have been much more likely to be designed to preserve ecosystems and habitat.

## Funding: Taxpayer Burdens, Hunter Benefits

A brief history of the Pittman–Robertson Act debunks the notion that hunters and hunting organizations fund contemporary government conservation. Legal intervention was deemed essential in 1937 to save hunter target species; Congress passed the "Federal Aid in Wildlife Restoration Act," commonly called the "Pittman–Robertson Act" (P–R), adding an 11 percent excise tax on rifle, shotgun, and ammunition sales, earmarked for use by state wildlife agencies to manipulate wildlife to increase hunter target species. States were required to match "grant funds with at least one dollar for every three federal dollars received" ('Pittman–Robertson' 2013, p. 3). Consequently, states that sold more guns received more Pittman–Robertson monies, and had more P–R funds available for wildlife manipulation. This legislation effectively tied the interests of state government wildlife agencies to the interests of state firearms industries.

As hunting became more and more of a pastime (as opposed to a source of necessary sustenance), hunter revenues declined. In 1970, the Pittman–Robertson Act was amended, adding "provisions for the deposit of the 10 percent tax on pistols and revolvers" ('Digest' n.d.). This amendment further tied government wildlife agencies to the firearms industry, and also radically altered *who* pays the P–R tax.

Today less than 5 percent of U.S. citizens hunt, but there are approximately 90 guns for every 100 citizens (Moss 2004; 'Hunters' n.d.). The most common reason for keeping a gun in the U.S. is now personal protection—usually a pistol or revolver. The second most common reason for owning a gun is target practice (Carroll 2005). Millions of women in New York, Boston, Los Angeles, Chicago, Detroit, San Francisco, and Philadelphia—who have never hunted and never will hunt—purchased handguns and paid the P–R tax. Some of these gun owners (like my sister) vehemently detest hunting, but nonetheless supported the P–R tax with their handgun purchase. Given that less than 5 percent of U.S. citizens hunt, and given that FWS P–R tax now draws largely from those who carry handguns for personal protection, government wildlife agencies *misappropriate* P–R funds by using these monies to manipulate wildlife and ecosystems on behalf of hunters.

Luckily, an alternative source of funding is available. Although hunters and hunting declined steadily over the last fifty years, wildlife watchers grew by one million *each year* between 1996 and 2001 (Robertson 2012). "The ratio of non-hunting outdoor enthusiasts to hunters grew more than 26 percent in the last ten years" so that there are now six times as many non-hunting wildlife enthusiasts as there are hunters (Robertson 2012). U.S. wildlife watchers currently contribute $46 billion to local, regional, and national economies. Non-hunters buy everything from "bird food to binoculars, from special footwear to camera equipment," spending large sums "to enjoy wildlife" ('Federal' n.d.). But if an excise tax on tents, cameras, hiking boots, climbing gear, sleeping bags, binoculars, backpacks, life preservers, skis, and canoes (and perhaps photography sales) were implemented to fund wildlife protection, federal and state wildlife agencies would need to shift focus from conservation on behalf of hunters and hunting to genuine preservation of wildlife and wild lands. Contemporary outdoor enthusiasts prefer to visit intact ecosystems, to watch wildlife on the hoof—they want to see deer and ducks alive on public lands, not dead in the back of a hunter's truck.

Given the decisive shift in public interest over the last thirty to fifty years, it is no longer just or sensible (if, indeed, it ever was) for wildlife agencies to cater specifically to hunters. FWS's ongoing alignment with hunting interests and firearms industries constitutes a flagrant breach of public trust and a gross misappropriation of public funds.

## "Public" Lands

Despite the requirement that public lands be "held in trust for the American people by the federal government," hunter interests hold sway. It is a breach of trust for federal agencies to exploit, manipulate, or damage public lands, or ecosystems and wildlife on public lands, on behalf of a tiny special interest group. Perhaps federal and state government wildlife agency oversight personnel are unaware that, with the drastic decline in hunters, hunter-friendly policies can no longer be considered management "in trust for the American people" ('Public' n.d.).

Government "management" policies for public lands are locked into the mindset of Americans who lived a century ago, when the majority of citizens hunted and public lands were perhaps rightly preserved for hunter interests—assuming hunter interests formed the majority. Today, U.S. citizens overwhelmingly choose to view rather than kill wildlife, and U.S. public lands ought to be managed accordingly—"in trust for the American people" ('Public' n.d.). Instead, public lands visitors are expected to acquire and wear fluorescent orange when entering public lands during long hunting seasons. At such times, hikers, bird watchers, backpackers, mountain bikers, and climbers—all other users—enter public lands knowing that they take their lives into their own hands—that they might be inadvertently shot by hunters.

There are several ways to justly manage public lands on behalf of the American people. For example, hunting can be restricted to a percentage of public lands reflecting the percentage of hunters—which is to say, public lands where hunting is allowed should be restricted to less than 5 percent of total public lands. This would leave 95 percent of public lands open for general use year-round for the 95 percent of the population that does not hunt. Alternatively, hunting could be restricted to a proportional percentage of time—5 percent, allowing the vast majority (the 95 percent) to feel safe on public lands 95 percent of the time. Based on the likelihood of yet further dwindling interests in hunting, perhaps hunters could be allowed on just 5 percent of public lands for just 5 percent of the time. Such changes in U.S. policies would hold U.S. public lands "in trust for the American people," rather than in trust for a mere 5 percent of the U.S. population.

## Wildlife "Refuges"

By definition, a refuge:

1 offers "shelter or protection from danger or distress,"
2 is "a place that provides shelter or protection,"
3 offers "recourse in difficulty" ('Refuge' n.d.).

Given this, we would expect a "wildlife refuge" to be a place where wild animals are sheltered and protected, but there are currently 556 National Wildlife Refuges in the United States, and they are decidedly dangerous places for wildlife.

U.S. wildlife refuges were established to "create well-stocked hunting grounds," as "a breeding ground for producing an outflow of game for sport hunting" (Kheel 2008). The home page of U.S. NWRS website announces, "Your Guide to Hunting on National Wildlife Refuges," and includes a search engine to helps users "Find the Perfect Hunt" ('Your' n.d.). The U.S. NWRS website also lists the agencies "Guiding Principles," including "wildlife management," defined as "active *manipulation* of habitats and populations, [as] necessary to achieve Refuge System and U.S. Fish and Wildlife Service missions" ('Guiding' n.d.). Hunting is listed first among the agencies "wildlife-dependent uses," followed by fishing ('Guiding' n.d.).

## Hunting as "Mercy Killing"

Contemporary hunters often assert that shooting wildlife is beneficial—even essential—for wilderness, ecosystems, and wildlife populations—no one wants Bambi to die a slow and painful death from starvation. But if deer are dangerously overpopulated, why is hunting limited to certain seasons? Why are hunting policies designed to *inflate* deer numbers? If deer (and other hunter target species) *are* overpopulated, wouldn't FWS stop eliminating (and start protecting) natural predators?

Those who claim that hunters prevent slow death by starvation also overlook the fact that hunters do not target thin, vulnerable, or sickly deer. Hunters tend to seek big bodies and big racks—the bigger the better. Why else would they attempt to attract target animals using bottled urine from does in heat (Heffernan 2012)? Rather than eliminate sickly deer, hunters strip "big game" populations of their strongest members, and their strongest genes. "This sort of discriminatory culling-of-the-fittest runs counter to natural selection and is effectively triggering a reversal of evolution by giving the unfit and defective a better shot at passing on their genes" (Robertson 2012).

Those who argue that hunting provides essential population control never seem to notice the hypocrisy—the absurdity—of human beings gunning down other species because we think they are overpopulated. We are the most dangerously overpopulated species: Do we really want to advocate shooting to kill as a reasonable and appropriate solution to overpopulation problems?

## Hunting as Procuring "Compassionate Meat"

With heightened awareness of the horrors of factory farming, some argue that hunting elk or rabbits or quail is preferable to buying beef or pork or turkey—that hunting is more compassionate than any factory farmed animal products: "Many hunters point out that death in the slaughterhouse is by no means more humane" than a bullet in the wild (Luke 1997).

Hunters cannot and do not shoot perfectly each time they pull the trigger, and hunters leave wounded animals to die slowly in the elements, or to be harassed and killed by other predators. Hunters wound animals who are not retrieved, animals who hemorrhage or die of infection—especially birds. Of the 200 million animals killed by hunters every year, 50 million are doves, 25 million are quails, and 20 million are pheasants (Gudorf and Huchingson 2010). Just three species of birds constitute nearly half of the 200 million animals killed annually, and fowl are still gunned down in the same way that my grandfather killed birds—with shotguns that scatter pellets, hoping to knock birds out of the sky, hopefully to be found and brought in by a "bird dog." Given this, it is not surprising that wounding when hunting is estimated to be as high as 30 percent (Gudorf and Huchingson 2010; 'Reducing' n.d.). Many wounded animals are left to die—those retrieved rarely die a painless death (Hatfield n.d.). Hunters very seldom kill quickly and cleanly with one shot. The average hunter is in the field only five to seven days each year—hardly an ample experience to determine a clear shot. Many novices, bent on achieving their legal limit, often "shoot from the hip" and injure rather than kill their target (Gudorf and Huchingson 2010).

Some hunters turn to bow hunting in the hopes of finding a measure of challenge. Bow hunting has a wounding rate of greater than 50 percent; bow hunters expend an average of 14 arrows per kill, and "[w]ounding and crippling losses are inevitable" (Hatfield n.d.). Each of 24 studies on the subject "concluded that for every deer legally killed by bowhunters, at least one or more is struck by a broadhead arrow, wounded, and not recovered. Studies of bowhunting reveal an average wounding rate of 54%" (Hatfield n.d.).

To determine whether or not hunting is cruel, we must know the definition of "cruel":

- "disposed to inflict pain or suffering";
- an act "causing or conducive to injury, grief, or pain" ('Cruel' n.d.). Hunting is cruel by definition.

To be fair, we must also figure in the physical and psychological pain and premature deaths of the many animals killed by Wildlife Services on behalf of hunters—those gassed, trapped, snared, and gunned down on behalf of hunters. Snaring and trapping are known to cause extensive, prolonged suffering not just for those who are killed, but for the larger community in which these animals live.

There is no nutritional need to consume flesh in twenty-first-century America—venison or beef or chicken. People who are genuinely committed to minimizing suffering must ask a broader question, do I need to eat animal products? Those who sincerely wish to reduce suffering (and protect ecosystems)—will hunt for potatoes and pickles in the garden, market, and pantry, rather than doves and ducks in the forest. For those who seek a compassionate diet, the answer is vegan, not venison.

## Hunting for Economy

Many hunters justify their sport as a means of low-cost sustenance, but when a prospective hunter queried online what it might cost to get into the game, replies noted the expense of a rifle/bow, ammunition, boots and clothes, license and permit—$500–$1500 just to get started ('How' n.d.). This assumes that one already owns a chest freezer and a truck, or some other way of transporting a large, bloody carcass. Ongoing expenses include ammunition, annual hunting licenses, the cost of transport, and gear upgrades and replacements. At $0.50–$1.50 for a single shot, ammunition is expensive (Gudorf and Huchingson 2010; Robertson 2012). (How many hunters are unwilling to burn up their money honing skills at a target range before they head for the woods?) Seasonal U.S. hunting licenses range from $20–$50 (depending on the state), with an additional license (in the range of $10–$30) required to shoot hunter target species, such as a deer, as well as for the privilege of killing a less common species (elk, antelope, moose). Hunting out-of-state generally costs at least four times as much … and hunting out-of-state is increasingly necessary as urban sprawl claims wild lands ('How' n.d.). Indeed, many hunters now drive long distances to reach areas open to hunters, and travel is therefore an increasingly significant portion of hunting expenses (along with owning, insuring, and maintaining an appropriate vehicle for rough terrain and for hauling a corpse home). Finally, one who hunts for food will need a chest freezer, and some will need to add the cost of professional butchering. This does not equate to a bargain buck (or even a discount duck).

Compare this with the cost of gardening, a wonderful family enterprise and bonding experience. Seeds and starters ($5–$15, depending on amount and variety) are the primary annual cost. Watering a garden entails no additional expense for those who already water a yard. For those who do not have a yard, community gardens are available in many locations (often free, sometimes $15–$50 for a season, depending on the community) ('Frequently' n.d.). A gardener can preserve produce by canning, or freezing, but gardeners can also choose to plant crops that keep well without refrigeration, such as winter squash and potatoes. Gardening for veggies is much cheaper than hunting for flesh, especially if one adds environmental costs and the cost to one's long-term health.

Another option, also much more economical than hunting, is to buy bulk staples. For example, it is pretty easy to find twenty pounds of rice for $10, and twenty-four pounds of navy or kidney beans for even less. Much of the world's population has depended on these, or similar grains and legume-based staples, for centuries—beans and rice, rice and lentils, potatoes and beans. With bulk grains and legumes and plants, it is easy to prepare delicious meals for one person on a budget of $20–$30 per week—including spices—now *that* is a markdown meal! See Ellen Jaffe Jones's *Eat Vegan on $4 a Day* for practical suggestions about this (Jones 2011).

Hunting is a comparatively expensive way to put food on the table.

> In reality, hunting is typically not a source of provision but actually drains family resources. Deer hunters, for example, spend on average over forty dollars per pound of venison acquired, once all the costs of equipment, licenses, transportation, unsuccessful hunts, and so forth, are calculated.
>
> (Luke 2007)

A hunter in *World Bowhunters Magazine* writes: "Nobody hunts just to put meat on the table because it's too expensive, time consuming, and extremely inconsistent"; hunters kill wildlife "because it's fun!" (Luke 2007). Honesty provides a breath of fresh air.

North Americans who hunt almost always hunt for pleasure, not for sustenance. Perhaps the best proof of this is that hunters kill some twenty-five million mourning doves, sixteen million squirrels, two million woodchucks, half a million prairie dogs, six hundred thousand crows, and sixty one thousand skunks every year ('Learn' n.d.). Have you ever been invited over for woodchuck soup or skunk stew?

## The "Eco-Hunter"

An environmentalist is

- "a person who is concerned with the maintenance of ecological balance and the conservation of the environment";
- "a person concerned with issues that affect the environment, such as pollution";
- "an expert on environmental problems";
- "a person who advocates or works" to protect "air, water, animals, plants," and other aspects of the natural environment ('Environmentalist' n.d.).

Hunters want bloated numbers of hunter target species and a dearth of carnivores competing to secure prey. Hunters enlist powerful government wildlife agencies to manipulate predators and ecosystems on their behalf—to bolster prey. Hunters have demonstrated over the course of more than a century that they are not interested in advocating for "the preservation, restoration, or improvement of the *natural* environment" ('Environmentalist' n.d.; emphasis added). Environmentalists are those interested in allowing predator/prey relationships to create and recreate their own natural balance, as they have for millions of years. Those who manipulate wildlife, altering natural ecosystems, do not qualify as environmentalists.

## Conclusion

On what legitimate grounds would an informed, sincere environmentalist align with hunter interests when hunter interests damage ecosystems? How can any informed, sincere environmentalist *fail* to take a stand *against* policies that destroy the integrity of ecosystems, which are generally supported by the gun lobby, hunters, and government conservation programs?

According to Aldo Leopold, a "thing is right when it tends to preserve the integrity, stability, and beauty of the biotic community. It is wrong when it tends otherwise" (Leopold 1968). Wildlife agencies damage the integrity and stability of ecosystems while manipulating biotic communities on behalf of hunters. Contemporary environmentalists in the U.S. who align with or support hunters and hunting do so at the expense of their mission—and therefore at the expense of their own integrity.

## Discussion Questions

1 Why do you suppose so many myths surround hunting? What social and cultural forces maintain hunting myths presented in this chapter?

2 Is it morally justifiable for an environmentalist to purchase a gun given that such a purchase pays for wildlife manipulation?

3 Should hunting be legal on public lands? Why or why not?

4 What are the advantages and disadvantages of holding hunters to 5 percent of public lands?

5 As long as we continue to allow hunting in wildlife refuges, should they be renamed? As a class, create a letter with suggested name changes to be submitted to Wildlife Services Department.

6 How do we, as one species among many, justly assess which species is most overpopulated?

7 Is it morally acceptable, when human population growth is creating the most significant and dangerous environmental problems, to control the populations of other animals (bears, elephants, or deer) and not our own?

## Essay Questions

1 Is hunting immoral? Is it natural? Is hunting outdated? Why?

2 Under what specific conditions might a hunter be an informed environmentalist and yet maintain integrity, if any? Put another way, are there any circumstances under which hunting might be morally justified for an environmentalist?

3 In what ways is wildlife manipulation beneficial for wildlife, environment, or hunters respectively?

4 Can you think of any population control methods that we might legitimately use both on ourselves and on other species? If you feel they cannot be used on humans, what does this indicate about such methods? When might freedoms and rights be breached, if ever, on behalf of the environment—on behalf of the life and safety of all that dwell herein?

## Suggested Further Reading

Fox, Camilla. 'Predators in Peril: The Federal Government's War on Wildlife.' Indiana Coyote Rescue Center. *Indiana Coyote Rescue Center Newsletter.* Winter 2009. Accessed 16 November 2012. http://www.coyoterescue.org/newsletters/winter2009.html.

Kemmerer, Lisa. *Eating Earth: Dietary Choice and Planetary Health*, Oxford University Press, 2014.

Kheel, Marti. *Nature Ethics: An Ecofeminist Perspective.* Lanham: Rowman and Littlefield Publishing Group, 2008.

Luke, Brian. "A Critical Analysis of Hunters' Ethics." *Environmental Ethics* 19:1, 1997: 25–44.

Luke, Brian. *Brutal: Manhood and the Exploitation of Animals.* Urbana: Illinois University Press, 2007.

## Note

For more on this subject see *Eating Earth*, Lisa Kemmerer, Oxford University Press, 2014.

## References

*AWI Quarterly* 2012, 'Caught in the Crosshairs: Effective Immunocontraception Faces Political Fire,' vol. 61, no. 2, pp. 22–24.

Carroll, J. 2005, 'Gun Ownership and Use in America: Women More Likely than Men to Use Guns for Protection,' *Gallup*, 22 November. Available from: http://www.gallup.com/poll/20098/gun-ownership-use-america.aspx. [Accessed 29 May 2014].

'Cruel,' *Merriam-Webster*, Available from: http://www.merriam-webster.com/dictionary/cruel. [Accessed 26 January 2013].

'Digest of Federal Resource Laws of Interest to the U.S. Fish and Wildlife Service: Federal Aid in Wildlife Restoration Act,' ('Digest') *U.S. Fish and Wildlife Service*, Available from: http://www.fws.gov/laws/lawsdigest/FAWILD.HTML. [Accessed 1 November 2013].

'Environmentalist,' *Dictionary Free Download*, Available from: http://www.thefreedictionary.com/environmentalist. [Accessed 19 November 2013].

'Federal Aid Division—The Pittman–Robertson Federal Aid in Wildlife Restoration Act,' ('Federal') *Conserving the Nature of America*, U.S. Fish and Wildlife: Southeast Region. Available from: http://www.fws.gov/southeast/federalaid/pittmanrobertson.html. [24 May 2012].

Fox, C.H. 2009, 'Carnivore Management in the U.S.: The Need for Reform,' *AWI Quarterly*, Fall vol. 58, no. 4, pp. 20–24.

Fox, C.H. 2011, 'Predators in Peril: The Federal Government's War on Wildlife,' Available from: http://www.projectcoyote.org/newsreleases/news_predatorinperil.html. [Accessed 14 October 2011].

'Frequently Asked Questions about Community Gardens,' ('Frequently') *Food Share Garden*, Available from: http://www.foodshare.net/files/www/Growing/Community_Garden_FAQ.pdf. [Accessed 5 June 2012].

Gudorf, C.E. and Huchingson, J.E. 2010, *Boundaries: A Casebook in Environmental Ethics*, Georgetown U. Press, Washington, DC.

'Guiding Principles,' ('Guiding') *National Wildlife Refuge System: U.S. Fish and Wildlife System*. Available from: http://www.fws.gov/refuges/about/mission.html. [Accessed 27 February 2012].

Hatfield, L. 'Report on Bowhunting,' Available from: http://www.bcdeerprotection.org/x-bowhuntingreport.pdf. [Accessed 13 January 2013].

Heffernan, T. 2012, 'The deer paradox,' *The Atlantic*, 24 October. Available from: http://www.theatlantic.com/magazine/archive/2012/11/the-deer-paradox/309104/. [Accessed 11 June 2014].

'How Much on Average Does Deer Hunting Cost? I Have Never been Hunting before and would Like to Start?' ('How'), Yahoo Answers. Available from: http://answers.yahoo.com/question/index?qid=20110503174847AA5QUQD. [Accessed 5 June 2012].

'Hunters out of the Whole Population in the U.S. What Percentage Actually Participates in Hunting?' ('Hunters') Yahoo Answers. Available from: http://answers.yahoo.com/question/index?qid=20080823102552AAnIkxv. [Accessed 5 June 2012].

Jones, E.J. 2011, *Eat Vegan on $4.00 a Day: A Game Plan for the Budget Conscious Cook*, Book Publishing Company, Summertown, TN.

Kheel, M. 2008 *Nature Ethics: An Ecofeminist Perspective*, Rowman and Littlefield Publishing Group, Lanham.

'Learn the Facts About Hunting,' ('Learn'), *The Humane Society of the United States*, HSUS Washington, DC.

Leopold, A. 1968, *A Sand County Almanac*, Oxford University Press, New York.

Luke, B. 1997, 'A Critical Analysis of Hunters' Ethics,' *Environmental Ethics*, vol. 19, no.1, pp. 25–44.

Luke, B. 2007, *Brutal: Manhood and the Exploitation of Animals*, Illinois University Press, Urbana.

Moss, D. 2004, 'E Word: Operation Prairie Storm' in *E Magazine: The Environmental Magazine* (July/Aug 2004), p. 6.

'Pittman–Robertson Wildlife Restoration' 2013 ('Pittman–Robertson'), *FY 2013 Budget Justification: Wildlife Restoration*, U.S. Fish and Wildlife Services, pp. 1–12. Available from: http://www.fws.gov/budget/2013/PDF%20Files%20FY%202013%20Greenbook/24.%20Wildlife%20Restoration.pdf. [Accessed 12 January 2013].

'Public Land' ('Public'), Wikipedia. Available from: http://en.wikipedia.org/wiki/Public_land. [Accessed 12 January 2013].

'Reducing Wounding Losses,' ('Reducing') *South Dakota Game, Fish and Parks*. Available from: http://gfp.sd.gov/hunting/waterfowl/wounding-losses.aspx. [Accessed 13 January 2013].

'Refuge,' *Merriam-Webster*, Available from: http://www.merriam-webster.com/dictionary/refuge. [Accessed 10 June 2012].

Robertson, J. 2012, *Exposing the Big Game: Living Targets of a Dying Sport*, Earth Books, Winchester, UK.

Yarrow, G. 2009, 'The Basics of Population Dynamics,' Available from: http://www.clemson.edu/extension/natural_resources/wildlife/publications/fs29_population_dynamics.html. [Accessed 1 May 2013].

'Your Guide to Hunting on National Wildlife Refuges,' ('Your') *National Wildlife Refuge System*, Available from: http://www.fws.gov/refuges/hunting/. [Accessed 27 February 2012].

# 7

# ESSENTIAL ELEMENTS FOR ELEPHANTS

## Problems and Solutions

*Valerie J. Chalcraft*

> They never do any harm unless provoked, even though they go about in herds, being of all animals the least solitary in habit.
>
> (Pliny the Elder, *Historia Naturalis*, 77 CE)

In my recent travels to Malaysia, Zambia, and Botswana, I heard stories about how much the landscape had changed in the last two decades. The forests are gone and there are few, if any, sightings of elephants where they were once abundant.

The elephants of Asia and Africa are in decline, not only in numbers but also with regard to the integrity of their species. The main cause of their decline is human exploitation, which fractures their family structure and destroys the lives of these individuals. Humans exploit elephants largely for two purposes: ivory and labor. Those who are lucky enough to survive these assaults on their person are further endangered by habitat loss, thanks to the pressures of human population growth.

During my travels in Zambia, I was told that a man's virility is judged by how many children he has. Villages are growing exponentially in Zambia, and the Miombo forest—the main source of fuel for cooking fires—also serves as primary habitat for elephants. In both urban and rural Zambia, baskets of charcoal are sold all along the roadways, and the smell of smoke is always in the air. Elephant habitat is also destroyed in both Africa and Asia to make way for farms to feed growing human populations. Individual and large-scale corporate farmers employ short-sighted "slash-and-burn" methods to clear-cut forests and make way for plantations, destroying already fragmented and damaged landscapes.

Elephants are fascinating, sensitive, complex individuals. For example, they mourn not only their dead (Poole 1996) but also bury and mourn dead humans (Siebert 2009). Elephants also play an important role in nature: They dig out water holes on which many other animals depend, and make paths through the forests, which work as fire breaks ('Asian Elephant' n.d.).

Elephants are also a "flagship" (Johnsingh and Joshua 1994) or "umbrella" species—their health coincides with the health of an ecosystem. Stable numbers, traditional herd structure, and the mental health of individual elephants are all essential to saving this species-at-risk. In the process of examining the problems and possible solutions facing elephants, this chapter simultaneously explores animal rights and environmental concerns, serving as an apt example of how much these two groups share in common, and why these two groups would benefit from joining forces—why they *must* join forces—to support existing projects designed to protect elephants and their forest homes.

## Crop-Raiding

Elephants require space, and their space is increasingly threatened and perpetually diminished by expanding human activities. Elephants are large-brained, large-bodied herbivores whose territories range from 15 $km^2$ to 11,000 $km^2$ (9.3 $mi^2$ to 6,835 $mi^2$) (Poole and Granli 2008). They consume more than 180 kg (400 lbs) every day (Handwerk 2005); depending on where they live, their food includes tree bark, roots, twigs, leaves, vines, and small stems. Asian elephants in the rain forests of Malaysia feed on 345 varieties of wild palm (Sukumar 1992).

In addition to palm, elephants partake in bananas, millet, sorghum, jackfruits, and sugarcane (Sukumar 1992). As human populations rise, we sprawl into elephant habitat, and our cultivated crops attract elephants, concurrently increasing human–elephant conflicts. Perhaps our crops are more tender and tasty than wild food sources, and they certainly offer higher nutritive value (Sukumar 1992). In any case, elephants find our intensive production of plants to be easy pickings. Bull elephants, in particular, are likely to raid crops, and may raid a field for several nights before moving on (Sukumar 1992). In an Indian study, crops generally form 9.3 percent of an adult bull's diet, but can reach as high as 20 percent (Sukumar 1992).

Since most crop-raiding elephants are bulls, it is not surprising that elephant–human conflicts tend to involve bulls. There is considerable individual variation in bull elephant aggression. Many of the highly aggressive bulls were previously injured by humans protecting crops (Sukumar 1992).

In the past, humans and elephants have co-existed peacefully. For example, in the early 1990s, Dr Cynthia Moss (of the African Elephant Conservation Trust) reported that elephants were not aggressive toward people in the Amboseli National Park, Kenya, where the Masai people and elephants lived without conflict (qtd. in Sukumar 1992). At that time, elephants were viewed as assets: Elephants opened up grassland, where Masai cattle grazed (Mangat 2008). However, the Masai's relationship with elephant herds has changed. The open land once used by cattle and elephants is now leased to farmers, and to protect their crops the Masai now spear elephants. A behavioral study shows that these elephants have developed a fear of the Masai, but that they do not fear researchers or tourists (Mangat 2008).

## Traditional Attempts at Deterrence

Even small losses can harm small farmers. Farmers who work the land in elephant territories work hard to prevent crop raids, and even retaliate for past raids, killing elephants with guns, poison-tipped spears, or poisoned food. Wealthier farmers put up expensive electric fences to deter elephants, while poorer farmers construct homemade electric fences (which kill elephants and other animals, including humans) (Sukumar 1992). For better or worse, elephants have been able to outwit electric fences by moving wires with the pads of their feet, or with their tusks—neither of which conduct electricity (Sukumar 1992). A guide based in Zambia told me that elephants knock down, drag, and toss trees onto electrical fences in order to reach tasty, nutritious crops on the other side (which qualifies, by the way, as "tool use"). Some farmers have gone so far as to dig trenches around crops, but this process is expensive, and elephants are smart enough—and powerful enough—to fill these ditches with dirt, then walk across to reach crops (Sukumar 1992).

Other attempts at deterrence have put humans in direct contact with elephants. India has established "elephant squads"—people who chase elephants from croplands using lights, firecrackers, and shotguns filled with birdshot (Theisen 2005). Although some farmers can afford expensive firecrackers or ammunition to shoot guns over elephants' heads, most farmers have only simple means to protect their crops: They shout and bang drums, shine flashlights and spotlights, or keep barking dogs. But faced with rows of lush produce, some elephants have inevitably learned to ignore these gentler methods. Furthermore, these techniques can disturb elephants, triggering an elephant rampage, in which both crops and humans may be trampled (Sukumar 1992). These human–elephant conflicts bring many negative repercussions, including significantly decreased free-living elephant populations.

Each year in Sri Lanka, more than 60 people and more than 200 elephants die as a result of elephant rampages (Haviland 2010a). In this little island nation, protests over ongoing, poorly managed human–elephant conflicts have emerged from an unexpected corner. In August, 2010, members of the Sri Lanka Wildlife Veterinarians' Association went on strike to protest the government's failure to prevent cattle ranchers from perpetually encroaching on elephant lands. In Sri Lanka, as elsewhere, cattle (rather than loggers) destroy forests, and compete with elephants for food sources (Haviland 2010b). Sri Lankan veterinarians need the support of both animal and Earth advocates if they are to effectively push the Sri Lankan government to create a sustainable, long-term plan to address the root of the problem—cattle ranching. Slash-and-burn logging to create pasture will only stop when consumers stop buying meat. Is this not a point on which animal and Earth advocates can agree?

## Captured for Work

It is generally more economical to capture and "break" young elephants than to breed them in captivity. For example, the logging industry does not breed elephants because a pregnant (then) lactating cow cannot work for 3 or 4 years, and her offspring will not be ready to work for 15 years ('ElefantAsia: Elephant baby bonus' 2014). Consequently, elephants are frequently captured, "broken," trained, and put to work by enterprising capitalists. Ironically, elephants forced to work within the timber industry are forced to destroy their own habitat: By removing trees from clear-cut forests, these displaced elephants further displace their wild counterparts. These most unfortunate captives, aside from losing their freedom and autonomy, and contributing to the demise of their own kind, are overworked and exhausted on a daily basis (World Wildlife Fund—Issues: Timber trade and illegal logging n.d.).

Before they are suitable to the tasks at hand, wild elephants must undergo "hard training." A *National Geographic* article describes traditional training, known as "The Crush," which is a "breaking ceremony" overseen by a shaman. The Crush requires that new captives be tied for 3 to 6 days, so they cannot move about freely, during which time they are beaten. These beatings are to instill fear. When they are untied, they are again tied and beaten ('Activists denounce Thailand's elephant "Crushing" ritual' 2002). Elephant capture, training, and forced labor cannot please either environmentalists, who are interested in protecting endangered species, or animal advocates, who are concerned about the welfare of the individual.

## Malaysia: Palm Oil and Translocation

Forests are also destroyed to create palm plantations. Malaysia is just one example of a country that is now in the process of rapidly decimating forests for timber, then replacing wild habitat with palm plantations. This process supplies Western consumers with a short-term supply of wood products and a longer-term supply of palm oil. Outside the Malaysian capital of Kuala Lumpur, palm plantations extend as far as the eye can see. Global palm oil production increased 50 percent between 2001 and 2006 as palm oil replaced now unpopular trans-fats (Fitzherbert et al. 2008), with Malaysia and Indonesia—holding 80 percent of Asia's remaining rainforests—producing 80 percent of the world's palm oil (Fitzherbert et al. 2008).

Elephants have no way to travel except through these newly-created enterprises. Large plantations can expect even more crop-raiding elephants because their long borders intersect with miles of elephant habitat. It is therefore not surprising that a herd of nine elephants reportedly destroyed a total of 4,400 hectares of palm oil trees and 180 hectares of rubber plantations in Malaysia over the course of 6 years (Sukumar 1992).

Destruction of forests for the palm industry is particularly disturbing because rainforests need not be destroyed to plant palm trees. Large tracts of abandoned land are available for such crops, but commercial palm planters prefer to subsidize their operations by clear-cutting forests and selling timber (Fitzherbert et al. 2008). Ironically, palm oil has also been promoted as a bio-fuel alternative, but since rainforests absorb carbon dioxide and produce oxygen, destroying these forests for bio-fuel will only increase global warming—and destroy wildlife, threaten species, and decimate ecosystems. Further, an EPA analysis shows that palm oil bio-fuels do not meet the minimum standards to qualify as a renewable fuel (Office of Transportation and Air Quality 2011).

While in Malaysia in 2007, I visited the Kuala Gandah Elephant Conservation Centre (KGECC). At KGECC I met crop-raiding elephants and infants orphaned when their crop-raiding mothers were killed. I assumed that a "conservation center" would be a place of sanctuary for elephants in need of relocation because they were raiding palm oil plantations, but I quickly realized that no elephant with any choice would opt to live at KGECC. The elephants had been brutally trained so that they would be suitable to entertain tourists with elephant rides. When chained for feeding, I noticed that the elephants immediately started swaying—the same stereotypic behavior among elephants chained in circuses ('Stereotypic behavior and behavioural disorders in elephants' n.d.). The elephants at KGECC are housed in a roofed holding area, from which I heard multiple trumpets: Elephants normally trumpet only when they are surprised, angry, or lost ('Some Facts about the Asian Elephant' n.d.).

As a tourist, I watched KGECC handlers control captive, trained elephants with bull hooks. Trainers ("mahouts") use bull hooks on elephants to establish physical dominance through pain and fear. Bull hooks are used to strike sensitive skin around the face, ears, and back of an elephant's legs (I have seen bull hook scars on elephants' legs in U.S. circuses). Several adult elephants at the KGECC, and one of the infants, had fresh injuries on their heads and necks. When I asked about the injuries, one staff member told me that the elephants had run their heads into "sticks"; another told me that the elephants had run their heads into "nails that stuck out from the walls." The wounds on these various elephants were located in the same place—on the forehead above the left eye—too coincidental to fit either the "stick" or the "nail" explanation. These wounds were caused by trainers with bull hooks, and the keepers had learned to lie to tourists about their brutal methods. Malaysia's Natural Resources and Environment Ministry admits in a statement that "the hooks were used to manage them when they misbehave or refuse to obey orders" (Asian Animal Protection Network n.d.).

KGECC holds orphaned elephants permanently at the facility, claiming that they may struggle to survive in the wild ('Malaysia Traveller: Kuala Gandah Elephant Sanctuary' 2014). In contrast, Dr Kate Evans (of Elephants for Africa) has documented cases of captive-reared elephants successfully adapting to life in the wild (Evans 2010). In truth, I felt very sorry for all of the elephants at KGECC, and could neither rationalize nor justify such brutal dominance and cruel exploitation. Unfortunately, the KGECC is not alone. For example, the staff at the Elephant Conservation Response Unit in Tangkahan, Indonesia use bull hooks to force elephants to perform tricks (Movahed 2014).

## Translocation, Overpopulation, and Culling

Most often, elephants at KGECC tote tourists around the facility. But these unfortunate elephants have also been trained to move other crop-raiding elephants from their native territories to reserves, a process called "translocation."

Translocation began in Malaysia in 1974, when lethal methods of elephant control sent Malaysian elephant populations into a hasty downward spiral. Instead of managing logging to accommodate elephants, Malaysia created the Elephant Capture and Translocation Unit (ECTU) in order to relocate displaced elephants who have become crop-raiders ('History of Kuala Gandah Elephant Sanctuary' 2011). While this program offers a measure of promise, translocation is not without serious problems: Not only are some elephants enslaved to work via ECTU and KGECC, but translocation too often results in elephant deaths, both en route and after release (Pinter-Wollman et al. 2009). Neither does translocation eliminate human–elephant conflicts. It is difficult to determine which elephants are the culprits—which elephants to move—and even if these individuals are identified and removed, other elephants step up to raid crops in their absence (Hoare 1999; 2001). Furthermore, elephant experts question the morality of removing bulls who are over 30 years of age from familiar environments and well-established territories. These older bulls are inclined to try to return to their home territories (Garai and Carr 2001).

In any event, there is no point in relocating yet more elephants to already depleted and diminished habitats. The ECTU has translocated 300–500 elephants to various small reserves throughout Malaysia (Elephantsanctuarymalaysia 2011), reserves that can only sustain a given number of elephants. Despite these losses, limitations, and the enslavement of some of the captured elephants, translocation is described by supporters as a "win–win" situation for both elephants and farmers (Elephantsanctuarymalaysia 2011). But a study of translocated elephants concluded that this practice increases mortality and human–elephant conflict, and that even if translocation helps reduce elephant conflicts in one area, the same (or more likely worse) conflict is created at another location (Fernando et al. 2012).

Ecosystem stress is also a serious problem in African parks that protect elephant herds, especially in areas where elephants eat tree bark, which can kill trees. A park guide at Chobe National Park in Botswana told me, "There are currently 185,000 elephants in the park—50,000 too many. As a result, there are increasing numbers of tall, dead trees which once provided food for giraffes—giraffe populations have declined." The Chobe guide blamed humans: "Tuberculosis used to kill many elephants at Chobe, but immunizations in the 1960s led to this artificially high population."

To reduce the overall number of elephants who are trapped within an artificial national park boundary, and to reduce the resultant environmental stress, several parks in southern Africa are "culling"—killing—elephants. South Africa's Kruger National Park resumed culling elephants in 2008 for the first time in 14 years (Timberg 2008). Although there are no fences around Chobe, heavy poaching in nearby Zimbabwe keeps elephants from wandering. Consequently, Chobe elephants also face culling. I asked the guide at Chobe about alternatives to culling, and he claimed that they had "tried everything" —sterilization surgery had led to sepsis and death, while contraception simply "didn't work."

Killing elephants to reduce artificially inflated numbers might, on first sight, seem acceptable to many environmentalists, who are more interested in populations than individuals. Such an approach is, of course, completely contrary to animal advocates. On closer scrutiny, this approach is rightly rejected by both camps. Killing elephants is a short-term fix that avoids dealing with the real problem: increased human populations and our destruction of elephant habitat.

## Ivory

One dangerous byproduct from culling is the accumulation of ivory stockpiles—tusks removed from the bodies of slaughtered elephants that happen to be worth a fortune. Until 1989, African elephant tusks could be traded internationally, and were traded primarily with buyers in Asia because African elephants were listed on Appendix II of the Convention on International Trade in Endangered Species of Wild Fauna and Flora (CITES). Appendix II indicates that the species is threatened, but controlled trade would still be allowed. In contrast, Asian elephants have been listed on Appendix I (endangered such that international trade is banned) since 1975 ('Wildlife trade: Elephant ivory trade FAQs' n.d.).

For African elephants, legal international trade combined with Asian demand lead to high prices and high rates of poaching. Older males with the biggest tusks were targeted first, followed by younger small-tusked elephants. Consequently, the African continent lost nearly 1 million elephants in the 1980s (Lemieux and Clarke 2009) and the CITES member countries voted to move African elephants from Appendix II to Appendix I. As a result, the bottom fell out of the market, and African elephants were safe from poaching (Sukumar 1992). Between 1989 and 2007, Africa's elephant populations increased by approximately 140,000 (Lemieux and Clarke 2009).

However, this population growth has not been uniform. The current ban only covers international trade—it does not ban ivory sales within any one country. Ten of 37 African countries with elephant populations have an internal ivory market serving tourists, and to a lesser extent, locals. Most of these domestic markets are located in Central Africa (Lemieux and Clarke 2009), where there is little regulation, and international buyers can easily tap into this market (Milliken et al. 2006; Courouble et al. 2003). As a result of these unregulated domestic markets, elephant numbers continue to decline within many African countries, especially in the Central African Republic, Zambia, and Angola (Lemieux and Clarke 2009). Even in countries that do not have a domestic market (for example, Zambia), elephant populations continue to decline if surrounding countries allow domestic trade in ivory. This is largely caused by lax border controls and the fact that elephants travel long distances, oblivious to international borders. Although Asia also has unregulated domestic markets, African domestic markets consume more ivory than Asian domestic markets (Hunter et al. 2004), perhaps because Asian female elephants do not grow tusks? African domestic markets alone cause the deaths of 4,000–12,000 elephants annually ('Crack down on domestic ivory markets in Africa needs major boost' 2005).

What should nations do with unsold "stockpiles" of tusks? Despite the fact that "there's no question that one-off sales promote poaching," several African countries pushed to lift the ban on international ivory sales, claiming that the money that ivory sales provided was essential for elephant conservation programs (Coghlan 2010). It makes no sense to attempt to save elephants by using profits from elephant body parts, and it is difficult to obtain credible reports that detail how resulting funds have been spent (Reeve et al. 2003). In any event, the CITES member countries Zimbabwe, Botswana, and Namibia were allowed an "experimental" sale of 50 tons of raw ivory to Japan in 1997; in 2007 Botswana, Namibia, and South Africa were also permitted to sell 60 tons of ivory to Japan. These sales were permitted with the understanding that they must maintain "adequate" methods to monitor poaching, and that Japan would not re-export the ivory. These countries now maintain this "regulated" market, and yet, their elephant populations continue to grow (Lemieux and Clarke 2009). That said, numbers alone do not predict the integrity of a herd, or the health and stability of elephant populations.

## Trauma and Breakdown: Post-Traumatic Stress Disorder

Poaching, culling and translocation have caused a breakdown in the social structure of many elephant communities. This breakdown threatens to destroy natural elephant behaviors. Increasingly, calves have only young, inexperienced mothers, without the benefits of guidance from elder males and females. Family units dominated by less-experienced females tend to have more conflicts with humans, perpetuating the cycle of violence. For example, inexperienced mothers have greater difficulty finding food and water (Allimadi 2007), so they are more apt to seek crops and villages in order to satisfy their basic needs. Making matters worse, when humans translocate orphans, these young elephants are more and more likely to join a group of youthful elephants that lacks the traditional social hierarchy of natural elephant herds.

Aggression has increased among elephants who lack stable family structure on South African reserves. For example, in the 1990s in Pilanesberg National Park, young male elephants killed both black and white rhinoceroses (Slotow and van Dyk 2001). Due to a lack of older males, young bulls were entering periods of sexual activity and aggression (musth) earlier than normal, filling the gap where older males used to stand. Hormone levels increase during musth, along with reproductive competition. In one experiment, introducing six older males suppressed early musth in younger males, and eliminated further rhino-directed aggression (Slotow et al. 2000). In Addo Elephant National Park, 90 percent of male elephant deaths are caused by bull elephant conflicts, in contrast with just 6 percent in more stable elephant communities (Highfield 2006). Bull elephant dominance conflicts rarely result in serious injuries in stable elephant herds, and when older bulls were introduced at Addo, aggression diminished. However, in May of 2010, a younger male who had been translocated from Kruger National Park killed a non-aggressive older male, also translocated from Kruger. An injury during translocation led the older male to lose a tusk (van Staaden 2010), which may have put him at a disadvantage.

India's remaining elephants also exhibit increased aggression. Villagers in southeast India aptly describe this change in herd behavior as "revenge." For example, 60 elephants rampaged through a village at Tingri, killing five people in an area where a train had killed a baby elephant ('World: South Asia—India Elephant rampage' 1998). Elephants in India have much reason to be angry—in North Bengal, 26 elephants have been killed by trains in a seven-year period on a 100 km stretch of land that crosses through four wildlife sanctuaries ('Forest Department struggles to guard elephant corridors' 2008). Unfortunately, in response to increasing human population, the Indian government is expanding a road in that same area to create a four-lane highway (Huggler 2006). While some villagers welcome the highway, others are delaying the process by protesting the government's demand for land on which to build this new road ('Villagers at NAHI door to give land: 153 families offer land for NH31D four-laning' 2013). If the road goes through, elephants are likely to protest predictable deaths when cars collide with members of pachyderm communities. And how will humanity respond to such retaliation?

Elephant "revenge" has also been reported in Uganda (Highfield 2006), where stress and the breakdown of elephant communities is, once again, the likely cause of aggression. Interestingly, similar stresses and breakdowns are also affecting Ugandan humans. In a fascinating and enlightening article, Dr Eve Abe compares hyper-aggression among orphaned male elephants with that of orphaned young male humans in her own Ugandan community, that of the Acholi people (Williams 2006).

The Acholi are trying to recover from 20 years of civil war. In Uganda's recent history, in order to obtain child soldiers, rebels killed the parents of young Acholi males or forced the boys to kill their own parents. As a result of such violence, there are over 200 camps of displaced Ugandans, many of whom grew up without parents or schools, and were raised by older children. These young males, like young bull elephants, tend to roam in groups and engage in violent, destructive behavior (Siebert 2006).

Such aggression is most likely a result of long-term, extreme stress, which affects brain regions associated with memory, emotions, and fear in animals (including humans). Psychologist Dr Gay Bradshaw describes how individuals (especially young individuals), who witness the brutal killing or capture of a family member, are left with psychological trauma (Bradshaw et al. 2005). Psychological trauma often manifests in post-traumatic stress disorder (PTSD), which entails an exaggerated startle response, unpredictable behavior, inattentive mothering, and hyper-aggression. Dr Bradshaw and Dr Lorin Linder observed that elephants in captivity exhibit these same symptoms if they have been violently separated from their mothers and trained with bull hooks (Bradshaw and Lindner n.d.). A traumatic stress reaction can be triggered by a sound, smell, or sight that reminds a victim of some past horror that they were forced to endure (Bradshaw et al. 2005). Such experiences, resulting in psychological trauma, likely account for deadly elephant rampages—both in the wild, and in circuses and zoos—just as they likely account for the violence of

Acholi youth. Since elephants learn from one another, they share their sense of fear and aggression (Handwerk 2005). This abnormal aggression—a perfectly normal response to the trauma of violence—further perpetuates the ongoing cycle of human–elephant conflict.

The fact that elephants are capable of exhibiting psychological trauma that identifiably parallels this same trauma in humans should be enough to stop elephant culling. Just as we do not "cull" humans to reduce overpopulation, we must not kill elephants, who, in truth, are not a species with an overpopulation problem—at least not compared with humanity.

When I asked the guide at Chobe National Park about the long-term impact of proposed culling, he knew of the cycle of violence that begins with elephant slaughter and noted that they would correct for this problem—entire herds would be killed, leaving no orphans; corpses would be burned so that no other elephant would see the remains. The Chobe guide added that the tusks would be burned in order to avoid fueling a market for ivory. At that time, I was unaware that Botswana—along with other nations—was legally selling ivory to Japan.

While Chobe's plan may reduce or eliminate long-term trauma, killing entire families and communities of sensitive, intelligent individuals cannot be considered an acceptable long-term solution for animal advocates. Furthermore, such a solution does nothing to curb human population growth and sprawl, or to prevent ongoing habitat destruction, and therefore cannot be considered an acceptable long-term solution for environmentalists.

## Healing

Although critical implications of translocation and culling are now widely documented, both practices continue as human populations and encroachment increase. The cycle of violence starts with humanity, is perpetuated by humanity, and must end with humanity. We must stop killing and exploiting elephants and instead foster healing. For both traumatized elephants and traumatized humans, healing requires that physical and social needs be met so that natural behaviors can replace violence brought on by trauma. Healing those traumatized by violence is a crucial step in promoting the health of an individual, family, community, *and* a species. A species can only thrive through the health of individuals. Consequently, several projects are now under way to heal and build bridges between humans and elephants in Uganda, Thailand, Kenya, and Sri Lanka.

In Uganda, Bradshaw and Abe are building a community center where it is hoped that both elephants and humans will recover from the horrific violence that was forced into their consciousness at a young age (Siebert 2006). These scholars have documented the symptoms of PTSD, as experienced by traumatized humans, in traumatized elephants, and are testing their theory: If the cause is the same, the cure is likely the same. They believe that both humans and elephants who have been traumatized by violence will heal if they are provided with a safe-haven where they have a sense of control and stable social attachments.

In Thailand, The Elephant Nature Foundation creates and maintains links with hill tribes, providing medicinal and educational assistance for both people and elephants ('Strategic goals' n.d.). In Kenya, the David Sheldrick Wildlife Trust provides sanctuary for orphaned and traumatized elephants, and has successfully rehabilitated and reintroduced more than 80 orphaned elephants into herds at Kenya's Tsavo National Park. Rehabilitating just one orphan takes 8 to 10 years ('Orphans Project' n.d.). Despite their brave effort, the lack of older herd members at the Sheldrick Wildlife Trust has made complete restoration of social dynamics impossible (Bradshaw et al. 2005). These projects in Uganda and Thailand deserve the support of both animal and Earth advocates.

Due to protests from wildlife veterinarians, Sri Lanka is reintroducing elephants who have been raised in orphanages back into natural habitats, which will help heal these traumatized, displaced, and too often exploited individuals (Haviland 2010b). Some Sri Lankan elephant orphanages are similar to the KGECC. For example, the Pinnawala Elephant Orphanage is a tourist attraction which breeds elephants, bringing young into a life of bondage for the sake of tourist elephant rides ('Pinnewala Elephant Orphanage in Sri Lanka' n.d.). Not surprisingly, a study at Pinnawala demonstrated that young residents display abnormal social behavior and stereotypical behaviors when chained for meals or during the night (Kurt and Garai 2001). Freedom should do a great deal to heal these individuals. However, steps must be taken to prevent human–elephant conflict once released.

## Solutions

### *Corridors*

If we are to protect the integrity of elephant communities, we must know where elephants travel. Toward this end, Elephant Voices has put together a catalogue of 260 adult elephants in Kenya to help track their movements in the Maasai Mara ('Introducing Elephant partners—Maasai Mara' 2010). Catalogues like this must be developed across all elephant territories to track elephant movements.

We must then establish large woodland corridors that connect and disperse elephants (and other species), reaching across international boundaries. Instead of clear-cutting forests indiscriminately, and killing or translocating elephants when they interfere with agriculture, human developments must be *planned around* elephant migratory routes, taking care not to destroy forests that provide passageway to other vital habitats, including water sources. Wildlife corridors must somehow circumnavigate human-made obstacles that fragment habitat, such as roads and railroad tracks. A study in central Africa (by the Wildlife Conservation Society and Save the Elephants) reported that roads act as prison bars, deterring elephants from crossing to reach habitat on the other side (Eccleston 2008). In Botswana, I heard elephants trumpet when vehicles were near, indicating distress. In Jharkhand, India, open-cast mines have damaged elephant migration paths, forcing elephants into previously unexplored areas, and into human settlements, threatening the safety and lives of both elephants and human beings. Families of those killed demand government compensation, but would it not be better for the government to invest in corridors to avoid such tragedies (Bhuiyan 2014)?

Elephant corridors can eliminate overpopulation in reserves by allowing elephants to travel between reserves, or into other traditional habitat areas. Such corridors are already being designed in India. The World Land Trust (WLT) and Wildlife Trust of India (WTI)—an Earth and an animal organization—are constructing wildlife corridors to accommodate traditional Indian elephant migration trails. As part of India's community-based elephant corridor plan, local residents must participate to make these large projects more feasible. WTI has identified 88 elephant corridors and has prioritized them according to their importance to elephants, and according to the feasibility of protecting a particular corridor. Some pastures and farmlands are left within elephant corridors, while other farmlands are purchased and replanted as forests. Due to high incidents of human–elephant conflict, a number of villagers and tribes must relocate. The WLT/WTI project is helping to provide more suitable sites for these villages, thereby returning remote areas to wildlife and wilderness, while also protecting people. This project aids and protects humans *and* elephants, as well as many other species, including highly endangered nonhuman primates who are just as likely to use the corridors ('Elephant corridors in India' 2011).

India's various states respond in diverse ways to human–elephant conflicts; several states in particular appear to be ready and waiting for help with finding and adopting effective measures to mitigate human–elephant conflicts. For example, the Indian states of Odisha and Pathrashole are desperately in need of corridors. Currently, locals try to block elephants from entering their region (not just from crops) with electric fences and water-filled trenches. But this causes greater elephant crop-raiding on the other side of the fence, leading to strife between people living in these different regions. Regional wildlife officials in the areas continue to use torches to try to scare elephants away from the crops, causing deadly elephant stampedes (Mukherjee 2014). Good news, however, comes from the Indian state of Tamil Nadu. In the wake of rapid development near forests inhabited by elephants, causing elephants to move into village areas, the Forest Department took a series of measures to prevent any further development in elephant habitat or corridors (Preetha 2013). Elephant corridors are expensive and require the support of the community and government, but they are a long-term strategy to protect humans and elephants. Corridors are truly "win–win" and should be supported and promoted by animal and Earth advocates alike.

## Industry Management

Effective industry management is also essential for protecting elephants and their habitat. In 2008, Malaysia palm oil companies announced that they are running out of forests to clear-cut, and that they therefore planned to expand their operations in Africa and South America ('Malaysia targets Africa and the Amazon for palm oil expansion' 2008). Sustainable land-use, like elephant corridors, requires community-based programs and international collaboration if we are to protect elephants ('Elephants in conflict' 2009).

One model is offered by the European Union, which put forward a bio-fuel policy regulating the development of palm oil plantations. This policy promotes palm oil plantations certified by the Roundtable on Sustainable Palm Oil (RSPO). Several loopholes have slowed the process, and certification does not secure compliance, but a few companies have, nonetheless, been certified. While such policies are headed in the right direction, they must have "teeth" to be effective. Additionally, plantation expansion must be limited to force farmers to grow palm trees on land that has already been cleared (Block 2009). Both animal and Earth advocates need to put pressure on the RSPO to close loopholes, limit plantation expansion, and enforce RSPO requirements. Such advocates might also organize boycotts against companies that refuse to join or adhere to such protective policies.

Effective logging industry management will also allow elephants and humans to co-exist more peacefully. A recent Indian study indicates that selective logging does not adversely affect elephant habitat, and when forests regenerate, selective cutting will again be possible. Given time, grasses and bamboo plants will grow into secondary rainforests across Asia, providing food for elephants and wood for human beings (Sukumar 1992).

## Replanting

Activists can help forests regenerate by planting trees. A group called Bring the Elephants Home (BTEH) has planted 350,000 trees for elephants in Thailand, Cambodia, and Borneo. They also offer educational programs for local villagers, rangers, and students ('Bring The Elephant Home: Help us to plant trees for elephants' n.d.).

The Indonesian Rainforest Foundation (IRF) has purchased or leased seven deforested regions in Indonesia. With the help of non-governmental organizations (NGOs) and government agencies (both local and central), IRF is in the process of replanting trees to expand remaining shards of forest in these devastated areas. They then protect these areas and seek to turn them into national parks. The IRF depends on public donations, and will eventually sell carbon credits to raise yet more funds ('Reforestation plan' n.d.). Working with local volunteers and other NGOs, and willing governments, this handful of earth and animal advocates have made a big difference.

## Alternative Livelihoods

Sustainable timber industries employ fewer people than large-scale slash-and-burn logging, so a reasonable forestry plan must also find alternative work for displaced loggers. In addition to designing corridors, the World Land Trust (WLT) and Wildlife Trust of India (WTI) work together to provide attractive alternative livelihoods ('Indian Elephant corridors appeal' n.d.). In the Chongwe District of Zambia, Community Markets for Conservation (COMACO) is mobilizing and training communities in skills for alternative livelihoods to the charcoal industry in order to help reverse alarming levels of deforestation. The program introduces such skills as fish farming, carpentry, and bee-keeping and trains others in conservation farming. The program also provides them with seeds for groundnuts and soybeans. After the training, farmers agree to comply with conservation requirements and steer clear of the charcoal industry ('Reversing trends of charcoal making' 2014).

Poachers can change careers, too. The CITES international ban has not deterred poachers, and violence has escalated between park rangers and poachers, with each side suffering casualties (Lemieux and Clarke 2009). But COMACO has enjoyed success through retraining poachers for a new career when they hand over their firearms. Candidates take a 6-week course in alternative trade skills, such as carpentry, bee-keeping, or metal work, and are then required to identify two or three other poachers with whom they will share these newly acquired skills. After COMACO staff verify that all requirements have been met, the group of initiates receives their start-up investment of tools and supplies, such as wood, beehives, and metal, so they can become a new business—begin their new careers—which COMACO continues to support while they become established. Follow-up is also essential, and COMACO staff make up to three visits every year to assess progress and success in alternative livelihoods, and to ensure that the workers are not poaching.

Retraining has proven to be a highly cost-effective way to reduce the threat of poaching in Zambia; only about 6 percent of participants return to poaching ('COMACO—Transform a Poacher' 2010). After 2 to 4 years, retrained poachers usually earn more in their new trade than they did as poachers. COMACO would like to expand their program in Zambia, and their retraining program could become a template for other countries that suffer from poaching—whether poachers are targeting elephants, bears, tigers, or sharks.

In Indonesia, indigenous communities living within the rainforests are very poor, usually feeding themselves and their families with just $1 a day. People living so frugally welcome even the meager sums that palm oil companies pay—even though these companies ultimately displace entire communities for the sake of palm plantation development (Quayle, pers. comm., 19 April 2012). As an incentive to keep local communities on their lands, and to provide them with enduring alternative income, the IRF provides training for locally-based sustainable forestry and agricultural programs, and offers education programs for children ('Reforestation responses' n.d.).

Ecotourism serves as another possible alternative livelihood, and could be a source of tax revenue for conservation. Incentives should be built in to encourage tourist camps and lodges to employ locals, even if they must train new employees. The IRF works with Indonesian landowners and indigenous populations to establish ecotourism lodges and educational centers for tourists that provide ongoing funds for conservation efforts in the replanted areas ('Reforestation responses' n.d.). Tourist facilities should also be encouraged to market the works of local arts and crafts to help provide legal incomes and prevent illegal activities (Lemieux and Clarke 2009; 'Rainforest art' n.d.; Spelman 2014).

Tourist camps and lodges must be carefully planned to avoid interrupting elephant migration. I visited a tourist camp in Botswana where an electric fence had been freshly erected to secure the camp. Some visitors reported elephant sightings within the camp just days before our arrival, and elephant footprints were clearly visible in the surrounding mud. I heard elephant trumpets near the electric fence in the middle of the night; it is likely that they had unexpectedly made contact with our electric wire, which had suddenly been erected in their territory. Animal and Earth advocates must promote and patronize responsible ecotourism.

Since sustainable logging also requires fewer elephant "workers," many of these enslaved individuals can be released from bondage. In Laos, ElefantAsia releases and rehabilitates elephants who have been exploited by the timber industry, and finds alternative sources of income for elephant handlers. ElefantAsia considers elephant rides, or so-called "elephant trekking," to be a legitimate alternative income because these elephants have already endured "The Crush" and carrying tourists is less stressful than logging: Females can continue to work while pregnant, and calves are desirable because they inevitably generate increased tourist income ('ElefantAsia—Elephant Baby Bonus' 2014). However, an elephant's spine is not designed to support tourists sitting on a wooden platform strapped to their backs. Rubbing causes skin sores and dangerous infections as elephants plod along carrying tourists, their youngest babies struggling to keep up. Under this stress, mothers sometimes cease to lactate leading to a high infant mortality ('Trekking' n.d.).

ElefantAsia is building a school to instruct Lao mahouts in animal-friendly training and handling (Maurer 2010, pers. comm., 8 December). While this is encouraging and I wish it to be true, only "protected-contact training" ('Protected contact' n.d.) behind a barrier is "animal-friendly" and would be incompatible with the close contact during trekking.

The end-goal for elephants must be release from human bondage, to live freely among their own kind. Consequently, many activists are exerting pressure to change the direction of ecotourism. For example, in response to such pressure from an international coalition of elephant advocates called "ElephAction group," Flight Centre (an international travel agency based in Australia) has withdrawn an ad depicting tourists riding on elephants, and Flight Centre now avoids depicting any form of elephant exploitation, including rides and trekking, elephants painting pictures, and elephants performing tricks. Instead, Flight Centre proactively promotes tours and volunteering opportunities with a heavy emphasis on sustainability and conservation (Orr 2010).

## Humane Deterrents

We must also develop ways to effectively, affordably, and humanely deter elephants from raiding crops. The "eleAlert," which won a National Science Foundation award in 2010, could also improve the effectiveness of crop fences. The eleAlert detects a breach of security, and helps identify problem areas, where humane deterrents can then be employed. A guide who directs tours in Zambia and Botswana told me about the newest elephant deterrent: chili pepper paste. When this paste is rubbed on twine and rags, then wrapped around wire fences, elephants stand back. In Zambia, gasoline-soaked rags serve the same purpose.

Another promising new deterrent exploits the African elephants' fear of bees. Lucy King found that African elephants fled when she played audio recordings of bees. They also fled when she played recordings of elephants who rumbled in response to encountering bees, demonstrating that elephants communicate danger to others in their herd. King suggests that farmers create "bee fences", hanging hives on poles roughly ten feet apart so that an elephant brushing up against a wire strung between these poles will cause the hive to shake, and the bees will swarm (Houreld 2010). Hives can also be strategically located to prevent elephants from stripping the bark from acacia trees, and bee-keeping can provide an alternative source of honey and extra income for locals. Needless to say, we cannot simply drive elephants away from crops without making sure that they have other viable food sources—we must have effective elephant corridors for both elephant travel and food sources.

## Consumer Choices and Political Pressure

Creating elephant corridors and alternative economies, managing industries, and improving crop deterrents are essential to the protection of elephants. As individuals, we can invest in these alternative industries and activist enterprises. We can also encourage international environmental organizations to form coalitions with organizations abroad in order to support, encourage, and sustain their work.

We should also encourage our governments to invest in long-term solutions and drop funding for ineffective short-term projects. In 1997 The United States government enacted the Asian Elephant Conservation Act (AsECA) and established the Asian Elephant Conservation Fund (administered by the U.S. Fish and Wildlife Service). Early on, the Asian Elephant Conservation Fund paid for electric fence improvements, as well as long-term solutions such as wildlife corridors and alternatives to grazing cattle in areas where they disrupt forest habitats (USFWS 2001). Unfortunately, more than 15 years later, the U.S. government seems unaware of the most effective ways of dealing with human–elephant conflicts. For example, in 2013 these funds supported elephant translocation in Malaysia and the creation of a 776–hectare-long (3 mile) electric fence in India. Only two projects focused on wildlife corridors—one in China and one in Thailand. Other worthy projects included tree planting in India and "improved land-use management" in Malaysia.

Among the many educational projects supported by U.S. funds, only one (in Sri Lanka) educated stakeholders on the "failure of traditional HEC [human–elephant conflict] mitigation and elephant conservation strategies and tools" and "the future direction of HEC mitigation" (USFWS 2013). Individuals can make a difference by writing letters to their government and governments abroad, requesting that funds support only projects that promote long-term solutions, including corridors and education, while discouraging funding for short-term ideas that create more problems than they solve, such as electric fences and translocation.

It is also important to write letters addressing ivory stockpile sales in markets across Africa; residents of non-African nations need to put pressure on African governments to shut down legal "unregulated" ivory markets (see Lemieux and Clarke 2009). Those living in CITES participant countries ought to urge their governments to place elephants in every country on Appendix I, to shut down legal, "regulated" markets, and to support alternative economies. Such comments have forced some countries to destroy ivory in symbolic gestures: In 2012, Gabon destroyed 4.8 tons of ivory. In 2013, the U.S. destroyed 6 tons of ivory while the Philippines destroyed 5 tons. In 2014, France destroyed 3 tons while China destroyed 6 tons of ivory. In the U.S. these symbolic efforts were supported by proposals for powerful and important new legislation: a "near-complete ban" on the import and export of ivory aimed at closing loopholes (Christy 2014). More proactive legislation in China is slower to materialize. China's government benefits financially by selling stockpiles to licensed carving factories, and the price of ivory has recently increased (Laing 2014). Despite the Chinese government's assurances that they are fighting illegal trade in ivory, 90 percent of ivory arrests at Kenyan airports are Chinese nationals headed to China or neighboring countries ('The illegal ivory trade threatening Africa's elephants' 2012).

Activists should urge governments to put trade sanctions on China. Instead of maintaining China's favored trade status, the U.S. government ought to pressure China to help protect endangered species (and honor human rights), and U.S. citizens should pressure their government to make these changes. We should also encourage funding that will bring lasting changes in China. While most of the programs of the African Elephant Conservation Fund focus on anti-poaching efforts across African nations, only one project (funded in 2012) focused on extensive media campaigns to raise awareness in China (USFWS 2012).

We must think about where our money goes not only as citizens and activists, but also as consumers. Do not buy ivory or elephant body parts in any form, in any country, ever. Ultimately, if we are to protect elephants—or our own species—we must reduce both our population (and our consumption). If human populations are significantly reduced, humans and elephants can co-exist peacefully, as they did for centuries prior to the human population explosion.

## Conclusion

Ecosystems all over the world suffer at the hands of humans; the disappearance of elephants and the degradation of their habitat are just two of the many effects of ongoing (seemingly endless) human expansion. As a flagship species—just one of many species desperately in need of fewer humans and more habitat—elephants draw our attention to the relations between individual, community, ecosystems, species, and Earth. Short-sighted approaches to the loss of elephants and their habitat—such as translocation, enslavement, and slaughter—must be replaced with methods that ultimately restore balance—both an ecological (species) and psychological (individual) balance. The collective efforts of animal and Earth advocates are essential if we are to save and protect elephants (and many other species and individuals at risk). Ultimately, animal and Earth advocates must work together to reduce human populations and their demands on the environment if we are to take our rightful place as just one species among many.

## Discussion Questions

1 Are humans invading elephant territory or are elephants invading human territory?

2 Section II begins with a poem, "Bison Sculptures." In what ways is the situation with bison in Montana similar to that of elephants in Asia?

3 Is there room on this human-dominated planet for elephants and bison, and other animals that require large amounts of territory or seclusion?

4 What light does Chalcraft's article shed on the moral problems that arise when we exploit elephants (and other wild animals) in circuses, zoos, and other forms of entertainment, such as television and movies? If zoos are cast as educational, does this change the moral equation?

5 What is the difference between training and torture? What is morally permissible between humans and other animals? Do we "train" children? When and how might it be okay to "train" another individual?

6 Does it make sense to attempt to save animals by selling their body parts?

7 Can you think of any examples of animals who have retaliated against individuals or communities? What does this indicate with regard to animal minds and emotions? Morally speaking, what does this indicate with regard to animals?

## Essay Questions

1 Do you read ingredients before making purchases? Make a list of items you can find in your home that contain palm oil. In what ways does palm oil consumption affect the environment and nonhuman animals? If you were taking action to reduce palm oil consumption in your community, where would you start and why?

2 If you visited Malaysia, would you ride an elephant just to see what it is like? Why or why not?

3 What steps would you take to decrease the value of ivory? What steps might you take to try to bring change in China on behalf of animals who are endangered by Traditional Chinese Medicines and/or the ivory trade?

## Suggested Further Reading

Anthony, L. (2009) *The Elephant Whisperer: My Life with the Herd in the African Wild*. New York: St. Martin's Press.

Bradshaw, G. A. (2009) *Elephants on the Edge: What Animals Teach Us about Humanity*. New Haven: Yale University Press.

Moss, C. J. (2000). *Elephant Memories: Thirteen Years in the Life of an Elephant Family*. Chicago: University of Chicago Press.

## References

'Activists denounce Thailand's elephant "Crushing" ritual' (2002, 16 October) *National Geographic Today*, http://news.nationalgeographic.com/news/2002/10/1016_021016_phajaan.html, accessed 25 April 2014.

Allimadi, M. (2007, 1 January) 'Profile: Dr. Eve Abe, Ethologist', *Black Star News*, http://www.blackstarnews.com/global-politics/others/profile-dr-eve-abe-ethologist.html, accessed 25 April 2014.

Asian Animal Protection Network (n.d.) https://groups.yahoo.com/neo/groups/aapn/conversations/topics/16914?source=1&var=1, accessed 10 October 2014.

'Asian elephant' (n.d.) *It's Nature: Endangered Species*, http://itsnature.org/endangered/asian-elephant/, accessed 25 April 2014.

Block, B. (2009, April 24) 'Can "sustainable" palm oil slow deforestation?' *Worldwatch Institute*, http://www.worldwatch.org/node/6082, accessed 25 April 2014.

Bhuiyah, M. (2014) 'Coal mining fuels the human-elephant conflict in Jharkhand', *Video Volunteers*, https://www.youtube.com/watch?v=O3J4JFwOHA4, accessed 25 April 2014.

Bradshaw, G. and Lindner, L. (n.d.) 'Post-traumatic stress and elephants in captivity,' http://www.elephants.com/joanna/Bradshaw&Lindner_PTSD-rev.pdf, accessed 25 April 2014.

Bradshaw, G. A., Schore, A. N., Brown, J. L., Poole, J. H., and Moss, C. M. (2005) 'Elephant breakdown,' *Nature*, vol. 433, p. 807, http://www.helpelephants.com/pdf/NATURE%20ELEPHANT%20BREAKDOWN.pdf , accessed 25 April 2014.

'Bring The Elephant Home: Help us to plant trees for elephants' (n.d.) http://www.bring-the-elephant-home.org/help-bteh/plant-trees-for-elephants/, accessed 3 November 2014.

Coghlan, A. (2010, 22 March) 'Corruption raises doubts over ivory sales,' *New Scientist*, http://www.newscientist.com/article/dn18654-corruption-raises-doubts-over-ivory-sales.html?full=true&print=true, accessed 25 April 2014.

'COMACO [Community Markets for Conservation]—Transform a poacher' (2010) http://www.itswild.org/transform-poacher, accessed 26 April 2014.

Courouble, M., Hurst, F., and Milliken, T. (2003) 'More ivory than elephants: Domestic ivory markets in the three West African countries,' Cambridge, UK: TRAFFIC

International, *TRAFFIC Online Report Series, no 8,* http://www.traffic.org/species-reports/traffic_species_mammals24.pdf, accessed 25 April 2014.
Christy, B. (2014, 11 February) 'United States tightens the noose on the ivory trade,' *National Geographic Daily News,* http://news.nationalgeographic.com/news/2014/02/140211-united-states-rules-wildlife-trafficking-ivory-science/, accessed 28 April 2014.
'Crack down on domestic ivory markets in Africa needs major boost' (2005) http://scienceinafrica.com/old/index.php?q=2005/july/ivory.htm, accessed 3 May 2014.
Eccleston, P. (2008, 28 October) 'African elephants afraid of roads because they mean danger.' *Daily Telegraph,* http://www.telegraph.co.uk/earth/earthnews/3353917/African-elephants-afraid-of-roads-because-they-mean-danger.html, accessed 25 April 2014.
'ElefantAsia: Elephant baby bonus' (2014), http://www.elefantasia.org/index.php/en/component/k2/item/94-elephant-baby-bonus, accessed 3 October 2014.
'Elephant corridors in India' (2011) http://www.outdoorconservation.eu/project-detail.cfm?projectid=24, accessed 3 May 2014.
Elephantsanctuarymalaysia (2011) 'History of Kuala Gandah Elephant Sanctuary,' http://elephantsanctuarymalaysia.blogspot.com/2011/10/.html, accessed 5 October 2014.
'Elephants in conflict' (2009, 17 June) http://www.elephantvoices.org/threats-to-elephants/-squeezed-into-conflict.html, accessed 25 April 2014.
Evans, K. (2010) 'Conservation and Welfare Implications of Releasing Captive-Raised African Elephants,' *Compassionate Conservation: Animal Welfare in Conservation Practice,* University of Oxford, Lady Margaret Hall, Oxford, UK, Lecture.
Fernando, P., Leimgruber, P., Prasad, T., and Pastorini, J. (2012, December) 'Problem-elephant Translocation: Translocating the problem and the elephant?' *Plos One,* vol. 7, no 12 , pp. 1–9, http://www.plosone.org/article/fetchObject.action?uri=info%3Adoi%2F10.1371%2Fjournal.pone.0050917&representation=PDF, accessed 3 May, 2014.
Fitzherbert, E. B., Struebig, M. J., Morel, A., Danielsen, F., Brühl, C. A., Donald, P. F., and Phalan, B. (2008, October) 'How will oil palm expansion affect biodiversity?' *Trends in Ecology and Evolution,* vol. 23, no 10, pp. 538–545, http://www.cell.com/trends/ecology-evolution/abstract/S0169-5347%2808%2900252-8, accessed 23 April 2014.
'Forest Dept struggles to guard elephant corridors' (2008, 17 July) *Express India,* http://www.expressindia.com/latest-news/forest-dept-struggles-to-guard-elephantcorridors/337231/, accessed 13 December 2010.
Garai, M. E. and Carr, R. D. (2001, July–December) 'Unsuccessful introductions of adult elephant bulls to confined areas in South Africa,',*Pachyderm,* vol. 31, pp. 52–57, http://www.african-elephant.org/pachy/pdfs/pachy31.pdf, accessed 26 April 2014.
Haviland, C. (2010a, 27 October) 'Sri Lanka's deadly clash with elephants', *BBC News South Asia,* http://www.bbc.com/news/world-south-asia-11596075, accessed 3 May 2014.
Haviland, C. (2010b, 17 August) 'Sri Lanka wildlife vets go on strike,' *BBC News South Asia,* http://www.bbc.co.uk/news/world-south-asia-10995118, accessed 26 April 2014.
Handwerk, B. (2005, 3 June) Quoting Joyce Pool of the Amboseli Elephant Research Project, in 'Elephants attack as humans turn up the pressure, *National Geographic News,* http://news.nationalgeographic.com/news/2005/06/0603_050603_elephants.html, accessed 26 April 2014.
Highfield, R. (2006, 17 February) 'Elephant rage: They never forgive, either,' *Sydney Morning Herald,* http://www.smh.com.au/articles/2006/02/16/1140064206413.html, accessed 26 April 2014.
Hoare, R. (1999) 'Assessing the evidence for the existence of habitual problem elephants,' *IUCN African Elephant Specialist Group Report,* http://www.african-elephant.org/hec/pdfs/hecprobele.pdf, accessed 26 April 2014.

Hoare, R. (2001) 'Management implications of new research on problem elephants,' *Pachyderm,* vol. 30, pp. 44–48, http://www.african-elephant.org/pachy/pdfs/pachy30.pdf, accessed 26 April 2014.

Houreld, K. (2010, 5 May) 'Buzz off! Elephants flee bees, new research shows,' *AP News,* http://www.marinij.com/news/ci_15025228?source=rss, accessed 26 April 2014.

Huggler, J. (2006, 12 October) 'Animal behavior: Rogue elephants,' *The Independent,* http://www.independent.co.uk/environment/animal-behaviour-rogue-elephants-419678.html, accessed 26 April 2014.

Hunter, N., Martin, E., and Milliken, T. (2004) 'Determining the number of elephants required to supply current unregulated ivory markets in Africa and Asia,' *Pachyderm,* vol. 36, pp. 116–128, http://www.african-elephant.org/pachy/pdfs/pachy36.pdf, accessed 26 April 2014.

'The illegal ivory trade threatening Africa's Elephants' (2012, 11 April) *BBC News World,* http://www.bbc.co.uk/news/world-17675816, accessed 26 April 2014.

'Indian Elephant corridors appeal' (n.d.) http://www.worldlandtrust.org/projects/india.htm, accessed 25 April 2014.

'Introducing elephant partners—Maasai Mara' (2010, 25 November) http://www.elephantvoices.org/news-media-a-reports/82-conservation/762-introducing-elephant-partners-maasai-mara.html, accessed 26 April 2014.

Johnsingh, A. J. T. and Joshua, J. (1994) 'Conserving Rajaji and Corbett National Parks: The elephant as a flagship species', *Oryx,* vol. 28, no 2, pp. 135–140.

Kurt, F. and Garai, M. (2001, June) 'Stereotypies in captive Asian elephants: A symptom of social isolation', *Scientific Progress Reports,* Vienna, pp. 57–63, http://www.helpelephants.com/pdf/KURT%20STEREOTYPY%20PAPER.pdf, accessed 26 April 2014.

Laing, A. (2014, 6 January) 'China destroys ivory stockpile in "significant symbolic step towards saving Africa's elephants"', *The Telegraph,* http://www.telegraph.co.uk/news/worldnews/asia/china/10553779/China-destroys-ivory-stockpile-in-significant-symbolic-step-towards-saving-Africas-elephants.html, accessed 28 April 2014.

Lemieux, A. M. and Clarke, R. V. (2009) 'The international ban on ivory sales and its effects on elephant poaching in Africa,' *British Journal of Criminology,* vol. 49, pp. 451–471, http://bjc.oxfordjournals.org/content/49/4/451.full, accessed 26 April 2014.

'Malaysia targets Africa and the Amazon for palm oil expansion' (2008, 25 August) http://news.mongabay.com/2008/0825-malaysia_palm_oil.html, accessed 26 April 2014.

'Malaysia Traveller: Kuala Gandah Elephant Sanctuary' (2014) http://www.malaysia-traveller.com/kuala-gandah.html, accessed 3 November 2014.

Mangat, R. (2008, 17 March) 'Amboseli: Dim future for the elephants,' *The East African,* http://www.theeastafrican.co.ke/magazine/-/434746/258418/-/14098ml/-/index.html, accessed 26 April 2014.

Milliken, T., Pole, A., and Huongo, A. (2006, April) 'No peace for elephants: Unregulated domestic ivory markets in Angola and Mozambique,' TRAFFIC International. *TRAFFIC Online Report Series, no 11,* Cambridge, UK, http://www.traffic.org/species-reports/traffic_species_mammals26.pdf, accessed 26 April 2014.

Movahed, N. (2014, 11 January) 'Reducing human–elephant conflict: But at what cost?' http://publicpolicyindonesia2014.wordpress.com/2014/01/11/reducing-human-elephant-conflict-but-at-what-cost/, accessed 27 April 2014.

Mukherjee, K. (2014, 25 January) 'Human, elephant conflict takes a new dimension,' *The Times of India,* http://timesofindia.indiatimes.com/home/environment/developmental-issues/Human-elephant-conflict-takes-a-new-dimension/articleshow/29328901.cms, accessed 26 April 2014.

Office of Transportation and Air Quality (2011, December) 'EPA Issues Notice of Data Availability Concerning Renewable Fuels Produced from Palm Oil under the RFS Program,' *EPA Regulatory Announcement*. EPA-420-F-11-046, http://www.epa.gov/oms/fuels/renewablefuels/documents/420f11046.pdf, accessed 26 April 2014.

'Orphans Project' (n.d.) David Sheldrick Wildlife Trust, http://www.sheldrickwildlifetrust.org/asp/orphans.asp, accessed 26 April 2014.

Orr, M. (2010, 27 December) 'Trunks up for Flight Centre,' http://www.facebook.com/#!/note.php?note_id=177329418954412&id=100000379405326, accessed 26 April 2014.

'Pinnewala Elephant Orphanage in Sri Lanka' (n.d.) http://www.elephant.se/location2.php?location_id=43&show=1, accessed 26 April 2014.

Pinter-Wollman, N., Isbell, L. A., and Hart, L. A. (2009) 'Assessing translocation outcome,' http://www.elephantconservation.org/programs/archive-of-projects/assessing-translocation-outcome/, accessed 3 May 2014.

Poole, J. (1996) *Coming of Age with Elephants: A Memoir*, Hyperion Press, New York.

Poole, J. and Granli, P. (2008) 'Mind and movement: Meeting the interests of elephants', http://elephantvoices.wildlifedirect.org/files/2009/04/chapter1_mindandmovement.pdf, 26 April 2014.

Preetha, M. S. (2013, 5 February) 'Forest Department measures to protect elephant corridors,' *The Hindu*, http://www.thehindu.com/todays-paper/tp-national/tp-tamilnadu/forest-department-measures-to-protect-elephant-corridors/article4380826.ece, accessed 25 April 2014.

'Protected contact' (n.d.) http://www.elephants.com/protected.php, accessed 4 May 2014.

'Rainforest art' (n.d.) http://www.indonesianrainforest.org/indonesian-rainforest/rainforest-art/, accessed 26 April 2014.

Reeve, R., Tuite, T., Gabriel, G., Bell, J, and Pueschel, P. (2003) 'The proposed sale of ivory from Botswana, Namibia and South Africa: Conditions and verification,' *Pachyderm,* vol. 35, July–December, pp. 115–131, http://www.african-elephant.org/pachy/pdfs/pachy35.pdf, accessed 25 April 2014.

'Reforestation plan', (n.d.) http://www.indonesianrainforest.org/indonesian-rainforest/reforestation-plan/, accessed 25 April 2014.

'Reforestation responses' (n.d.) http://www.indonesianrainforest.org/indonesian-rainforest/reforestation-responses/, accessed 25 April 2014.

'Reversing trends of charcoal making: A growing challenge for COMACO' (2014) http://www.itswild.org/n0608/reversing-trends-charcoal-making, accessed 25 April 2014.

Siebert, C. (2006, 08 October) 'An elephant crackup?' *The New York Times*, http://www.nytimes.com/2006/10/08/magazine/08elephant.html?_r=1&pagewanted=, accessed 25 April 2014.

Siebert, C. (2009) *The Wauchula Woods Accord: Toward a New Understanding of Animals*, Scribner, New York.

Slotow, R. and van Dyk, G. (2001) 'Role of delinquent young "orphan" male elephants in high mortality of white rhinoceros in Pilanesburg National Park, South Africa,' *Koedoe*, vol. 44, no 1, pp. 85–94, http://www.koedoe.co.za/index.php/koedoe/article/view/188, accessed 25 April 2014.

Slotow, R., van Dyk, G., Poole, J., Page, B., and Klocke, A. (2000) 'Older bull elephants control young males', *Nature*, vol. 408, pp. 425–426.

'Some Facts about the Asian Elephant ' (n.d.) http://www.trekkingcentrallaos.com/html/elephants_info.html, accessed 5 October 2014.

Spelman, K. (2014) 'Together, we save forests: Rainforest protection must include bottom-up approaches that engage the people living in the forests', http://ensia.com/voices/together-we-save-forests/, accessed 23 April 2014.

'Stereotypic behavior and behavioural disorders in elephants' (n.d.) http://www.upali.ch/stereotypicbehaviour_en.html, accessed 3 May 2014.
'Strategic goals' (n.d.) http://www.elephantnaturefoundationuk.org/?s=strategic+goals&search submit=, accessed 26 April 2014.
Sukumar, R. (1992) *The Asian Elephant: Ecology and Management*, Cambridge University Press, Cambridge UK.
Theisen, W. (2005, 5 June) 'Elephant rage', *National Geographic Explorer Blog*, https://groups.yahoo.com/neo/groups/nhnenews/conversations/topics/9325, accessed 26 April 2014.
Timberg, C. (2008, 26 February) 'South Africa to resume elephant culling,' *Washington Post*, http://www.washingtonpost.com/wp-dyn/content/article/2008/02/25/AR2008022500970.html, accessed 26 April 2014.
'Trekking' (n.d.) http://www.elemotion.org/elephant-tourism/trekking-2/, accessed 25 April 2014.
US Fish and Wildlife Service (2001) Asian Elephant Conservation Act: *Summary report 1999–2001*, https://www.fws.gov/international/pdf/summary-report-asian-elephant-1999-2000.pdf, accessed 27 April 2014.
US Fish and Wildlife Service (2012) Wildlife Without Borders: *African Elephant Conservation Fund*, https://www.fws.gov/international/pdf/project-summaries-african-elephant-2012.pdf, accessed 28 April 2014.
US Fish and Wildlife Service (2013) *Asian Elephant Conservation Fund FY2013*, https://www.fws.gov/international/pdf/project-summaries-asian-elephant-2013.pdf, accessed 27 April 2014.
van Staaden, H. (2010, 22 May) 'Tourists watch in horror as elephant gored', *News24.com*, http://www.news24.com/SouthAfrica/News/Tourists-watch-in-horror-as-elephant-gored-20100502, accessed 26 April 2014.
'Villagers at NAHI door to give land: 153 families offer land for NH31D four-laning' (2013, November 19) *The Telegraph*, http://www.telegraphindia.com/1131119/jsp/siliguri/story_17585632.jsp#.U2VX8FcWmIA, accessed 3 May 2014.
'Wildlife trade: Elephant ivory trade FAQs' (n.d.) http://www.worldwildlife.org/what/globalmarkets/wildlifetrade/faqs-elephant.html, accessed 26 April 2014.
Williams, C. (2006) 'Elephants on the edge fight back,' *New Scientist*, vol. 189, no 2539, p. 39, http://www.elephantvoices.org/multimedia-resources/document-download-center/category/100-articles-interviews-popular-media.html?download=128:elephants-on-the-edge-fight-back, accessed 3 May 2014.
'World: South Asia—India elephant rampage' (1998, 24 December), BBC News, http://news.bbc.co.uk/2/hi/south_asia/241781.stm, accessed 26 April 2014.
World Wildlife Fund—Issues: Timber trade and illegal logging (n.d.) http://wwf.panda.org/what_we_do/endangered_species/elephants/asian_elephants/areas/issues/habitat_loss_fragmentation/timber_trade_illegal_logging/, accessed 3 November 2014.

# 8

# TRAPPED

## Individuals, Species, and Ecosystems at Risk

*Lisa Kemmerer and Anja Heister*

"I don't want the cheese, I just want out of the trap."

(Spanish Proverb)

### Skiing in Montana—Anja's Story

Something dark between the limbs of a nearby fir tree caught my eye when I was skiing with my partner, Dave, and his son, admiring mountains and snow-covered trees in Montana public lands. We swerved for a closer look, and found a small marten hanging from a trap by her front leg. The visual brought to mind old photos of Ku Klux Klan lynchings: head fallen to the side, lifeless body swaying lightly in the breeze. Her dark eyes looked frightened—she would never have willingly allowed a human to come this close to her, but she was utterly helpless.

Dave wrapped his coat around the pine marten while I released the chain that secured the trap to the tree's trunk. Once free, Dave placed the pine marten gently on the snow's crusty surface, and we worked frantically to free her leg. It took us about five minutes to figure out how to release the spring-loaded jaws. The marten remained motionless after that, and we feared the worst, but in the next instant she was on her feet, hobbling away, dragging her injured leg. She paused for a moment, then took flight again, and did not stop until safely beyond our reach. Then she stopped to look back at us, as if to say thanks. The three of us cried as we watched, pondering the incredible pain and suffering she had endured, shuddering at the thought of what would have happened if we had not found her, and wondering how she might possibly survive with such a damaged leg.

I have never shaken the image of that skinny young marten held fast in a steel-jawed foothold trap, dangling from a tree alongside a trail on public lands—perhaps for several days and nights—in an otherwise lovely Montana National Forest. Because of her, I co-founded Footloose Montana to spearhead an initiative to end recreational and commercial trapping on Montana's public lands, where I have worked for the past five years.

## Montana's Furbearer Trapping Program

The use of body-gripping traps and snares for commercial and recreational trapping is banned on public lands in California, Colorado, Massachusetts, and Arizona, but most states still allow trapping for profit and pleasure, including Montana (Reed 1999a). Montana's Furbearer Trapping Program (MFTP), administered by the Montana Department of Fish Wildlife and Parks (MDFWP), offers an apt example of U.S. trapping regulations and methods—and of the problems and moral issues that arise wherever trapping is legal.

There are somewhere between 250,000 and 750,000 trappers in the U.S., fluctuating with demand and the commercial value of pelts. For example, a growing demand for fur in Russia, Europe, and Asia likely spiked trapper licenses for 2007–2008; 1,200 additional Montana trapping licenses were sold when the price of bobcat pelts rose to nearly $450 (Giddings 2008; Babcock 2007); trapping spiked again when Montana added a trapping season for gray wolves (*Canis lupus*) in 2012. Even so, less than half of one percent of Montanans purchase a Furbearer Trapping License in any given year (Reed 1999b).

A Montana conservation/furbearer trapping license for a Montana resident (12 years and older) costs $28 and a wolf trapping license costs $28, insignificant in comparison with the value of even one pelt ("Furbearer: Montana" 2013). Fur prices were as follows in 2012–2013: Bobcat $589, wolverine $235, fisher $145, marten $84, badger $25, otter $113, coyote $94, raccoon $27, beaver $31, red fox $66, mink $29, muskrat $12, weasel $3, and skunk $4 ("Average" 2013).

Trapper target species are placed in one of three categories:

- "Furbearer" is an official U.S. Fish and Wildlife Services (USFWS) designation for species with commercially valuable fur, includes only ten species: beaver, otter, muskrat, mink, marten, fisher, wolverine, bobcats, swift fox, and lynx.
- "Predators" include only the coyote, weasel, skunk, and civet cat (spotted skunk).
- Wolves are generally also considered predators, but wolves lost their endangered species protection in Montana in 2011. Consequently, until the MDFWP decides that they will be reclassified as either "game" or a furbearer species, wolves are categorized separately as a species "in need of management."
- "Nongame Wildlife" includes species without other legal classification, such as badger, raccoons, and red fox ("Montana Furbearer" 2013 and "Furbearer: Montana" 2013).

Regulations and licensing only apply to Furbearers and wolves. *Montana residents can trap Predators and Nongame Wildlife free of charge.* There are no regulations, no quotas, and no licenses for Predators or Nongame Wildlife—excluding wolves ("Furbearer: Montana" 2013).

A high percentage of Furbearers are rare (lynx and otters, for example), others dangerously rare and likely to be listed as Threatened in the near future (such as wolverines, fishers, and martens). Five Furbearers—bobcat, fisher, otter, swift fox and wolverine—are already "protected" by quotas, which are established to limit "take," thereby protecting populations and species from extirpation or extinction. Montana's 2013–2014 quotas were as follows: 7 fishers, 30 swift foxes, 95 otters, and 1,895 bobcats. One district's bobcat quota (now closed) exceeded the limit by 25 individuals—and these are only *reported* kills ("Furbearer Trapping Guide" 2013).

State regulations do not apply to species protected federally: Species listed as Threatened/Endangered under the U.S. Endangered Species Act (ESA) are regulated at federal (USFWS) and international levels. For example, to protect endangered species, CITES (Convention on International Trade in Endangered Species) requires that quotas be set by member nations for the killing of any species-at-risk, and that a federal permit accompany any exported pelt belonging to an endangered species—or any species similar in appearance. The U.S. ESA lists lynx as Threatened, and because bobcats resemble lynx, CITES requires that each of these species be regulated, and that their furs travel with a government permit.

State regulations require federally protected animals to be released *when possible*. Injuries caused by neck snares frequently "lead to death" after an animal is released (Iossa et al. 2007, 344). Seemingly minor injuries to these predators, such as a broken tooth (from biting at steel traps or metal wires), can be life-threatening because they reduce hunting efficiency. Threatened/Endangered hunters who are missing teeth or limping, have little chance of survival. If a trapper decides that the victim is unlikely to survive, s/he must notify MDFWP immediately to determine "disposition and/or collection of the animal" ("Furbearer: Montana" 2013).

State regulations also require trappers to notify MDFWP within 24 hours after killing any of the six species regulated by quotas. When a quota has been reached, MDFWP closes the season within 48 hours ("Furbearer: Montana" 2013, 5). Additionally, MDFWP requires trappers to present pelts taken from any one of the six species that have quotas, as well as those of martens, to be tagged. Finally, trappers are required to present trapped fishers, otters, or wolverines, and the skulls of any bobcats or martens for data collection ("Furbearer: Montana" 2013). Using these remains, MDFWP biologists determine the age, sex, and diet of the deceased (in the hope of better understanding population dynamics).

With avid trapper Brian Giddings at the helm, Montana's Furbearer Trapping Program establishes trapping regulations, including quotas and seasons. Regulations tend to protect trapper interests rather than ecosystems or the interests and needs of wildlife. There are no restrictions on the number of people who may purchase a Montana trapping license, on the number of traps they may set, or on the length of traplines[1]—some trappers operate 100- to 200-mile traplines ("Minutes" 2008, 12). Regulations note that "traps *should* be checked at least once every 48 hours," but this recommendation carries no force ("Furbearer: Montana" 2013; emphasis added). (Forty-eight hours is a very long time, indeed, to wait with body parts crushed by a metal device, deprived of food, water, and shelter). While trappers frequently claim that pelts degenerate quickly if traps are not checked often, this is certainly not the case when temperatures are below zero—trapping season is throughout winter. That said, how could any trap-check time requirement be enforced?

Quotas, generally designed to protect species-at-risk, permit species-at-risk to be targeted by trappers. *Common sense suggests that any species in need of quota protection ought to be fully protected from trapping and snaring.* Instead, quotas are set and trapping continues … until the federal government steps in, listing a particular species as Threatened/Endangered, thereby protecting that particular species from being targeted for as long as it remains listed. Even when listed by the federal government, trapping for *other* species is still permitted in core habitat areas for Endangered/Threatened species—which is to say, these animals are still caught and killed in traps and snares.

Voluntary MDFWP surveys over the course of ten years indicate that an average of 50,000 animals are *reported* as trapped in Montana every year ("State" 2005, table, p. 2); MDFWP recorded a total of 66,919 wild animals killed by trappers in Montana in 2011–2012 ("Furbearer Program" 2013). It is important to note that the return rate of completed trapper surveys is a mere 35 percent ("Furbearer Program" 2013). Trappers, who have a vested interest in keeping trapping seasons open and quotas high, are the primary source of data used to determine trapping regulations (Dickson 2006). The effectiveness of regulations *depends* on trapper reports. How can government agencies adequately regulate trapping or monitor species numbers with such sparse and biased data?

It is impossible to know what percentage of trappers comply with regulations. Montana is a big state with vast wilderness. How would anyone know if a trapper failed to present pelts for tagging, or continued to target a quota species long after the quota had been met and the season closed? How many trappers bring in the body of an endangered species *knowing* that the pelt (sometimes worth hundreds of dollars) will be confiscated? Trapping regulations are notoriously difficult to enforce, and it is impossible to know how many animals, or which species, are *actually* trapped each year.

Trappers constitute less than one percent of Montana's population. While trappers walk away with the profits, taxpayers pick up the tab for damage done. For example, swift foxes were trapped out of Montana and reintroduced at taxpayer expense. Once reintroduced, *despite taxpayer investment in every extant swift fox in Montana,* MDFWP opened a trapping season on swift foxes, whose pelts currently sell for $50–$125 online ("Prices" n.d.; "Swift Fox Skins" n.d.). Wolves were similarly extirpated from Montana (Government hunters reported the last wolf ill in the Yellowstone area in 1940), and reintroduced at taxpayer expense, and as soon as they were delisted from the U.S. ESA trapping was permitted. At the 2014 Fur Harvesters Auction, gray wolf pelts marketed for $164–$625 ("NAFA" 2014). Why are trappers—largely responsible for their extirpation—permitted to profit from even one swift fox or wolf pelt when taxpayers have paid the bill for reintroduction?

Even researchers only guess at current wolverine (*Gulo gulo*) populations (partly because a high percentage of radio-collared wolverines are killed in traps, and partly because this species is very reclusive). A minimum of 400 breeding *pairs* are essential to long-term genetic viability, yet it appears that there are only between 250–300 wolverines remaining in the lower 48 states (Schwartz et al. 2009). Current estimates place Montana's wolverine population at a mere 200, *with a breeding population of only 34 individuals* (Cegelski et al. 2006, 197). Trappers reduced the wolverine population in Montana's Pioneer Mountains by an estimated 50 percent between 2003 and 2005 (Squires et al. 2007). An unnerving 77 percent of mature females trapped in Montana were pregnant (Anderson and Aune 2008). Despite ongoing debate over whether or not to list wolverines as Threatened, MDFWP maintained a trapping quota of 5 until 2012 ("Wolverine" n.d.). Though they are not currently targeted due to legal action aimed at getting wolverines protected, as wide-ranging predators, wolverines are extremely vulnerable to *any* traps set in their habitat (Wuerthner 2013).

Trapping is responsible for 86 percent of lynx (*Lynx canadensis*) deaths. Lynx, like wolverines, are far-ranging predators with a slow reproductive rate. MDFWP maintained a trapping season for lynx until threatened with a lawsuit in 1999 (Fox 2004a, 26). In February of 2000, USFWS listed the Canada lynx as Threatened ("Montana Field Guide: Canada Lynx" n.d.), preventing trappers from targeting these big cats. Nonetheless, trapping continues in core habitat, and at least nine lynx have been trapped in Montana since 2000 ("Lawsuit" 2013). Consequently, a coalition of conservation organizations filed suit in 2013 against MDFWP on behalf of lynx.

The swift fox (*Vulpes velox*), North America's fastest and smallest (5–7 pounds/2.2–3.2 kg) native canine, was once abundant across the Great Plains. Victims of ubiquitous trapping and predator poisoning in the 1900s, they are believed to have become extinct in Montana by 1969 ("Montana Field Guide: Swift Fox" n.d.). They were reintroduced in 1998 at the 1.5 million-acre Blackfeet Reservation in northern Montana (Stephens and Anderson 2005). In 2006, yet more swift foxes were released at the Fort Peck Reservation in Eastern Montana ("Basic" n.d.). MDFWP hastily opened a swift fox trapping season with a quota of 20. Trappers refer to swift fox as "naïve" (easy to trap). Official records for the 2012–2013 trapping season report 21 swift foxes trapped—even *reported* kills were above the established quota ("Furbearer Trapping Guide" 2013).

In 2003 a MDFWP wildlife biologist noted that fishers (*Martes pennanti*) are "one of the lowest density carnivores in the state," such that fieldwork across three winters yielded only 11 incidences of verifiable evidence of fisher presence (Vinkey et al. 2006, 61). Fishers were trapped out of Montana before the 1930s, reintroduced (from British Columbia and Minnesota) in 1959, and again between 1988 and 1991 ("Endangered" 2011). Despite these efforts, the only significant fisher population in the Northern Rockies lives along the Montana–Idaho border (Bitterroot and Clearwater regions). Nonetheless, MDFWP maintains a fisher trapping quota of seven. While the west coast population was accorded federal protection in April 2004, fishers in the Northern Rockies were denied ESA protection in 2011—despite their rarity and susceptibility to trapping ("Fisher" n.d.). In Idaho, where fishers are listed as Critically Imperiled, a 2012 nonprofit Freedom of Information

(FOIA) request revealed that traps set for wolves impact fishers heavily (Cole 2013). As long as it is legal to trap fishers—or even place traps and snares in critical habitat—this rare and slow-to-reproduce predator will continue to decline.

For more than 40 years, researchers have emphasized the importance of protecting martens (*Martes americana*). They identify trapping as a "major cause for local decline" and recognize this species as at risk of a threat for "extirpation" (Buskirk and Ruggiero 1994, 26, 29). Nonetheless, martens are only listed by MDFWP as in need of moderate conservation. According to avid trapper and head of MDFTP, Brian Giddings, "this popular furbearer is the key species in western Montana mountain trapping and has been attracting more interest in recent years. Trappers take roughly 1,000 marten each year. Pelts are valuable, bringing about $25 to $40 on the fur market" (Fraley 2002).

Trappers extirpated river otters (*Lontra canadensis*) from most of their U.S. habitat. Since 1976 more than 4,000 otters have been reintroduced in 21 states ("IUCN" 2011). They remain a "sensitive species," listed in Appendix II of the Convention on the International Trade in Endangered Species (CITES), yet MDFWP only lists river otters as a species with moderate conservation needs. Though their populations appear to be declining (again), Montana quotas allow trappers to take a whopping 95 river otters every season ("Montana Field Guide: River Otter" n.d.).

Grey wolves *(Canis lupus),* once numbering about 2 million in North America, were nearly extirpated (on behalf of ranchers and hunters) by the U.S. government's (ongoing) Predator Control Program. In the 1960s, only a smattering of wolves remained (in a portion of Minnesota and Michigan). In 1995, at taxpayer expense, wolves were reintroduced to Yellowstone National Park, and in April of 2011, a rider attached to a fast-track budget bill, stripped Montana wolves of their ESA protected status—the first time that a species was delisted by politics (rancher, trapper, and hunter interests) rather than via scientific data ("Restoring" n.d.). Immediately after delisting, MDFWP classified wolves as "in need of management" and opened a hunting season; hunters killed 166 wolves in 2011 (Zuckerman 2012). In 2012 an additional leg-hold trap season was initiated *without any quota* (King 2013). Official records indicate that hunters and trappers killed 225 grey wolves in 2012–2013 ("FWP" 2013) in addition to the 104 wolves killed by the federal government's ongoing Predator Control Program. Among the dead were seven wolves wearing GPS research collars who were killed right outside Yellowstone National Park (Brown 2007; Maughan 2012).

## Trapping for U.S. Fish and Wildlife Services (USFWS) Predator Control Program

U.S. government wildlife agencies continue to operate on the misguided and outdated belief that certain species pose a significant economic threat to ranchers and the interests of hunters, that these interests are worth protecting *even at the expense of ecosystems*, and that killing great numbers from certain species is a viable solution to this supposed problem. The U.S. government Predator Control Program, run by Wildlife Services, spends "over $100 million annually" to poison, trap, and shoot "several million animals" every year, including "black and grizzly bears, beavers, mountain lions, coyotes, and wolves" (Keefover-Ring 2009, i). Thanks to the U.S. government's Predator Control Program, trappers (as well as hunters and poison patrol) are employed and engaged each year at taxpayer expense.

## Trapping for Science

MDFWP promotes trapping as "important" for "wildlife research," conflating trapping for science with trapping for profit and recreation ("Wildlife" 2010). Trapping for science, especially live-trapping, is not contentious—trapping for recreation and profit *is* contentious. Scientific trapping is organized and executed by wildlife "professionals," presumably on behalf of wildlife populations. For example, live traps used to relocate garbage-can-raiding bears or to reintroduce extirpated species. By conflating personal self-interests (pleasure and profits) with scientific wildlife agendas, trappers and trapping organizations attempt to garner public support to safeguard contentious recreational and commercial trapping.

## Trapping on "Public" Lands

Thirty-five percent of Montana's lands are public, presumably held by the government "in trust for the American people," but trapping is legal on public lands; trap and snare casualties occur *primarily* on Montana's public lands ("Public Lands" n.d.). During the 2012–2013 trapping season, distressed citizens reported more than 50 dogs caught in traps (most of which were set on public lands). Two of these dogs died in traps. In the same year, two cats lost limbs because of traps.

Furbearer trapping regulations only require that traps be placed 30 feet from the centerline of any public road, 50 feet from roads and hiking trails on public lands, and 300 feet from designated trailheads ("Furbearer: Montana" 2013, 3). There are no regulations whatsoever for those trapping Predators and Nongame Wildlife. Although trappers constitute less than half of one percent of Montana residents, they endanger all who venture onto public lands ("Public Lands" n.d.). Why are trappers allowed to profit from and destroy wildlife held in public trust at risk to other citizens and their companion animals? Inasmuch as the vast majority of citizens would prefer to enjoy public lands as natural ecosystems, and as safe places for children and companion animals, trapping ought to be banned. I trust that someday an informed public will ban traps from public lands.

## Traps and Snares

Traps and snares were designed by trappers for trappers. They are not designed to be humane; they are designed to protect pelts and thereby maximize trapper profits.

## Steel-Jaw Foothold/Leghold Traps

On average, trappers own about 205 traps, foothold traps being the majority (59%), followed by body-gripping traps (22%) and snares (11%) (Fox 2004a). Steel-jawed foothold (or leghold) traps are the most common. They are designed to clamp onto an animal's leg when a foot lands on a steel plate, triggering two spring-loaded metal jaws. There are two types of foothold traps: long-spring and coil-spring; the latter is generally preferred because the grip is stronger. They come in various sizes, but are perhaps most frequently set to catch coyotes, bobcats, raccoons, and skunks. Foothold traps can also be set underwater to drown muskrats, otters, mink, and beavers (Fox 2004a).

Foothold traps cause severe bruising, deep lacerations, and fractures. They dislocate joints and damage teeth and gums (because animals caught and in extreme pain tend to desperately attack and bite traps). Animals caught in steel-jawed traps sometimes die of exposure because they are immobilized in extreme winter weather without access to food, water, or shelter. They can also perish when predators attack. Some animals escape traps by twisting in circles, dislocating their joint, then chewing through skin, flesh, ligaments, and tendons, leaving their foot behind in the trap in order to escape (Stone 2010). They have little hope of survival. This desperate act is called "wring off." For those who walk in Montana, it is not uncommon to come upon a trap gripping only a paw.

Foothold traps fail to meet the most basic animal welfare standards, and though they remain legal, they have been condemned as inhumane by the American Veterinary Medical Association, the American Animal Hospital Association, the World Veterinary Association, and the National Animal Control Association (Fox 2004a).

## Conibear Traps

The body-gripping Conibear trap, named after Canadian inventor and trapper Frank Conibear, "consists of two metal frames hinged at the center point and powered by two torsion springs to create a scissor-like action" (Fox 2004a, 34). Their baited trigger "is designed to snap shut on an animal's spinal column at the base of the skull, causing a fatal blow," but in order for this to happen, the appropriate size animal must enter the trap at the optimal speed and proper angle (Fox 2004a, 34). Releasing this trap requires both knowledge and considerable strength—usually two people. Consequently, when a dog is caught in a Conibear trap, there is little that terrified, desperate onlookers can do... except watch their dog die.

Conibear traps are commonly set underwater to drown semi-aquatic species such as muskrats, mink, beaver, and river otters. Drowning is not considered a humane death in veterinary medicine or lab animal research, for obvious reasons (Ludders et al. 1999). For species that are physiologically adapted to aquatic life, drowning is a slow process. A beaver can remain underwater for 15 minutes, a muskrat for 17 minutes, and an otter for 22 minutes. A beaver is therefore likely to struggle for at least 10–13 minutes before losing consciousness, muskrats and otters for 15–20 minutes (Gilbert and Gofton 1982). Mink caught in underwater traps "struggle frantically prior to loss of consciousness, an indication of extreme trauma" (Fox 2004b,

1172)—as one would expect of any drowning. Because most animals trapped in aquatic sets struggle for more than three minutes before losing consciousness, some researchers have concluded that such traps do not "meet basic animal welfare standards," and therefore cannot "be considered humane" (Fox 2004b, 1172).

## Snares

A snare is a wire noose designed to catch a passing animal and tighten when he or she struggles to escape. "Snares are generally categorized as neck, body, or leg snares" (also called '"foot"' snares), depending upon which area of the animal's body the trap is designed to target (Fox, 2004a, 37–38). Neck snares are set vertically (above ground) to catch and strangulate canids (coyotes, foxes, and wolves) and cats (bobcats and lynx). Leg snares are usually set horizontally, and are designed to catch and restrain rather than kill, in the process cutting off circulation and cutting through flesh to bone. Body snares are also used to drown aquatic animals, including beaver and otters.

## Killing Trapped or Snared Animals

Trapped or snared animals who do not drown or suffocate or die from exposure or attacking predators, must wait—exposed to the elements, without water or food, with body parts mangled and strangled—until the trapper shows up. No regulations guide the killing of trapped animals, so trappers are free to use whatever method they prefer—which means low-cost methods that preserve pelts. Magazines for trappers "often advocate suffocation, drowning, gassing and hitting with clubs… to minimize pelt damage" (Iossa et al. 2007, 347). The *Trapper Education Manual* advises trappers to stand "on the animal's chest to compress organs, which leads to death" ("Trapper" 2005, 98). For young and inexperienced trappers, the *Trapper Education Manual* recommends striking "smaller furbearers such as raccoon, opossum, and fox hard at the base of the skull with a heavy wooden or metal tool to kill or render them unconscious" (p. 98). They also suggest that young trappers place a foot "over the heart and chest area," where compressing organs "will lead to death" (p. 98). Alternatively, they are counseled to use "submersion techniques or kill-type traps" in the hope of avoiding the necessity of dispatching a trapped, struggling, wounded, and terrified animal (p. 98). Given well-documented links between violence toward animals and violence toward people, it would make a lot more sense to discourage adults from trapping with children, or better yet, simply ban trapping for those under eighteen.

## Environmental Ethics: Deceit and Suffering

"Cruel" is defined as "causing or conducive to injury, grief, or pain" ("Cruel" n.d.). Trapping is inherently cruel. Nonetheless, the International Organization for Standardization (ISO) outlined international humane standards for lethal traps in 1999. Though "humane trapping" is clearly an oxymoron, some traps are even worse than others.

It is important to note that *in assessing traps and snares, the ISO's only concern was physical suffering—not psychological trauma or premature death.* But psychological distress (i.e., fear), physiological stress and pain can also be observed and assessed in trapped animals through behavioral analyses and stress-related hormonal and blood-cell measurements" (Fox 2004b, 1171). Trapped animals show "significant changes in hormones, enzymes and electrolytes, as well as muscle pH [and] have increased levels of serum cortisol, indicating a stress response" (Iossa et al. 2007, 344). As one would expect, they also exhibit an increased heart rate and elevated body temperature—also stress indicators (Kreeger et al. 1990).

In a remarkable show of callousness (that ought to have disqualified them from attempting to assess humane methods), ISO researchers tested how long a trapped animal struggled before lapsing into unconsciousness.[2] Further undermining their study, the ISO decided that trap criteria for animal welfare ought to differ according to species, and should be determined by timing death for 70 percent of victims. What of the other 30 percent? Is it acceptable if they linger for hours—or days—in traps designed to kill quickly?

Their criteria were as follows:

- short-tailed weasels: 70 percent of animals tested must be unconscious within 60 seconds;
- American pine martens, Canadian lynx, and fishers: 70 percent within 120 seconds;
- all other animals, 70 percent within 180 seconds (Iossa et al. 2007).

Not surprisingly, a more extensive and less arbitrary review of traps (both those designed to restrain and those designed to kill) concluded that many trapping practices—including those previously approved by the ISO—are inhumane, and that current trapping legislation "fails to ensure an acceptable level of welfare for a large number of captured animals" (Iossa et al. 2007, 335). Nonetheless, the international fur trade is alive and well, and the U.S. is a major exporter of "wild-caught fur," sending pelts to Russia, the European Union, China and other Asian nations.

Aside from the suffering, environmental ethics philosopher Paul Taylor objects to those who knowingly take advantage "of an unsuspecting animal" (Taylor 1986, 186). He argues that such deception, when used "to bring advantage to the deceiver, reveals the deceiver's view that the animal has either no inherent worth at all or a value lower than the deceiver" (186). With regard to trappers, the former is probable, the latter is certain. Trappers pride themselves in their ability to deceive and trick animals, baiting and concealing traps and snares on well-traveled animal paths in the hope of luring and ensnaring wildlife. Traps lie outside any experience that wildlife has evolved to anticipate in their natural environment. They cannot adapt to traps because there is rarely a future for those caught.

## Species and Ecosystems

*The environmental impact of trapping has never been systematically assessed.* Montana trapper surveys indicate that more than half a million animals were trapped in the past ten years—*and these are only the official figures, i.e. those reported.* No official records document how many "protected" species are caught and/or killed in traps (perhaps to avoid inevitable lawsuits), but unintended casualties *necessarily* include threatened and endangered species (as well as beloved companion animals). Dick Randall, former government trapper and fur dealer, testified before Congress that for each target animal he secured, he trapped two non-target animals. Randall's ratio, when applied to the *reported* number of trapped animals, suggests that trappers have eliminated a staggering one-and-a-half million animals from Montana's ecosystems in the past ten years. This figure does not include dependent young, who waited in vain for a trapped parent to return to the den with food. In the Bitterroot Valley, for example, two dead cubs were found next to a mountain lion who had suffocated in a snare.

Despite obvious links between trapping and Threatened/Endangered/extirpated species, and despite the removal of tens of thousands of predators from local ecosystems, many environmental organizations support trappers and trapping—presumably because this attracts a few more paying members. For example, a Defenders of Wildlife representative in Bozeman, Montana six years ago stated that Defenders "opposes trapping only in cases where it's harming rare wildlife or where it's not done in a responsible and sustainable way" (Gallagher 2008). Does this Defenders representative understand that trapping is indiscriminate, that traps capture *any* animal (usually predators) who steps into their life-wrecking grip—including threatened and endangered species? Has this Defenders of Wildlife representative ever seen the numerous online photos of random animals caught in traps—eagles and fawns and dogs? Trapping is *inherently* irresponsible, unsustainable, and harmful to many species-at-risk.

Trapping reduces biodiversity—the backbone of healthy ecosystems. Aldo Leopold wrote: "a thing is right when it tends to preserve the beauty, integrity, and stability of the biotic community. It is wrong when it tends otherwise" (1968, 262). Trapping compromises the beauty, integrity, and stability of ecosystems. According to Leopold's environmental assessment, trapping is therefore "wrong."

## Taking Action for Animals and Ecosystems

As consumers, educators, lobbyists, and activists, each of us can help reduce the suffering, premature death, and environmental devastation caused by trapping. First and most fundamentally, consumers must read labels in order to avoid putting money down for products that contain fur, whether a cat's toy, bedroom slippers, winter coat, or perfume containing extracts from beaver castor sacs. Second, if someone you know is considering conventional traps in hope of resolving a human–animal conflict—whether with a mouse, beaver, raccoon, or coyote—contact The Humane Society of the United States, In Defense of Animals, People for the Ethical Treatment of Animals, or Footloose Montana to learn about humane methods for resolving human–animal conflicts. Third, take political action against trapping: Write to elected officials to let them know how you feel about traps and snares—and why. Encourage policy-makers to end recreational and commercial trapping on public lands, or at a minimum, to regulate trapping more closely. For example, ask legislators to:

- Require that trap placements be clearly posted whenever traps are placed on public lands.
- Ban traps in heavy-use areas of public lands.
- Ban traps within 1,000 feet of public lands trails, trailheads, and campsites.
- Require trappers to check traps at least once every 24-hours (with stiff penalties for infringements).
- Require licenses for trapping Predators and Non-game Wildlife.
- Require trappers to keep wildlife officials informed (using a map) regarding the exact whereabouts of their traps.
- Keep records of all trapped or snared Endangered/Threatened species.

Finally, start a citizen's initiative to ban recreational and commercial trapping, or to ban recreational and commercial traps on public lands (see Born Free U.S.A's "How to Organize a Successful Anti-Trapping Campaign" at http://www.bancrueltraps.com/c1_getting_started.php).

## Conclusion

Montana's Furbearer Trapping Program exemplifies the many problems and negative outcomes associated with trapping:

- Trapping alters ecosystems, removing tens of thousands of predators each year.
- Traps are indiscriminate, catching two unintended victims for every target animal.
- Traps are indiscriminate, catching Threatened/Endangered species and companion animals.
- Conibear traps cannot be released without specialized knowledge and considerable strength, which makes these traps extremely dangerous (usually deadly) to unintended victims.
- Due to extirpations and species depletion, trapping is expensive for the taxpaying public.
- Trappers and trapping are notoriously difficult to regulate.
- Trapping is extremely cruel.

In short, trapping is bad for animals, bad for ecosystems, and bad for citizens. Goethe wrote, "Knowing is not enough; we must apply. Willing is not enough; we must do" (Jensen n.d.).

## Discussion Questions

1 What is our obligation with regard to laws that, on careful and thorough consideration, we find contradict other laws? What is our obligation with regard to laws that, under careful consideration, we find to be immoral?

2 Does it make sense that commercial and recreational trapping is legal in nations that have enacted and maintain animal welfare laws? Why or why not?

3 Animal cruelty laws prohibit neglecting a dog (for example not having a water bowl available), while it is legal for trappers to drown, strangle, bludgeon or stomp a coyote to death. How is this discrepancy likely to be justified by those who support current laws?

4 Are compassion and empathy desirable attributes for those working in wildlife conservation? If so, how might environmental policies reflect these fundamental human values?

## Essay Questions

1 Legally, traps are protected as the property of trappers. For example, in Montana, "state law prohibits people from disturbing traps or trapped animals" ("Trapper Guidelines" n.d.). If you were walking on Montana's public lands and came upon a steel-jawed trap holding a prairie dog by his or her leg, what would you do? What if you found a member of an endangered species caught in the same trap? What if you found your dog caught in this trap, instead? Given the above law, what would you do if *you* were caught in the trap?

2 Research your state's trapping laws. Is it legal to trap on public lands? If so, is it legal for you to release your dog or cat from a trap on public lands? What are your thoughts regarding your state's trapping laws—in what ways do they strike you as reasonable, immoral, necessary, unjustifiable, and so on?

3 Research your state's trapping laws and trapper target species. Write a letter to your legislators requesting specific changes in your states trapping laws based on local environmental concerns, such as the need to protect a specific fragile ecosystem or a particular species-at-risk, problems inherent in removing predators, concerns about indiscriminate methods, and so on.

## Suggested Further Reading

Bekoff, M. and Pierce, J. (2009) *Wild Justice: The Moral Lives of Animals*, The University of Chicago Press, Chicago and London.

Fox, C. H. and Papouchis, C. M. (eds) (2004) *Cull of the Wild: A Contemporary Analysis of Wildlife Trapping in the United States*, Brainerd: Bang Publishing.

Paquet, P. C. and Darimont, C. T. (2010) 'Wildlife Conservation and Animal Welfare: Two Sides of the Same Coin,' *Animal Welfare*, vol. 19, pp. 177–190.

Ripple, W. J., Estes, J. A., Beschta, R. L., Wilmers, C. C., Ritchie, E. G., Hebblewhite, M., Berger, J., Elmhagen, B., Letnic, M., Nelson, M. P., Schmitz, O. J., Smith, D. W., Wallach, A. D., and Wirsing, A. J. (2014) 'Status and Ecological Effects of the World's Largest Carnivores,' *Science*, vol. 343, pp. 151–162.

## Notes

1 A trapline is a geographic line along which traps are set.

2 Trapping research is usually undertaken to protect trapper interests. These horrific experiments are just one of the many unmeasured (and usually unmentioned) effects of legalized trapping.

## References

Anderson, N. J. and K. E. Aune. (2008) "Fecundity of Female Wolverine in Montana." *Intermountain Journal of Sciences,* vol. 14, no 1–3, pp. 17–30.

"Average Pelt Price Values." (2013) *Montana Fish, Wildlife and Parks: State Furbearer Program Newsletter.* Spring 2013 [PDF] Wildlife Division, Helena. Accessed Mar. 26, 2013. http://fwp.mt.gov/hunting/trapping/.

Babcock, M. (2007) "Number of Trapping Licenses Sold has been Climbing," *Great Falls Tribune, 4 Dec.*

"Basic Facts about Swift Foxes." *Defenders of Wildlife.* Accessed Apr. 28, 2013. http://www.defenders.org/swift-fox/basic-facts.

Brown, M.. (2013) "Montana Hunters, Trappers Kill at Least 223 Wolves." *Billings Gazette,* 28 Feb. Accessed Mar. 13, 2013. http://billingsgazette.com/news/state-and-regional/montana/montana-hunters-trappers-kill-at-least-wolves/article_a7cd046c-69ee-5e9b-b1b2-50ba37ea422b.html.

Buskirk, S. W. and Ruggiero, L. F. (1994) "American Marten." *The Scientific Basis for Conserving Forest Carnivores: American Marten, Fisher, Lynx, and Wolverine in the Western United States*, L.F. Ruggiero, K.B. Aubry, S.W. Buskirk, L.J. Lyon, and W.J. Zielinski (eds). U.S. Department of Agriculture, Forest Service, Rocky Mountain Forest and Range Experiment Station. Fort Collins, CO, 7–37.

Cegelski, C., Waits, L. P., Anderson, N. J., Flagstad, O., Strobeck, C, and Kyle, C. J. (2006) "Genetic Diversity and Population Structure of Wolverine (*Gulo gulo*) Populations at the Southern Edge of their Current Distribution in North America with Implications for Genetic Viability," *Conservation Genetics*, vol. 7, pp. 197–211.

Cole, K. (2013) "State Public Records Request Shows Widespread Capture and Mortality of Non-Target Animals Related to Idaho Wolf Trapping During 2011/2012 Trapping Season," *The Wildlife News*, 14 Feb. Accessed Feb. 16, 2013. http://www.thewildlifenews.com/2013/02/14/state-public-records-request-shows-widespread-capture-and-mortality-of-non-target-animals-related-to-idaho-wolf-trapping-during-20112012-trapping-season/.

"Cruel." (n.d.) *Merriam-Webster.* M-W.com. Accessed Jan 26, 2013. http://www.merriam-webster.com/dictionary/cruel.

Dickson, T. (2006) "Maintaining a Buckskin Lifestyle in a Polar Fleece World," *Montana Outdoors: The Magazine of Montana Fish, Wildlife and Parks*, Nov.–Dec. Accessed Sept. 10. http://fwp.mt.gov/mtoutdoors/HTML/articles/2006/Trapping.htm.

"Endangered and Threatened Wildlife and Plants; 12-Month Finding on a Petition To List a Distinct Population Segment of the Fisher in Its United States Northern Rocky Mountain Range as Endangered or Threatened With Critical Habitat." *Federal Register. USFWS: Proposed Rules* 76.126. June 30, 2011. Accessed Apr. 10, 2013. http://www.fws.gov/policy/library/2011/2011-16349.pdf.

"Fisher." *United States Fish and Wildlife Service: Endangered Species*. Accessed Feb. 12, 2013. http://www.fws.gov/mountain-prairie/species/mammals/fisher/.

Fox, C. H. (2004a) 'Trapping Devices, Methods, and Research', in C.H. Fox. and C.M. Papouchis (eds) *Cull of the Wild: A Contemporary Analysis of Wildlife Trapping in the United States*, Brainerd: Bang Publishing, 2004: 31–59.

Fox, C. H. (2004b) 'Welfare, Well-being, and Pain: Wildlife Trapping, Behavior and Welfare', *Encyclopedia of Animal Behavior*. Ed. M. Bekoff. Santa Barbara: Greenwood Publishing Group, pp. 1170–1175.

Fraley, J. (2002) "Montana Outdoors Portrait: Martens," *Montana Outdoors. A Magazine of Montana Fish, Wildlife and Parks*. Accessed Feb. 27, 2013. http://fwp.mt.gov/mtoutdoors/HTML/articles/portraits/marten.htm.

"Furbearer: Montana Trapping and Hunting Regulations." (2013) *Montana Fish, Wildlife and Parks*. Helena. Accessed May 1, 2014. http://fwp.mt.gov/eBook/hunting/regulations/2013/furbearer/index.html.

"Furbearer Program: Statewide Harvest and Management Report 2011–12 Montana." (2013) *Montana Fish, Wildlife and Parks*. Helena, 2009. Accessed May 5, 2014. http://fwp.mt.gov/hunting/trapping/.

"Furbearer Trapping Guide." (2013) *Montana Fish, Wildlife and Parks*. Helena. Accessed Apr. 5, 2013. http://fwp.mt.gov/hunting/planahunt/huntingGuides/furbearer/default.html?harvestTabHeader.

"FWP Reports Wolf Hunting and Trapping Season Results." (2013) *Montana Fish, Wildlife and Parks*. Helena. Accessed May 1, 2014. http://fwp.mt.gov/news/newsReleases/headlines/nr_4069.html.

Gallagher, S. (2008) "Groups Seek Endangered Status for Wolverines," *San Francisco Chronicle*. SFGate, *24* August. Accessed April 16, 2013. http://www.sfgate.com/news/article/Groups-seek-endangered-status-for-wolverines-3198203.php.

Giddings, B. Personal e-mail to Heister. June 23, 2008.

Gilbert, F. and Gofton, N. (1982) "Terminal Dives in Mink, Muskrat and Beaver," *Physiology and Behavior*, vol 28, no 5, pp. 835–840.

Iossa, G., Soulsbury, C.D. and Harris, S. (2007) "Mammal Trapping: A Review of Animal Welfare Standards of Killing and Restraining Traps," *Animal Welfare,* vol 16, pp. 335–352.

"IUCN Otter Specialist Group." *International Union for Conservation of Nature and Natural Resources*. Aug. 9, 2011. Accessed Apr. 4, 2013. http://www.otterspecialistgroup.org/Species/Lontra_canadensis.html.

Jensen, Anthony K. "Johann Wolfgang von Goethe (1749–1832)." *Internet Encyclopedia of Philosophy: A Peer-Reviewed Academic Resource*. Accessed Apr. 4, 2013. http://www.iep.utm.edu/goethe/.

Keefover-Ring, W. (2009) "War on Wildlife: The U.S. Department of Agriculture's Wildlife Services. A Report to President Barack Obama and Congress," *WildEarth Guardians: A Force for Nature*. Feb.. Accessed Aug. 20, 2012. http://www.wildearthguardians.org/Portals/0/support_docs/report-war-on-wildlife-june-09-lo.pdf.

King, J. (2013) "Montana's Final Wolf Harvest Numbers Revealed." *KGVO.com*. Accessed Apr. 10, 2013. http://newstalkkgvo.com/montanas-final-wolf-harvest-numbers-revealed/.

Kreeger, T. J., White, P. J., Seal, U. S. and Tester, J. R. (1990) "Pathological Responses of Red Foxes to Foothold Traps," *The Journal of Wildlife Management*, vol. 54, no 1, pp. 147–160.

"Lawsuit filed to protect lynx from trapping." (2013) *Independent Record*. Accessed March 21, 2013. http://helenair.com/news/local/lawsuit-filed-to-protect-lynx-from-trapping/article_eb576912-927f-11e2-b9e4-0019bb2963f4.html.

Leopold, A. (1968) *A Sand County Almanac*, Oxford University Press, New York.

Ludders, J. W., Schmidt, R. H., Dein, F. J. and Klein, P. N. (1999) "Drowning is not Euthanasia," *Wildlife Society Bulletin*, vol. 27, no 3, pp. 666–670.

Maughan, R. (2012) "All Yellowstone Park Wolf GPS Collared Wolves Were Killed in the Wolf Hunt." *The Wildlife News*. Dec. 12, 2012. Accessed Apr. 1, 2013. http://www.thewildlifenews.com/2012/12/12/all-yellowstone-park-wolf-gps-collared-wolves-were-killed-in-the-wolf-hunt/.

"Minutes." (2008) *Montana Fish, Wildlife and Parks Commission Meeting*. June 12, 2008. 1–18. Accessed July 18, 2008. http://fwpiis.mt.gov/content/getItem.aspx?id=34458.

"Montana Field Guide: Canada Lynx." Accessed Jan. 30, 2013. http://fieldguide.mt.gov/detail_AMAJH03010.aspx.

"Montana Field Guide: River Otter." Accessed Apr. 14, 2013. http://fieldguide.mt.gov/speciesDetail.aspx?elcode=AMAJF10010.

"Montana Field Guide: Swift Fox." Accessed Apr. 14, 2013. http://fieldguide.mt.gov/detail_AMAJA03030.aspx.

"Montana Furbearer." *Never Miss*: *Never Miss a Special Permit Application Deadline Again!* Accessed Dec 29, 2013. http://nevermisstags.com/mtspecies/furbearer.html.

"NAFA March 2014 Fur Auction Results." *Trapping Today*. Mar. 3, 2014. Accessed May 1, 2014. http://www.nafa.ca/wp-content/uploads/2014/02/NAFA_2014-02-WF-USA.pdf.

"Prices." *Peterson Furs: Your Friendly, Country, Fur Buyer*. Accessed Mar. 27, 2013. http://www.petersonfurs.com/furs.htm.

"Public Land." *Wikipedia*. Accessed Jan. 12, 2013. http://en.wikipedia.org/wiki/Public_land.

Reed, T. "Trapping in the United States." (1999a) *High Country News: For People Who Care About the West*. . Accessed Jan. 27, 2013. https://www.hcn.org/issues/152/4918.

Reed, T. "Is Trapping Doomed?" *High Country News: For People Who Care About the West*. (1999b). Accessed Jan. 27, 2013. https://www.hcn.org/issues/152/4917.

"Restoring the Gray Wolf." *Center for Biological Diversity*. Accessed Apr. 19, 2013. http://www.biologicaldiversity.org/campaigns/gray_wolves/.

Schwartz, M. K., Copeland, J. F., Anderson, N. J., Squires, N. J., Inman, R. M., McKelvey, K. S., Pilgrim, K. L., Waits, K. L. and Cushman, S. A. (2009) "Wolverine gene flow across a narrow climatic niche," *Ecology*, vol. 90, no 11, pp. 3222–3232.

Squires, J. R., Copeland, J. P., Ulizio, T. J., Schwartz, M. K. and Ruggiero, L. F. (2007) "Sources and Patterns of Wolverine Mortality in Western Montana," *The Journal of Wildlife Management*, vol. 71, no 7, pp. 2213–2220.

"State Furbearer Program Newsletter." (2005) *Montana Fish, Wildlife and Parks*. Helena. Accessed Aug. 19, 2011. http://fwpiis.mt.gov/content/getItem.aspx?id=31419.

Stephens, R. M. and Anderson, S. H. (2005) "Swift Fox (*Vulpes velox*): A Technical Conservation Assessment," *USDA Forest Service, Rocky Mountain Region*, January 21. Accessed Apr. 19, 2013. http://www.fs.fed.us/r2/projects/scp/assessments/swiftfox.pdf

Stone, B. Personal phone conversation with Heister. January 20, 2010.

"Swift Fox Skins." *Chichester Inc. International Merchants for Exotic Natural Products Since 1992.* Accessed Feb. 2, 2013. http://www.chichesterinc.com/SwiftFoxSkins.htm.

Taylor, P. (1986) *Respect for Nature: A Theory of Environmental Ethics*, Princeton University Press, Princeton.

"Trapper Educational Manual: A Guide for Trappers in the United States." (2005). *International Association of Fish and Wildlife Agencies.* http://www.fishwildlife.org/files/Trapping_Student_Manual_2005.pdf.

"Trapper Guidelines for Placement of Ground Sets." *Montana Fish, Wildlife, and Parks.* Accessed June 13, 2014. *fwp.mt.gov/fwpDoc.html?id=32308.*

Vinkey, R. S., Schwartz, M. K., McKelvey, K. S., Foresman, K. R., Pilgrim, K. L., Giddings, B. J. and LoFroth, E. C. (2006) "When Reintroductions are Augmentations: the Genetic Legacy of Fishers (*Martes pennanti*) in Montana," *Journal of Mammalogy*, vol. 87, no 2, pp. 265–271.

"Wildlife Management and Regulated Trapping in Montana." (2010) *Montana Fish, Wildlife and Parks.* Helena..Accessed Mar. 19, 2013. http://fwp.mt.gov/hunting/trapping/.

"Wolverine." *United States Fish and Wildlife Services: Endangered Species.* Accessed Mar. 28, 2013. http://www.fws.gov/mountain-prairie/species/mammals/wolverine/.

Wuerthner, G. (2013) "Montana Wolverine Trapping Ignores Ethics and Biology," *The Wildlife News*, 9 Jan.. Accessed Jan. 10, 2013. http://www.thewildlifenews.com/2013/01/09/montana-wolverine-trapping-ignores-ethics-and-biology/.

Zuckerman, L. (2012) "Montana Wolves Killed in Hunting Season, but Population Still Grew," *Reuters. Huffington Post.* Accessed Apr. 12, 2013. http://www.huffingtonpost.com/2012/03/02/montana-wolves-killed-hunting-season_n_1315383.html.

# 9

# SEEING IS BELIEVING

## Nature Films in a Patriarchal Culture

*Melanie J. Martin*

Filming nonhuman animals in their natural habitats may seem an innocent or even noble pursuit—indeed, it has the potential to be. Some nature films attempt to teach viewers to behave more responsibly toward the natural world—the *Planet Earth* series, for instance, strives to show nonhumans behaving naturally, in their natural setting. Their filming crews purportedly disturb their subjects as little as possible. However, filmmakers who unapologetically disturb wildlife and invade habitats promote the idea that humans of Western cultures[1] should have complete dominion over the rest of the natural world. In doing so, many nature films follow a longstanding tradition of viewing individual animals as mere "specimens," asserting Western mastery of faraway places. Like American and British safari hunters of the late 1800s and early 1900s, contemporary "camera hunters" (filmmakers) travel to "exotic" lands to capture rare or elusive nonhumans. By successfully navigating remote terrains and finding an animal of a particular species, nature filmmakers attempt to prove that no nonhuman or landscape is off-limits to humans of Western culture. Analysis indicates that such filmmakers attempt to claim these animals and their habitats through the act of filming, by bringing home lasting proof of their mastery of the natural world. In the process, filmmakers ignore how enacting this apparent superiority complex has, for centuries, destroyed ecosystems and caused great suffering to nonhuman animals. Paradoxically, some of these filmmakers attempt to intrigue and educate viewers about other animals and ecosystems, with the apparent goal of building support for conservation. The Western, patriarchal inclination toward dominion seems to be subconsciously engrained in the minds of filmmakers, often promoting a disrespectful and destructive attitude toward other animals and nature—even when filmmakers may believe they are promoting a responsible attitude toward wildlife. Most often, our longstanding desire to conquer the natural world, enacted by filmmakers, teaches viewers that nonhuman animals and their habitats are mere objects to be exploited and subjugated for human purposes.

## Western Dominance: Safari Hunting and Filmmaking

Not all cultures kill for sport, entertainment, and self-gratification. In fact, many safari hunters of the late 1800s to early 1900s comment that indigenous guides seem uninterested—even unwilling—to search for "exotic" nonhumans. In his narrative *Hunting the Elephant in Africa*, safari hunter C.H. Stigand criticizes his indigenous African guides for having no desire to hunt nonhuman animals for sport (1986, p.19). According to his narrative, the guides perceive his obsessive quest for the elusive situtunga to be ridiculous, and they are reluctant to "penetrate" the interior swamps where this animal lives (1986, pp.22–23). This land was their home; they felt no need to conquer it.

Historically, the desire to hunt and film "exotic" nonhumans is largely a white, Western, male phenomenon. Westerners have sought nonhumans in faraway lands as a means of domination, since individual wild animals are often viewed as symbolizing the lands where they live (Kaplan, 1997, p.61). Thus, for hundreds of years Westerners have amassed "collections" of both living and dead nonhuman animals to "prove" that they have conquered "exotic" places and powerful nonhumans. In doing so, they also subjugate humans of other cultures, claiming dominion over distant lands and resources.

This mindset of domination is evident in nonfiction safari hunting narratives such as Theodore Roosevelt's "The Game of the High Peaks: The White Goat," Peter Capstick's "The Black Death," C.H. Stigand's *Hunting the Elephant in Africa,* and Frank Buck's *Bring 'Em Back Alive.* Numerous twenty-first century wildlife documentaries, such as Animal Planet's *Spy on the Wild,* the *Crocodile Hunter* series, and *The Jeff Corwin Experience*, continue the tradition of amassing collections of nonhumans. Though filmmakers do not attempt to kill nonhumans, they often terrorize these individuals by pursuing and/or capturing them. The narrator frequently laughs at the nonhuman's obvious distress after capture, suggesting to viewers that the suffering of other animals is irrelevant. This is one of the most dangerous messages carried in nature films: Individual nonhumans are mere objects, not beings who feel and think. These films render very real feelings irrelevant in the eyes of viewers. Certainly some filmmakers appear to have worthy objectives, such as educating viewers about the variety, intricacies, and splendor of the nonhuman animal world, but their films tend to uphold cultural disrespect for other species in both subtle and overt ways.

Modern-day camera hunting is rooted in the "Nature Man" movement of the later 1800s and early 1900s. During this time, men no longer spent their days working the fields or hunting for sustenance; they had lost one of their primary social roles, and presumably, along with it, their supposed mastery over nature. This drew many men to seek to reclaim dominance over nature by hunting, causing sport hunting among white men to escalate, especially during the Roosevelt era. Killing the rarest, most elusive wild animals in unfamiliar territories offered men the sense of conquering nature that they craved (Alaimo, 2000, p.97). Unfortunately, at a time when humans might have forged a new, healthier relationship with nature based on respecting all life, they sought new ways of exploitation and entertainment through violence.

Thus, camera hunting, which had begun in the 1850s, continued to escalate along with sport hunting. It was viewed from the start as an expression of masculinity (Gandy et al., 2000, p.208). Men shot wild animals with both cameras and guns, often at the same time. Photos allowed them to bring home lasting evidence of their manly prowess. Arthur Radclyffe Dugmore, an early nature photographer, commented that nature filmmaking was so "manly" a hobby that it made traditional hunting seem a "boy's sport" (qtd. in Philo and Wilbert, 2000, p.214).

The desire for expeditions into the wilderness was revived in the 1960s, partly as a result of programs like *Wild Kingdom* and Jacques Cousteau's documentaries. These programs bore a striking resemblance to earlier safari hunting narratives, in which man was dominant and deadly with regard to nature. In both cases, "hunters" were guided by the same patriarchal, imperialistic, and anthropocentric mindset that caused their initial alienation from nature. As a consequence, men had to keep conquering more nonhumans and more wild places, never satisfied.

The idea of conquering other animals with a camera may at first seem absurd, but in Western culture, people claim power through the act of looking. Governments oftentimes establish their power by letting citizens know they can be watching at any time (Foucault, 1979, pp.201–17). "Big Brother is watching you," people often say. "Regular folks" thereby learn to claim power in the same way, especially white men, who assume that they, too, have the right to exert power over others visually. The "privileged eye" of white masculine power seeks to objectify and claim dominion over other animals and habitats through unlimited observation (Davies, 2000, p.251). Thus, capturing individual nonhumans on film provides filmmakers with an enormous sense of power over other animals. Filmed nonhumans remain prisoners of our gaze, a testament to human superiority, instilling in viewers the idea that other animals are under our power, and are objects for our enjoyment.

## Feminization of the Land

Our perception of nature governs our perception of nonhuman animals. Westerners tend to maintain a feminized conception of the land, which affects how we treat the land and nonhumans therein. For example, the term "virgin" is often used to describe allegedly untouched wilderness—a landscape available for men to "have their way" with. In the stolen land known as America, male settlers saw the vast expanses of forest as "virgin" land waiting to be claimed. Land is equated with fertility, and men are traditionally the ones to "plant the seeds" (Kolodny, 1975, p.xiii). Settlers spoke of nature as a "*Paradise* with all her virgin beauties" (Kolodny, 1975, p.12). Colonist Thomas Morton described American nature in 1632 as "Like a faire virgin, longing to be sped,/And meete her lover in a Nuptiall bed"—yearning, that is, for men to "plant their seeds" (Kolodny, 1975, p.12)—never mind the millions of indigenous people who were or would be massacred, leaving others displaced and disillusioned for centuries. Their presence had to be erased, either physically or figuratively, through language, for Western dominion to take root.

Likewise, society often defines women as unplowed land; "plow her" is a sexual term that males sometimes use to refer to intercourse. In other words, Westerners define environments as passive, just as women have been historically labeled as passive—awaiting male action (Plumwood, 1993, p.4). Westerners typically view environments as backdrops for *our* activities, rather than as dynamic, active communities of individuals. Thus, Western men have seen environments, like women, as something to claim and exploit for their purposes before someone else does so.[2]

## Masculinization of "Wild" Animals

While presenting the land as feminine, contemporary filmmakers tend to present nonhumans as highly masculine. Qualities typically associated with masculinity in Western culture include strength, fortitude, and virility. An individual nonhuman who is perceived as possessing such qualities has often been viewed in Western cultures as the ideal animal of the species—just as the man represents the ideal human (Adams, 2004, p.40). Filmmakers masculinize nonhumans because they have come to view their prey as dangerous and threatening to their own masculinity. In Western culture, hunters (including camera hunters) tend to view other animals as competitors, individuals who might claim dominance over the land—and thus, over the hunter—by escaping. This plays into the camera hunter's view of filming wild animals as a dangerous competition in which the filmmaker seeks to assert his supremacy over the natural world. Filmmakers therefore tend to focus on the allegedly dangerous nature of other animals, highlighting characteristics typically viewed as masculine. In the popular *Crocodile Hunter* series, for instance, Steve Irwin wrestled with crocodiles and other nonhumans such as snakes, often exclaiming, "Danger, danger!" Meanwhile, the crocodile was in terror for his or her life, and rightly viewed the man as posing a deadly threat.

Not surprisingly, National Geographic advertises its program *Dangerous Encounters with Brady Barr* (on its website) with the words: "Get up close and personal with some of the most feared predators on the planet and find out who reigns supreme." In truth, Barr harasses nonhuman animals and invades their habitats, scrambling into an aardvark burrow, provoking snakes to attack, wrestling crocodiles, and unleashing a one-man reign of terror on countless other animals. Instead of admitting that Barr is terrorizing wild animals, the website portrays the victims—the nonhumans—as dangerous and threatening. For example, Barr begins narrating an episode called "Dragon Hunt" (about the "deadly" Komodo dragon): "I'm Dr Brady Barr. As a herpetologist, I work with many deadly reptiles. Up 'til now, I've always had a good understanding of how each animal can kill me using venom, constriction, power, or teeth" ("Dr Brady," 2010). Apparently this "beast" is a particularly dangerous one to Barr—at least he wants to be sure that *we* think so. He proceeds to lasso a Komodo dragon, then straddles him or her to subdue his unhappy victim while two male assistants help restrain the dragon in the background, and another holds the lasso attached to the dragon's neck. Barr apparently gives no thought to the dragon's terror as he or she struggles desperately to break free, in dread of meeting an untimely death. Instead, Barr exclaims gleefully, "I caught a dragon!"

Eventually the slow-moving dragon lies completely inert, having given up, unable to fight back or escape. Barr proceeds to strap a camera and radio transmitter onto the Komodo dragon, and Barr releases his victim so that he can observe how the dragon kills prey (presumably, one might discern, with those many sharp teeth). As in most nature films, the individual nonhuman's feelings of terror are ignored in order to portray men as brave, and because animals are assumed to be irrelevant in light of human curiosity and the revered quest for "scientific" knowledge. Such films propagate the harmful idea that humans—at least white, Western males— are entitled to provoke, exploit, terrorize, and even endanger and harm other animals for their purposes, however vague, frivolous, and capitalistic.

Throughout the 1990s, violence increased greatly within the natural history film genre, refocusing on the allegedly dangerous nature of nonhumans (Burt, 2002, p.48). The series *When Animals Attack* portrays wild animals attacking humans, as if without cause. Animal Planet's *Human Prey* and *I'm Alive* do the same. By casting nonhuman animals as inherently violent and dangerous, nature documentaries follow trends set by horror movies like *Jaws*, *Anaconda*, and *Lake Placid*. Such programs portray nonhumans as dangerous and unpredictable, without admitting that humans as a whole are far more dangerous to wild animals. Viewing these films uncritically increases our society's desire to conquer such animals by killing, capturing, and harassing them.

An episode of *I'm Alive* (2009), in which a man is attacked by a leopard, begins like a scene in a horror movie. An overhead, angular view reveals a man sitting in the dark. A fast-moving series of clips then shows a leaping leopard, a screaming woman, and a man lying on the ground, howling in pain. The camera then appears to fall, so the viewer cannot clearly see what is happening. Through the bouncing vision of the moving camera, which disorients viewers, the man is rushed to surgery, creating an intense feeling of danger and panic. Shortly thereafter, while the man's wife is describing her reaction to her husband's near death at the hands of a terrifying wild beast, the camera flashes to repeated images of the man being attacked by the wild leopard. This flick, like so many wildlife shows, relies on shock and terror to captivate viewers. By airing such clips, Animal Planet is unapologetically teaching viewers that other animals are dangerous, deadly—the enemies of humanity.

"Dangerous" is relative. Any animal—human or nonhuman—can be dangerous. By using "dangerous" to describe wild animals, the films fail to fairly portray the complexity of both humans and nonhumans, as well as the complexity of interspecies interactions.

For filmmakers, as for safari hunters, the perceived danger of nonhumans lies in their ability to outwit Western humans, eluding their grasp and threatening their masculine dominance (and their masculinity in general). "The male and the white subject need to be in control: They assume power lies with them, in their gaze, and are uneasy when there is an entity that seems to elude their control" (Kaplan, 1997, p.62). Men imagine that evasive nonhumans assume the privileged eye that white men have historically claimed. Their "danger" lies in their ability to watch hunters while remaining unseen, a common theme in countless horror movies featuring masculinized nonhumans.

The filmmaker's tendency to masculinize wild animals is most fundamentally evident in an overt focus on filming male animals. The male animal's story is portrayed as *the ultimate* story, while the female (if filmed) is presented as a passive vessel for procreation (Crowther, 1995, pp.128–31). Corwin says of the Alaskan bull moose, "bulls will spar until one gives up and one claims possession of the cow" ("Alaska," 2001). His words depict his own patriarchal view of the female as male property. He never questions whether the cow might *choose* the stronger of the two bulls as a protective and reproductive strategy. In his vision, the bull is the competition, and the cow is the prize, an object to be claimed by the male who reigns supreme.

## Camera Hunting as a Man's Pursuit

With the land cast as feminine, and wild animals as masculine, filmmaking becomes the rightful pursuit of men, and men alone.

"The privileging of males, and the preoccupation with the strong male in particular, underlies another typical wildlife television story, the naturalist as hero, a man with a quest" (Crowther, 1997, p.131). Since the 1960s, nature films have centered on a quest for a particular animal, undertaken by an "intrepid naturalist host" who represents the ideal man (Chris, 2006, p.55). By tracking down the idealized male animal, the naturalist hero proves that he not only matches this animal's prowess, but is yet more powerful. *He* is the ultimate hero.

Meanwhile, women are often seen as objects, like nonhumans. When they aren't excluded completely, they are given only minor roles. "The early twentieth-century motion pictures … establish wildlife filmmaking as a largely masculine project, one that assigns particular tasks to women as help-mates" rather than portraying women as equals (Chris, 2006, p.3). Men lead; women follow. Rugged "nature men" such as Jeff Corwin and Steve Irwin seek, capture, and/or wrestle with "manly" animals, while women, if they appear at all, serve as assistants. Terri Irwin (the wife of Steve Irwin) has a significant role in many of the *Crocodile Hunter* films, but Steve is the star, the one who wrangles with and overcomes the nonhuman animals, who are portrayed as overtly dangerous. When the camera is rolling, Terri Irwin is merely an assistant. In fact, Terri admits, "I'm the straight man who plays off of the wild man. I'm Jane, he's Tarzan. It's always been like that" (Berrettini, 2010). When given voice, she overtly "marvels at his bravery and possibly reckless behavior" without expressing any agency of her own, or any will to change, stop, or take part in her husband's adventures (Berrettini, 2010). With Steve cast as the stereotypical nature man, Steve and Terri both lead viewers to believe that he is in his element in situations where Terri, who has a background in wildlife rehabilitation, would never dare to tread. Through this positioning, Steve Irwin—and many other white male "celebrity conservationists" like him—assume a stance of belonging in and claiming dominion over a place that was not their homeland historically (Huggan, 2013, pp.206–207).

Men are portrayed in nature documentaries as able to "penetrate" the world of wild animals because of their skill and knowledge (Crowther, 1997, p.294). Men know. Men teach. Men conquer. In subtle ways, these ideas become engrained in the minds of viewers. Men are often captivated by manly prowess. Others, though they may not know why, cannot identify with the manly "nature guide," and may feel marginalized by this presentation (294). This serves to encourage men to exploit nonhumans and their environment, while encouraging women to remain on the sidelines.

Many children watch manly nature programs, which condition them to accept culturally defined sex roles from a young age. Girls are unlikely to identify with the male stars of nature documentaries, leading them to unconsciously accept the idea that nature, especially the vast wilderness that is rightly the realm of the manly television explorers, is not their place. Meanwhile, these documentaries teach boys to see nature as something to conquer and dominate. This may be why boys in America are far more likely than girls to kill or abuse animals and insects for fun—they have learned that "real men" dominate nature (Herzog, 2007, p.14). They have not learned to see the fragile life of other living beings, or to see caring for other living beings as a sign of strength. By indoctrinating youth, filmmakers encourage them to uphold traditional roles that harm humans, nonhumans, and habitats.

## Conquering and Marginalizing through Objectification

A "nature guide" inevitably explains a few facts about a particular wild animal, presenting that individual as an object to be examined, like a car or a table. In the grip of the nature guide, animals are specimens rather than individuals. The nature guide presents animals as easily defined, thereby oversimplifying what it means to be a particular bat, snake, or frog.

Believing that we understand "other" animals often makes us feel superior. We collect knowledge of nonhumans like we collect trophies of wild animals. "What we know about them is an index of our power," and thus, an index of what we've conquered (Berger, 1992, p.38). A film clip and a brief list of facts serve as a trophy for the filmmaker. Arrogant pride concerning our society's technological advancements leads us to believe that the camera shows us every aspect of the animals' lives. This "penetrating technology," enhanced by the nature guide's concise list of facts, makes other animals seem easily understandable—simple in contrast to the human being (Kaplan, 1997, p.62). In an episode titled "Into the Heart of Darkness" (2001), Corwin searches for what he terms "the very elusive, very rare black caiman" in the rainforests of Ecuador. The episode's title speaks to a Westerner's view of faraway jungles as dark, forbidding mysteries. Corwin must travel many miles down rivers and through forests to find the caiman. He finds many other animals along the way, but he must locate an elusive caiman to prove he can "penetrate" the dense jungles and conquer the nonhuman of his choice. He at last comes upon a caiman, who he beams into his white male world with a headlamp. He then snatches the caiman from the water, wrestles the rightly terrified animal into submission, and raises the subdued caiman, victorious. He proceeds to describe the conquered caiman in typical nature guide style, with a brief list of facts. Mystery solved, journey over, wild animal conquered and released.

Westerners often imagine themselves to be infinitely complex, but feel comfortable defining nonhumans in a few sentences. Those who subjugate wild animals usually portray them in simple terms, leading viewers to see them as inferior in their simplicity (Spivak, 2001, p.2203). Filmmakers rarely speculate about how a nonhuman animal might feel, or experience the world; they focus on facts, limiting wild animals to the status of object, thus affirming their own superiority.

Indeed, our society has learned to view the feelings of nonhuman animals as irrelevant, beyond knowing, and expendable on behalf of our own interests. We test atomic bombs in ocean ecosystems, destroying many nonhuman individuals and their homes. We use sonar in ocean habitats, disorienting and hurting cetaceans such as whales and dolphins. We likewise destroy the habitat of countless nonhuman animals in the tropical rainforests in the name of palm oil and for the sake of eating hamburger, simultaneously forcing endangered primates and many other nonhuman animals into deadly confrontations with humans. If nature films wish to inform viewers in a way that promotes respectful, responsible interactions with other species, they will portray nonhumans as individuals whose physical, mental, and emotional wellbeing hold priority. Only then will we begin to reverse our indifference to and destruction of the environment and other animals. How different nature films would be if the nature guide asked questions he or she couldn't answer, expressing a sense of wonder and mystery about the world around us.

## Penetrating Private Spaces

Programs such as Animal Planet's *Meerkat Manor* and *Orangutan Island* offer viewers a chance to surreptitiously watch nonhumans over a prolonged period of time. "Fly-on-the-wall" documentaries, like Animal Planet's *Spy on the Wild,* are increasingly popular (Davies, 1998, p.214).

In *Spy on the Wild,* filmmakers "mount" tiny cameras on wild animals, which serve as a prosthetic "privileged eye" that allows viewers to continuously observe the private lives of wild animals, and to give the human eye access anywhere wildlife goes. One camera, attached to a bee, carries viewers through his or her daily flight. In another episode, a camera that is hidden in an underground cave allows viewers to follow a troop of monkeys into their private dwelling, a place previously off-limits to the human eye. In Brady Barr's "Dragon Hunt," the narrator aptly refers to his "spy camera." In another episode of *Spy on the Wild*, a golden eagle is released with two small cameras strapped to her back so that we might observe where she goes and what she sees during her flight—an aerial view to which humans typically don't have access ("An Extra Wing," 2006). As the eagle begins her flight, the narrator informs viewers: "Every detail of Tilly's superb aerodynamic design is *captured* by miniature camera technology." Viewers then see a brief clip of the land below, including several shots from a variety of angles, to provide a somewhat panoramic view. These clips are brief, disorienting, and shaky, and it is unlikely that they provide any real sense of what flying must be like for an eagle—the camera is trying to capture what lies beyond our grasp.

In another episode, focusing on gannets, filmmakers again explain their gadgetry in glowing terms, noting that they use "a global positioning satellite receiver [that] records speed and position every ten seconds" ("Gannets," 2006). This "excellorometer" "measures every wingbeat above and below the water," and is fixed to a gannet's tail feathers. The bird is then released for his or her morning flight ("Gannets," 2006). The commenter proudly notes: "Beyond this point, they [gannets] used to be completely off the radar. No one knew where they went, or what they did"—and under the influence of Western science, *we can't not know* ("Gannets," 2006). These programs teach us that no place, and no nonhuman, is off-limits to humans of Western culture. We hold dominion over the rest of the animal kingdom, and over every cave and flower.

In the popular series *Orangutan Island,* viewers have extended access to the daily lives of a community of orangutans on an island in central Borneo. This group of young orangutans have all been orphaned—their mothers were killed, most likely by poachers who kidnapped their babies for the exotic pet trade, by palm plantation owners who kill nonhuman primates to avoid any damage to their profits, or for the bushmeat trade—for those who eat orangutans. These unfortunate orphans are brought to this protected island in hopes, according to the program, that they will learn to survive on their own in the wild. Although the program may inspire people to care more about these endangered primates, it also mis-educates. Because they live on an island, these orphaned orangutans must live in crowded *Survivor*-style mixed group company. Such a social structure would never exist for orangutans in the wild. Furthermore, they often walk upright, which is abnormal and very dangerous for a primate that normally swings through the trees and rarely (if ever) descends to the ground ("Orangutans," 2008). Walking on the ground leaves orangutans vulnerable to predators and parasites, and to human disease (*Guidebook*, 2009, pp.38–9). The program claims to be preparing these orangutans for possible release, while encouraging behavior that will bring a death sentence in the wild. Nonetheless, the orangutans' humanlike behavior is lauded rather than lamented by the narrator, as if other primates are to be valued in relation to their similarity to human beings ("Orangutans," 2008).

The individuals on Orangutan Island are anthropomorphized in other ways as well. They are given names—and character stereotypes—that make them seem humanlike and thus familiar, while greatly simplifying and denying who they are—orangutans, not human beings. On the program's webpage "Meet the Orangutans," for instance, viewers can read about "Cha Cha, the Social Butterfly" and "Jasmine, the femme fatale," giving viewers an unrealistic sense of familiarity with orangutans. Some of the anthropomorphized descriptions are quite forced. For example, in an episode on tool-use, the narrator says in a tongue-in-cheek tone: "Fiber-rich leaves make a good snack in-between fruit meals. But they can be a little boring, so Milo … made a green leaf club sandwich. Hold the Mayo!" ("Inventive," 2008). And in an episode featuring interactions between the human filmmakers and orangutans, the narrator says: "Jordan [the orangutan] is a budding filmmaker … maybe Chen Chen has his eye on a Hollywood career" ("Filming," 2008). As this episode shows, filmmakers on Orangutan Island make no attempt to keep a respectful distance from the orphaned orangutans who are trapped on this island, almost invariably orphaned by violent humans. The orangutans examine filming equipment, play catch with the producer, and hug the filming crew. Their lives are thoroughly edited, not just on film but in their unnatural, contrived existence—rejects from the exploitative world of humanity, they are exploited by filmmakers.

Rather than appreciate the intelligence and grace of orangutans in the natural world, viewers watch them in a virtual community that is tailored to human comforts and interests. The orangutans' unnatural behaviors, and their ongoing association with humanity, is dangerous: These orangutans are expected to be released into the wild. They are unlikely to survive walking upright, or approaching human beings, who too often aim to kill with guns or machetes in such instances, whether out of fear, to protect crops, because they have acquired a taste for primate flesh, or in order to kidnap orangutan babies for the "exotic pet" trade.

Such programs are also part of the reason that tourists sometimes get too close to wild animals. For instance, visitors at the Bukit Lawang rainforest tourism site in Sumatra often feed orangutans fruits, touching or even holding the orangutans while other tourists take photos (*Guidebook*, 2009, p.32). After visiting Bukit Lawang, some people post videos of themselves feeding or hugging orangutans on YouTube ("Hugging an orangutan", 2008; Dellatore, 2010; "Mum and Jackie the Orangutan," 2009). These activities are strictly forbidden, mainly to protect the orangutans: There is a serious risk of spreading human diseases to orangutan communities, where diseases like colds or the flu can be life-threatening, especially to infants (Hadisiswoyo, 2010). When filmmakers air footage of "wild" animals interacting with human beings, as they do on Orangutan Island, they encourage disrespectful and dangerous interactions with wild animals.

Anyone with a computer can watch continual footage of the private lives of nonhumans trapped in many zoos, such as the National Zoo in Washington, D.C. ("Animal Cams"). By airing such footage, zoos foster the attitude that humans are superior—we have the power to control other creatures, including the comparatively new ability to expose every aspect of their lives for our entertainment or "knowledge" (Malamud, 1998, p.58). Continual surveillance bolsters our feeling of superiority. Like the animals in *Spy on the Wild,* zoo animals cannot escape the camera's gaze. Electronic zoos "are not about the animals themselves, but about what we can do with them" (264).

Constant surveillance asserts that nonhumans can be controlled for our purposes, and studied so that no mystery remains to these simplified non-individuals (Haggerty and Trottier, 2013, pp.6–7). In "solving" the mystery of wild animals, we define and limit who they are—we conquer them.

## The Power of Consumers

How we perceive wild animals and their habitats has a lasting effect on how we treat nonhumans and environments. Films teach us to invade so-called sanctuaries and nature reserves with the brash demand to see nonhumans when and where we choose, invading their privacy, terrorizing them, and damaging and destroying their homes. Many so-called nature tourism ventures promise visitors a guaranteed close-up sighting of certain nonhumans, though they disturb nonhumans and sensitive ecosystems in the process.

An attitude that nonhumans and their habitats can and should be conquered and controlled also governs how we treat local animals and ecosystems—whether we support trapping and gunning down wildlife from helicopters or inhumane factory farming, and whether or not we advocate for the environment or willfully degrade our surroundings. A film tradition that celebrates humans as conquerors keeps us trapped in the cycle of conquering. In doing so, films encourage us to unquestioningly partake in practices that oppress and harm nonhumans and the environment.

By thinking critically about films, we can keep them from negatively affecting our lives and minds. By choosing the programs we watch carefully, and by submitting feedback when we see something that disturbs us, we can influence nature films—which, in turn, affect our larger culture. We can also engage others in conversations about nature films that we watch, thereby encouraging more people to think critically about this film genre, and perhaps even to avoid programs that promote harmful attitudes.

A critical analysis of nature films is part of a larger need to ponder consumption, whether of food and bath products or water and energy. Once informed, we might choose to engage in conversations and write letters to the editor about important issues, helping to raise awareness and encouraging dialogue. We might also choose to spend more time outside, simply admiring the mysteries of the natural world, whether in the form of a mountain, a leaf, or an ant. As individuals, we can work to change exploitative forms of domination that discount or diminish the natural world and the many wild individuals who dwell therein, whether expressed through nature films or packaged animal products. But we must first train ourselves to recognize domination in all of its insidious forms.

## Conclusion

Nature films are in a unique and powerful position to promote respectful attitudes toward nonhumans and their habitats, and to build support for conservation. Unfortunately, many of today's documentaries maintain and present the archaic view of nature as feminine, of nonhuman animals as competitors, and of those who go out into the wilderness as heroes who conquer and prevail over wild nature.

Films are just one of the ways that we acquire attitudes about nature and nonhumans. Because of the prevalence and power of television—especially for children—nature films must now be recognized as a primary means of promoting violence toward nonhumans and nature more generally. Along with a media that presents meat-eating as the norm, earth and animal activists as terrorists, and people who disregard the suffering of nonhumans, films help to create a culture of violence and apathy toward nonhumans and the environment. Consumers must actively work to dismantle the system of Western human superiority by rejecting expressions of human superiority, whether in the form of novels and movies about "killer animals," or products that come from factory farms.

Nature filmmakers have a powerful influence. In light of extinctions and global climate change, responsible filmmakers must make a conscious choice to diverge from the archaic Nature Man narrative that focuses on the exploitation of individual nonhumans and the environment. Viewers can take an active role in bringing about this change. We can request that nature films ask at least as many questions as they answer. We can express our desire that filmmakers weave into their narratives a sense of wonder and compassion for the natural world, and a poetic speculation about what it *may* be like to be one of these individual animals. We can also request that films include possibilities as to how we might promote broad-scale social change that will be reflected in our relationships with nature as a whole, producing films that might serve as tools for change (Packwood Freeman and Jarvis, 2013, p.265). This shift in nature films can help people to understand that these many other marvelous creatures exist alongside us, as our equals, and that many aspects of nature—including many aspects of individual nonhumans—are beyond our comprehension. In this way, viewers and filmmakers can work together to recreate a visual representation that more closely aligns with our rightful place in the world of animals, and our place on the larger planet. In the process, we will foster a more responsible society that strives to respect and protect both wild lands and wildlife.

## Discussion Questions

1 What TV shows did you watch as a child that helped shape your understanding of nature? How does this essay help you to revisit and reassess these shows?

2 In what ways do men project masculine constructs onto nonhuman animals?

3 Morally speaking, can we film nonhuman animals for our entertainment? If so, which species/individuals and under what conditions?

4 People visit zoos, circuses, and aquariums primarily as forms of entertainment. How is this better or worse than the film industry's exploitation of animals?

5 Do other animals have the right to privacy? Would it be morally acceptable/justifiable/reasonable for human beings alone to be granted this right?

6 Have you seen any nature films recently that promote masculinist or imperialistic attitudes described in this essay? Describe aspects of the film that promote masculinist or imperialistic attitudes.

7 Do you think female filmmakers tend to portray wilderness and wildlife in a different manner from male filmmakers? In what ways? Why might this be the case?

## Essay Questions

1 Offer a feminist critique of one of the TV shows you watched as a child.

2 If you were to create a nature or wildlife film, how might your own internalized patriarchy or privilege influence the lens through which you present the world?

3 Envision a nature film that successfully steers clear of masculinist and imperialistic attitudes. Describe the storyline, portrayal of subjects, filming, narration, etc.

4 Have you watched a movie or TV show that embodies or promotes both speciesism and racism? Explain how you see speciesism and racism as analogous in this particular film.

5 Have you ever seen a film that presents seemingly contrasting ways of viewing the world? How and why does it do this?

6 Watch *The Hottentot Venus* and/or *The Elephant Man* and describe and critique power over otherness represented in the movie.

## Suggested Further Reading

Alaimo, S., 2000. *Undomesticated ground: Recasting nature as feminist space*. Ithica and London: Cornell UP.

Crowther, B., 1995. Toward a feminist critique of television natural history programs. In P. Florence and D. Reynolds, eds *Feminist subjects, multi-media: Cultural methodologies*. Manchester UP, pp.127–48.

Gandy, M., Stallybrass, P., White, A., and Ryan, J. R., 2000. Hunting with the camera: Photography, wildlife and colonialism in Africa. In C. Philo and C. Wilbert, eds *Animal spaces, beastly places: New geographies of human/animal relations*. London: Routledge, 203–21.

Kaplan, E. A., 1997. *Looking for the other: Feminism, film, and the imperial gaze*. NY and London: Routledge.

Malamud, R., 1998. *Reading zoos: Representations of animals and captivity*. NY: New York UP.

Mitman, G., 2013. *Reel nature: America's romance with wildlife on film*. Cambridge: Harvard UP.

## Notes

1 For the purposes of this essay, references to Western culture refer primarily to Euro-American and British culture even though the attitudes discussed may apply to other Westernized societies as well.

2 While patriarchal societies provide men with considerable control over women, men are also oppressed by this system of social conditioning. They are encouraged, if not compelled, to fulfill an

ideal of masculinity that may not fit their particular persona, and they may be looked down on, or ostracized, if they are not appropriately masculine.

## References


Adams, C., 2004. *The pornography of meat.* NY: The Continuum International Publishing Group, Inc.

Alaimo, S., 2000. *Undomesticated ground: Recasting nature as feminist space.* Ithica and London: Cornell UP.

"Alaska," 2001. *The Jeff Corwin Experience,* Season 1 Episode 4, 4 February 2001. [TV Programme] Animal Planet. Discovery Communications, Inc.

Berger, J., 1992. *About looking.* NY: Vintage.

Berrettini, M. L., June. 2010. Danger! Danger! Danger! or When animals might attack: Adventure activism and wildlife film and television. *Scope* 17, [online] Available at: http://pdxscholar.library.pdx.edu/cgi/viewcontent.cgi?article=1006&context=ta_fac. [Accessed 27 May 2014].

Burt, J., 2002. *Animals in film.* London: Reaktion Books.

Chris, C., 2006. *Watching wildlife.* Minneapolis: Minnesota UP.

*The Crocodile Hunter,* 1997–2004 [TV Programme] Animal Planet. The Best Picture Show Company.

Crowther, B., 1995. Toward a feminist critique of television natural history programs. In P. Florence and D. Reynolds, eds *Feminist subjects, multi-media: Cultural methodologies.* Manchester UP, pp.127–48.

Crowther, B., 1997. Viewing what comes naturally: A television approach to natural history. *Women's Studies International Forum,* 20(2), pp.289–300.

*Dangerous Encounters with Brady Barr,* 2 January 2010, [Online] National Geographic. Accessed September 2010. http://natgeotv.com/uk/dangerous_encounters.

*Dangerous Encounters with Brady Barr, "Dr Brady,"* 28 January 2010, [TV Programme] National Geographic. Available at: http://www.youtube.com/watch?v=nqcAKcdMK1w.

Davies, G., 1998. *Networks of nature: Stories of natural history film-making from the BBC.* Ph.D. University College London.

Davies, G., 2000. Virtual animals in electronic zoos. In C. Philo and C. Wilbert, eds *Animal spaces, beastly places: New geographies of human/animal relations.* London: Routledge. pp.243–67.

Dellatore, D. 26 July 2010. Sumatran Orangutan Society. Personal Interview. Medan, Sumatra.

Foucault, M., 1979. Panopticism. *Discipline and punish: The birth of the prison.* NY: Vintage Books.

Gandy, M., Stallybrass, P., White, A., and Ryan, J. R., 2000. Hunting with the camera: Photography, wildlife and colonialism in Africa. In C. Philo and C. Wilbert, eds *Animal spaces, beastly places: New geographies of human/animal relations.* London: Routledge, 203–21.

*Guidebook to the Gunung Leuser National Park,* 2009. Medan, Indonesia: Yayasan Orangutan Sumatera Lestari-Orangutan Information Center, 38–9.

Hadisiswoyo, P., 25 July 2010. *Orangutan Information Center.* Personal Interview. Haliban, Sumatra.

Haggerty, K. D. and Trottier, D., 2013. Surveillance and/of nature: Monitoring beyond the human. *Society and Animals,* pp.1–20.

Herzog, H. A., 2007. Gender differences in animal-human interactions: A review, *Anthrozoös.* 20.1, 7–21.

Huggan, G., 2013. *Nature's saviours: Celebrity conservationists in the television age.* Abingdon: Routledge.

"Hugging an Orangutan," 1 April 2008 [Video online] Available at: http://il.youtube.com/watch?v=m0c6CZu-ses&p=70F07AC1C4D070F6&index=7 [Accessed 26 Sept. 2010].

*I'm Alive,* "Man Survives Leopard Attack," 2009, [TV Programme] Animal Planet. Gurney Productions. Available at: http://animal.discovery.com/videos/im-alive-videos/.

"Into the Heart of Darkness," 2001 *The Jeff Corwin Experience,* Season 1, Episode 8, 2001, [TV Programme] Animal Planet, Discovery Communications, Inc.

Kaplan, E. A., 1997. *Looking for the other: Feminism, film, and the imperial gaze.* NY and London: Routledge.

Kolodny, A., 1975. *The lay of the land.* North Carolina UP.

Malamud, R., 1998. *Reading zoos: Representations of animals and captivity.* NY: New York UP.

"Mum and Jackie the Orangutan," 9 February 2009, [Video online] http://il.youtube.com/watch?v=5HWQvvB_11M&feature=related. [Accessed 26 September 2010].

The National Zoo. "Animal Cams." Accessed September 26, 2010. http://nationalzoo.si.edu/Animals/WebCams/default.cfm.

*Orangutan Island*, "Filming," 7 December 2008 [TV Programme] Animal Planet. Discovery Communications, Inc. Available at: http://animal.discovery.com/videos/orangutan-island-filming-orangutans.html.

*Orangutan Island*, "Inventive," 8 July 2008, [TV Programme] Animal Planet. Discovery Communications, Inc. 8 July 2008. Available at: http://animal.discovery.com/videos/orangutan-island-webisodes-inventive-orangutans.html.

*Orangutan Island*, "Meet the Orangutans." [online] Available at: http://archive.today/dLptU Accessed 1 June 2014.

*Orangutan Island*, "Orangutans," 7 December 2008, [TV Programme] Animal Planet. Discovery Communications, Inc. Available at: http://animal.discovery.com/videos/orangutan-island-orangutans-walking-upright.html.

*Orangutan Island,* [TV Programme] Animal Planet, 2008. Available at: http://www.animalplanet.com/tv-shows/other/videos/orangutan-island.htm.

Packwood Freeman, C., and Jarvis, J. L., 2013. Consuming nature: The cultural politics of animals and the environment in the mass media. In Marc Bekoff, ed. *Ignoring nature no more: The case for compassionate conservation.* The University of Chicago Press.

Philo, C, and Wilbert, C., 2000. Animal spaces, beastly places: An introduction. In *Animal spaces, beastly places: New geographies of human–animal relations.* London; NY: Routledge, pp.1–33.

Plumwood, V. 1993. *Feminism and the mastery of nature.* London: Routledge.

Spivak, G. C., 2001. From "A critique of postcolonial reason" in *Norton Anthology of Theory and Criticism.* W. Cain, L. A. Finke, B. Johnson, and V. Leitch, eds. NY: W.W. Norton, pp.2197–208.

*Spy on the Wild*, "An Extra Wing," 2006. [TV Programme] BBC Worldwide Ltd. and Discovery Communications, Inc.

*Spy on the Wild*, "Gannets," 2006. [TV Programme] BBC Worldwide Ltd. and Discovery Communications, Inc.

Stigand, C. H., 1986. *Hunting the elephant in Africa.* NY: St. Martin's Press.

# SECTION III

# Common Ground

## Dietary Choice

**Poem:** Bernard Quetchenbach

### The Cicada Killer

She looks
the part, a hornet
                    magnified,
sculling in the grass
below an oak.

I missed the first
                    attack
but guess it falconlike,
the giant tumbled by extended
                                        legs—

dangerous, I suppose,
as any
      heavy prey.
They fall together from the leaves.

The killer's wings are thin and
                                        tinted
orange
and lift
the great cicada from the ground.

"The Cicada Killer" first appeared in *Owen Wister Review* and is included in *Everything As It Happens*, a poetry collection from FootHills Publishing.

# 10

# SO YOU WANT TO STOP DEVOURING ECOSYSTEMS? DO THE MATH!

*John M. Halley*

## Introduction—Human Impact

Humans use more resources than any other species. By 1986 we were already either consuming, diverting or otherwise interfering with 40 percent of the total annual biomass (mass added by growth or reproduction of living organisms) produced on land every year (Vitousek et al., 1986). In the seas, we "only" appropriate 8 percent of aquatic productivity (Pauly and Christensen, 1995). Humanity also intervenes dangerously in major biogeochemical cycles. For example, 39 percent of nitrogen entering the global ecosystem every year is introduced by human beings, mostly as fertilizers (Falkowski et al., 2000). People appropriate more than half of the planet's usable, flowing freshwater. For example, 98 percent of US rivers are impeded, mostly by dams (Vitousek et al., 1997). During growing seasons, many of the great river systems of the world, such as the Indus, actually run dry before they find the sea. This is caused by irrigation. Human emissions have inflated the atmospheric carbon dioxide reservoir by 35 percent, up from 280 ppm in 1,800 to 379 ppm in 2005 (Le Treut et al., 2007). Such excessive emissions have serious consequences, including climate change. It is no longer statistically credible to attribute the global warming we have observed since 1850 to natural causes (Halley, 2009).

Another measure of our enormous and dangerous impact is the sheer biomass of humanity and that of domesticated animals. For example, the biomass of Great Britain's 800 million chickens is now over *100 times* greater than the total biomass of Britain's wild birds. A 1999 study estimated that the biomass of wild birds in Great Britain had fallen from 14,200 tonnes in 1968 to 12,742 tonnes in 1988, a decrease of 10 percent (Dolton and Brooke, 1999). Given the continuing trend in wild-bird biomass, this number should have fallen to 11,468 tonnes by 2008. But there are now more than 800 million chickens in the U.K. If each living chicken weighs 1.5 kg then their total biomass is 1.2 million tonnes. The ratio of these two masses is 104.64—chickens in the U.K. weigh 105 times more than the total biomass of Britain's wild birds. Similar ratios prevail in other industrialized countries.

In 1979, the biomass of humans—when there were only 4.4 billion of us—was *three times* that of *all* the planet's wild mammals, which of course includes such massive animals as whales, elephants, buffalo, and bears (Atjay et al., 1979). Since the late sixties, the rate of human population growth has decreased, and by the middle of the twenty-first century, our population should start to decline (Cohen, 2008). Unfortunately, this good news is offset by new and disturbing trends, most notably, increased per-capita consumption, especially of animal products. Since this increased demand for animal products is met primarily through factory farming, this is a catastrophe for animal welfare. It also portends a major acceleration of species extinction.

As humans continue to multiply and consume, everyone else's space is reduced. Currently, the main factors driving species extinction are overhunting, species introduction, climate change, and habitat loss. Of these, the most important is now habitat loss, especially through conversion of wildlands to agriculture (though in the future, climate change may take the lead; Pimm, 2008). Although it is impossible to be precise about annual extinctions caused by habitat loss, a conservative estimate for the Amazon alone (where most habitat loss is caused by conversion of forests to grasslands for grazing animals) is 3,000 species per year. While debates continue over exact numbers, current rates of extinction are hundreds—perhaps thousands—of times greater than pre-human extinction rates (Pimm et al., 1995).

Species extinction rates are estimated by species-area curves, which show how species numbers (on islands, for example) vary in relation to square feet of habitat (Brooks et al., 1999). Similar rules apply for "islands" of various types—not just for those surrounded by water, but for any isolated habitat fragments, including those that have been divided and isolated by human development. Each "island" can support a limited biodiversity, and as the island's size decreases, so does the number of species that can coexist. This means that the conversion of most of a rainforest to cultivation will not only lose the species that once dwelled in the cultivated areas, but also many of species in the remaining forest, because they cannot be sustained in such diminished areas, and therefore go extinct. Thus, because this preference drives others to clear rainforests in order to graze cattle and raise crops to feed cattle (and chickens and pigs), people's preference for meat-eating leads directly to the extinction of species (and also, of course, to the death of most of the individual wild animals in the affected areas).

The purpose of this essay is to explore how human dietary preferences affect habitat loss and species extinction using quantitative methods. Toward this end, I first calculate the annual food-energy distribution for a sample of people in Europe (a "standard" diet). I then estimate the area of cultivation required for such a diet, and compare this area to that required for a vegetarian or vegan diet. Finally, I discuss the advantage of a meat-free diet with regard to habitat loss and species extinction.

Quantitative analysis is limited. In addition to the technical difficulties described later (see "Limitations," p. 158), this approach compresses all the individual stories of humans and other animals into featureless numbers. When it comes to changing our lives or our lifestyle, the quantitative approach alone is entirely inadequate. Nonetheless, a mathematical approach can play an important role: In Western cultures, numbers tend to "pack a punch." For example, the fact that the mass of (factory-farmed) chickens exceeds all natural birds one hundred times over is a powerful testament to the imbalance we have created. In this spirit, this article uses numbers to quantify the cost (in biodiversity) of our dietary choice. Combined with other information, this mathematical analysis provides incentive for informed, concerned citizens to think more carefully about what they choose to eat.

## Quantifying the Impact of Dietary Choice

In terms of reduced environmental impact, various forms of vegetarianism and veganism compare favorably with meat-eating. For example, the environmental impact of humans using soy protein is somewhere between 4.4 and 100 times less than that of humans consuming animal protein (Reijnders and Soret, 2003). More recently, scientists have started to analyze the U.S. diet in terms of global warming, noting that the average American diet throws 1.5 tonnes more $CO_2$ into the atmosphere, per person per year, than does a vegan diet (Eshel and Martin, 2006). This is equivalent to the difference between driving a high efficiency car and an SUV.

The Food and Agriculture Organization's 2006 landmark report, *Livestock's Long Shadow*, highlights the environmental impact of an animal-intensive diet (Steinfeld et al., 2006). This report notes that 18 percent of our annual greenhouse-gas emissions stem from animal production. Other authors, such as Goodland and Anhang, argue that more than half of our greenhouses gases (51 percent) stem from animal farming (Goodland and Anhang, 2009). The interesting and controversial aspect of Goodland and Anhang's analysis is that they include in their figures the direct respiration of livestock (Bell et al., 2014; Herrero et al., 2011). As explained by Barnosky, natural respiration would also have occurred in the vast natural herds of megafauna (such as mammoths and mastodons, who were probably exterminated by humans), but such great beasts existed in considerably smaller numbers than current domestic herds, which are maintained above natural carrying capacity by petrochemicals (Barnosky, 2008).

As an applied ecology teacher, given that our diet affects the environment, I needed to develop some method by which students of applied ecology could calculate the environmental impact of their personal dietary choices (using the methods in Brewer and McCann's *Laboratory and Field Manual of Ecology*, 1982). To discover their dietary "ecological footprint," each student keeps a three-day record of their consumption of common food categories. With the help of a handy-reference table, students look up the calorific value of food types. Then, using this, they calculate the calorific value of foods they consume over a three-day period, and by extrapolation, estimate their annual caloric consumption. On the basis of well-known agronomic relations, students also calculate the area of cultivation that is required to support each food category consumed. For example a square meter can supply 1,000 kcal per year in the form of oranges, but only 40 kcal in the form of cheese.

Based on information gathered from this exercise (between 2007–09), I have assembled a database, from which I have randomly extracted the results for 25 students in order to estimate average consumption (Figure 10.1).

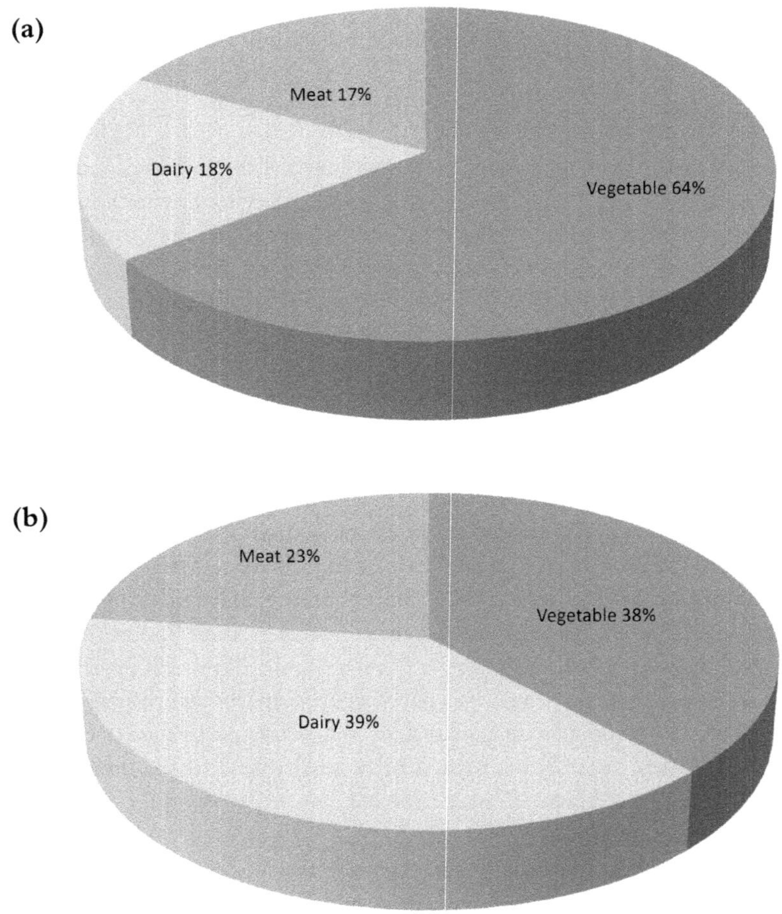


**FIGURE 10.1** "Share of (a) energy intake derived from meat, other animal products (dairy and eggs), and vegetables compared with (b) corresponding areas needed to support the components of this diet (b) for a random sampling of Greek students in Applied Ecology at Ioannina. Greek people tend to eat a high proportion of fish relative to Americans, but their consumption of fish is still not particularly high, and since footprint calculations involving fish are somewhat different, fish are not included in this figure. This omission slightly reduces the apparent consumption of animal products.

With an overall caloric intake of 0.894 million kcal/year, students typically receive 64 percent of their caloric intake from vegetable sources, 17 percent from meat sources, and 18 percent from dairy products or eggs. This is a typical pattern in Western countries, a pattern which differs substantively from other areas of the world, in which vegetables still generally supply 90 percent of caloric intake. When we examine land area required to support common food categories for Western consumers, we find that, on average, a total of 1.107 acres are required. We also establish the percentage of land necessary for the various food groups: 38 percent for vegetables versus 23 percent for meat sources and 39 percent for other animal products. Animal products require considerably more land per kilocalorie because this land must support not only humans but also farmed animals.

On the basis of patterns of consumption recorded by students, I constructed two dietary options: the *Partial Vegetarian* and the *Partial Vegan*. The first is likely to represent someone on their way to becoming a full vegetarian, who regularly foregoes a fixed portion of meat intake, and makes up for lost caloric energy by consuming more vegetables *and* more dairy products. The *Partial Vegan* is likely to represent someone on their way to becoming a full-fledged vegan, who eschews a fixed percentage of meat *and* dairy, increasing only vegetable consumption to meet energy requirements.

I calculated the required area of cultivation for each of these proposed diets—the result is shown in the Figure 10.2. This figure assumes that each kcal of meat removed from the partial vegetarian's diet will be replaced with vegetables *and* non-meat animal products, in a ratio of 18:64 (as is the case with students, shown in Figure 10.1). In this scenario, 22 percent of replacement foods come from dairy and eggs, while 78 percent come from vegetable products. If consumers fill the meat-gap with a *greater* proportion of dairy and eggs (rather than vegetables), *there will be no reduction in land area required—and it might even increase*. While vegetarianism usually requires less land than a meat-based diet, this land-use reduction is rooted in the shift to vegetables. In contrast, vegetarians who increase their consumption of dairy and egg products, similarly increase their "ecological footprint" because dairy and egg products require farmed animals, who require more land because calories are cycled through cattle and chickens. Therefore, simply removing meat leads to a 6.2 percent reduction of land use (at best) if dairy and egg consumption expands with a reduction in meat. In contrast, this same improvement occurs with just a 15 percent commitment to a vegan diet. However, some authors (Haddad and Tanzman, 2003) argue that most vegetarians typically turn to vegetable alternatives, substituting less than 15 percent of lost flesh calories with dairy and eggs. Such a vegetarian would require only 0.918 agricultural acres, offering a 17 percent reduction over the standard meat-dairy-and-eggs diet. Vegetarians who employed a 90:10 replacement ratio (favoring vegetable-derived products) would need only 0.787 acres, representing a 29 percent land area reduction in comparison with a standard Western diet.

Figure 10.2 shows that reducing the amount of meat in one's diet, without reducing other animal products, leads only to a marginal reduction in land area

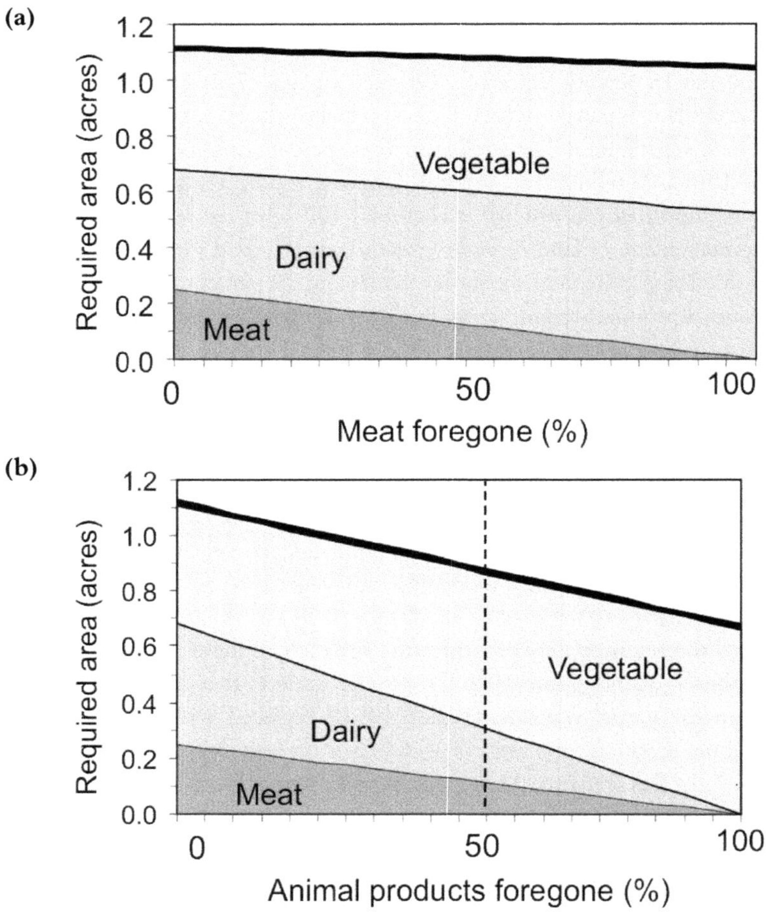


**FIGURE 10.2** This graph shows the amount of land required to support (a) partial vegetarians and (b) partial vegans respectively.

The partial vegetarian (a) reduces meat consumption, and replaces this with vegetable and dairy consumption. The partial vegan (b) replaces animal products "across the board" with vegetable products. The horizontal axis represents the degree of replacement; the total vegetarian and the total vegan are therefore represented on the far right of each graph. Accordingly, a 50 percent vegan is indicated by the dotted line on graph (b). A 50 percent vegan has cut consumption of animal products by half, replacing half of their animal calories (meat, milk, and eggs) with vegetable calories.

The bold line (on top) defines the drop in total land area required as consumers shift from meat to dairy and vegetables (vegetarian, graph a) or from meat and dairy to vegetables (vegan, graph b). For example, the 50 percent vegan (dotted line, right panel) will require a total of 0.885 acres to produce the food he or she consumes (0.125 acres for meat consumption, 0.216 acres for other animal products consumed, and the remaining 0.544 acres for vegetables).

For the purposes of this graph, each person is assumed to have a caloric requirement of 0.894 million kcal/year (as is the case for my Greek students), and a ratio of 18:64 is assumed for meat replacement with vegetables and non-meat animal products.

required (left panel). Thus switching to a vegetarian diet has little effect on our ecological footprint. In contrast, reducing our intake of animal products overall greatly reduces land use, and therefore is highly beneficial to the environment (right panel). The vegan diet requires just 0.662 acres of cultivated land, compared to 1.107 acres for the standard Western diet, rich in animal products.

## Discussion

The most important insight provided by this statistical analysis is that a reduction in our intake of animal products is rewarded with a major reduction in environmental impact. Another important insight is that merely removing meat achieves very little—the whole spectrum of animal products must be reduced if we are to reduce our ecological footprint. This is not surprising because raising farmed animals requires much more land than does raising vegetables, and farmed animals are required if we are to have eggs or dairy products.

Figure 10.3 provides a more detailed view of the environmental cost of various foods. This barchart shows the acreage needed to produce the main components of the standard Western (Greek) diet, excluding fish. Note that Figure 10.3 demonstrates that the energy cost of a square meter of land is considerably higher for animal products than for vegetables. For example, an acre of land planted with potatoes could support eight times the number of people supported by a square meter of land devoted to animal products, excluding milk. Potatoes can support 3.81 times as many people as can milk products. Note that cheese has a strikingly low caloric yield in relation to land use. Cheese requires a very large area of land in exchange for very few consumable calories.

Why is there only a 40 percent overall gain if we remove *all* animal products from our diet? It is because, in contrast to the analysis of others (such as Reijnders and Soret, 2003), I am comparing complete diets rather than focusing on a single element (such as protein). This means that vegan and vegetarian diets are assessed based on the current spectrum of consumption for each sector, and are being compared with a meat-based diet that already includes 64 percent vegetable products. Thus, there is only a 36 percent increase in vegetable consumption possible. Furthermore, high-yield-per-acre vegetable products, like potatoes and beans, are seldom prominent in the standard vegan or vegetarian diet, while comparatively lower-yield-per-acre products are more common, such as vegetable oils (especially in Greece).

Environmental and animal activist agendas both point to a vegan diet, but these two agendas do not always coincide. For example, when we compare yields *within* the meat sector, we find that students eat nearly four times as much poultry as goat or lamb, but that poultry contributes only twice as much land use. But poultry farming entails far greater animal welfare problems. So environmentalists may sometimes say "eat more chicken" while animal advocates say "avoid eating chicken." Nonetheless, as environmental factors beyond land use are considered (such as the affects of nitrogen from chicken manure), much of this conflict is likely to disappear (Steinfeld et al., 2006).

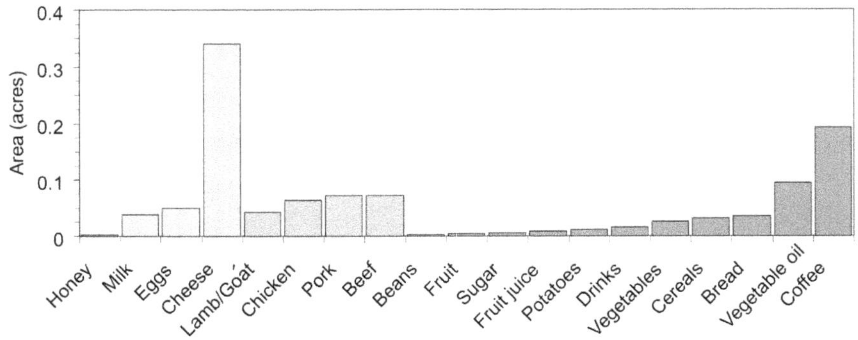


**FIGURE 10.3** Land area required for the principal components of the standard Greek diet (excluding fish) as represented by a random sample from applied ecology students (as noted above). The sum of the height of all bars totaled adds up to 1.107 acres. This, then, is the average area required to support student diets, as reported in questionnaires.

Note: "Drinks" include wine, beer, and soft drinks, but not milk or coffee.

In any event, environmental and animal activists have more in common with regard to dietary concerns than most people realize. If 300 million people (approximately equal to the population of the U.S.) were to become even 50 percent vegan, the environmental results would be significant, saving approximately 0.223 acres per person. The total area moved out of production and back to habitat would be over 270,000 km$^2$. This is larger than the area of the Amazon currently lost each year through deforestation (about 23,000 km$^2$), which not only exploits farmed animals, but also causes the extinction of 3,000 species annually. Clearly there is much that environmentalists and animal advocates can agree on with regard to diet. Whether we care about the planet, species, or individual animals, we need to move toward consuming more vegetable products, and away from meat, dairy, and eggs.

## Limitations

I have analyzed the relation between diet, basic nutritional energetics (calories in versus calories out) and land use. This analysis is only one small part of the larger environmental picture with regard to diet. Modern animal agriculture is linked to a truly staggering number of environmental problems: natural-habitat destruction, greenhouse gas emissions, soil degradation, aquifer depletion, and various kinds of pollution—including the creation of dead-zones (Steinfeld et al., 2006). Industrialized animal production (factory farming) also entails animal abuse, massive social costs (many small farmers are replaced by a few mega-units), an increase in zoonotic illnesses (like bird and swine flu), and the misuse (and consequent ineffectiveness) of antibiotics (Steinfeld et al., 2006). While all of these issues demand equal time and consideration, the above analysis serves to show the direct link between environmental and animal welfare agendas through the prism of personal dietary preferences and land use.

It is also important to note that nutritional energetics is only one part of dietary assessment, which must ultimately assess food qualities, such as protein and fiber content. There is a rich and growing body of literature that examines the challenges and benefits of a vegetable-based diet in comparison with a diet rooted in animal products (Campbell and Campbell, 2005; Eshel and Martin, 2006; "Position of the American Dietetic Association," 2009) .

A land-based assessment also ignores the energy input required for agricultural production, cooking, storage, and processing various food types. Again, as discussed by Eshel and Martin, evidence suggests that these factors, when considered, point to the environmental efficacy of the vegetarian/vegan dietary choice (Eshel and Martin, 2006).

## Conclusions

Because of the enormous mass of humanity currently on the planet, even small changes in lifestyles, such as consuming fewer animal products, can have an enormous environmental impact. A diet rich in animal products leads to an environmentally devastating chain of events: greater land use, conversion of natural environments to farmland, displacement and death of wild animals living in these regions, and reduced wild populations in remaining wildlands (which leaves these populations more prone to extinction). Thus, the consumption of animal products leads to habitat destruction and species extinction. Humans can help end this disastrous spiral by consuming less meat, dairy, and eggs.

From an environmental point of view, if one is concerned about deforestation and species loss, we ought to seriously reduce our consumption of animal products—not only meat but all animal products. Merely moving towards a diet rich in non-meat animal products (dairy and eggs) achieves virtually nothing. Thus the environmentalist is led to a conclusion similar to the animal activist: removing meat from one's diet is not enough; it is much better to be quarter-vegan than 100 percent vegetarian.

## Appendix: Methods of Calculation

The energy intake of most Western consumers is composed of a vegetable component $V$, a dairy and egg component $D$ and a meat component $M$. The sum of these must meet required energy $E$.

$$E = M + D + V$$

Changing to a vegetarian diet can be done by reducing the meat component of our diet while increasing the other two sectors, leading to new proportions:

$$E = \alpha M + \beta(D + V)$$

Since we need total energy intake to remain the same, we can estimate the increase of dairy and vegetable foods ($\beta$) when these foods replace meat intake:

$$\beta = \frac{E - \alpha M}{D + V}$$

Our total land "footprint" as consumers, (the area of cultivation required), is the sum of the areas required for each food type. For example, suppose the total area needed to support a standard diet (1.107 acres) consists of meat, dairy, and vegetable portions $A_M$, $A_D$ and $A_V$. Provided that the only changes made are *between* sectors, but not *within* sectors, the land areas needed will change proportionately.

For the above three sectors, the total area needed is:

$$A_1 = \alpha A_M + \beta(A_D + A_V)$$

Thus, the area required to support a partial vegetarian will be:

$$A_1 = \alpha A_M + \frac{E - \alpha M}{D + V}\left(A_D + A_V\right)$$

Similarly, the area required by a partial vegan will be:

$$A_2 = \alpha\left(A_M + A_D\right) + \frac{E - \alpha(M + D)}{V} A_V$$

This analysis demonstrates that we only need a single parameter ($\alpha$) to describe the change of areas needed for cultivation as we change the ratios between sectors in our diet. Thus, there is a direct and simple link between reduction of animal products in our diet and land area saved.

## Discussion Questions

1 As a class, how many interrelated factors can you list between the cattle industry and the mass extinction of animals in the Amazon? (To be included in the list, someone must be able to articulate relevant interrelated factors.) How many interrelated factors can you list between hog/poultry industries and mass extinctions in the Amazon?

2 How do the planetary effects of an omnivorous, vegetarian, and vegan diet differ? With regard to animal suffering, how do the effects of an omnivorous, vegetarian, and vegan diet differ? When you decide what to consume for lunch, which seems more important to you, and why?

3 Given what you know about diet and the planet, why do you suppose we are not all vegans? What are the most important factors blocking people from shifting to a plant based diet, and what creative strategies might help overcome these blocks?

4 Section III begins with a poem titled, "The Cicada Killer." How do you feel about the killing ways of other flesh-eaters? Do you react differently depending on predator and prey? Do you feel differently if a hawk kills a snake or a baby bunny? What if a snake kills a baby bird or a shrew? Is there any value to what seem to be unfounded, emotional reactions/attachments with regard to predator/prey encounters?

## Essay Questions

1 Halley comments that some 3,000 species disappear every year from the Amazon. Use WWF's "Footprint Calculator" (see suggested further reading below) to estimate your ecological footprint ("How many planets do you need?"). If you are/were an omnivore, how much would your footprint change if you were vegan? If you are/were a vegetarian, how much would your footprint change if you were vegan?

2 The respiration of humans and farmed animals are not currently included in the IPCC estimates of anthropogenic greenhouse gas production. Why do you suppose they are excluded from this critical calculation? Consider contrasting opinions on the subject (articles by Steinfeld et al. 2006; Goodland and Anhang 2009; and Herraro et al. 2011) and take a stand as to whether or not such respirations should be included in IPCC estimates of anthropogenic greenhouse gas production. Defend your position.

## Suggested Further Reading

Brewer, R., McCann, M.T., 1982. *Laboratory and field manual of ecology*. Saunders College Publishing.

Goodland, R., Anhang, J., 2009. Livestock and climate change: what if the key actors in climate change are… cows, pigs, and chickens. *World Watch*, November/December 2009, 10–19.

Herrero, M., Gerber, P., Vellinga, T., Garnett, T., Leip, A., Opio, C., Westhoek, H.J., Thornton, P.K., Olesen, J., Hutchings, N., 2011. Livestock and greenhouse gas emissions: The importance of getting the numbers right. *Animal Feed and Science Technology* 166, 779–782.

Steinfeld, H., Gerber, P., Wassenaar, T.D., Castel, V., Rosales, M., de Haan, C., 2006. *Livestock's long shadow: environmental issues and options*. Food and Agriculture Org.

WWF Footprint Calculator (http://footprint.wwf.org.uk/).

## References

Atjay, G., Ketner, P., Duvigneaud, P., 1979. Terrestrial primary production and phytomass, in *The Global Carbon Cycle*. John Wiley, Chichester, UK, pp. 129–181.

Barnosky, A.D., 2008. Megafauna biomass tradeoff as a driver of Quaternary and future extinctions. *Proceedings of the National Academy of Sciences* 105, 11543–11548.

Bell, M.J., Cloy, J.M., Rees, R.M., 2014. The true extent of agriculture's contribution to national greenhouse gas emissions. *Environmental Science and Policy* 39, 1–12.

Brewer, R., McCann, M.T., 1982. *Laboratory and field manual of ecology*. Saunders College Publishing.

Brooks, T.M., Pimm, S.L., Oyugi, J.O., 1999. Time lag between deforestation and bird extinction in tropical forest fragments. *Conservation Biology* 13, 1140–1150.

Campbell, T.C., Campbell, T.M.(II), 2005. *The China study: The most comprehensive study of nutrition ever conducted and the startling implications for diet, weight loss and long-term health.* BenBella Books, Inc.

Cohen, J.E., 2008. *How many people can the Earth support?* Paw Prints.

Dolton, C.S., Brooke, M. de L., 1999. Changes in the biomass of birds breeding in Great Britain, 1968–88. *Bird Study* 46, 274–278.

Eshel, G., Martin, P.A., 2006. Diet, Energy, and Global Warming. *Earth Interact.* 10.

Falkowski, P., Scholes, R.J., Boyle, E., Canadell, J., Canfield, D., Elser, J., Gruber, N., Hibbard, K., Högberg, P., Linder, S., 2000. The global carbon cycle: a test of our knowledge of earth as a system. *Science* 290, 291–296.

Goodland, R., Anhang, J., 2009. Livestock and climate change: what if the key actors in climate change are... cows, pigs, and chickens? *World Watch*, November/December 2009, 10–19..

Haddad, E.H., Tanzman, J.S., 2003. What do vegetarians in the United States eat? *American Journal of Clinical Nutrition* 78, 626S–632S.

Halley, J.M., 2009. Using models with long-term persistence to interpret the rapid increase of Earth's temperature. *Physica A* 388, 2492–2502.

Herrero, M., Gerber, P., Vellinga, T., Garnett, T., Leip, A., Opio, C., Westhoek, H.J., Thornton, P.K., Olesen, J., Hutchings, N., 2011. Livestock and greenhouse gas emissions: The importance of getting the numbers right. *Animal Feed and Science Technology*. 166, 779–782.

Le Treut, H., Somerville, R., Cubasch, U., Ding, Y., Mauritzen, C., Mokssit, A., Peterson, T., Prather, M., 2007. Chapter 1: Historical Overview of Climate Change, in Solomon, S., Qin, D., Manning, M., Chen, Z., Marquis, M., Averyt, K., Tignor, M., Miller, H., Eds (Eds), *Climate Change 2007: The Physical Science Basis.* Contribution of Working Group I to the Fourth Assessment Report of the Intergovernmental Panel on Climate Change. Cambridge University Press.

Pauly, D., Christensen, V., 1995. Primary Production Required to Sustain Global Fisheries. *Nature* 374, 255–257.

Pimm, S.L., 2008. Biodiversity: Climate Change or Habitat Loss—Which Will Kill More Species? *Current Biology* 18, R117–R119.

Pimm, S.L., Russell, G.J., Gittleman, J.L., Brooks, T.M., 1995. The Future of Biodiversity. *Science* 269, 347–350.

Position of the American Dietetic Association: Vegetarian Diets, 2009. *Journal of American Dietetic Association* 109, 1266–1282. doi:10.1016/j.jada.2009.05.027.

Reijnders, L., Soret, S., 2003. Quantification of the environmental impact of different dietary protein choices. *American Journal of Clinical Nutrition*. 78, 664S–668S.

Steinfeld, H., Gerber, P., Wassenaar, T.D., Castel, V., Rosales, M., de Haan, C., 2006. *Livestock's long shadow: environmental issues and options.* Food and Agriculture Org.

Vitousek, P.M., Ehrlich, A.H., Ehrlich, P.R., Matson, P.A., 1986. Human Appropriation of the Products of Photosynthesis. *Bioscience* 36, 368–373.

Vitousek, P.M., Mooney, H.A., Lubchenco, J., Melillo, J.M., 1997. Human domination of Earth's ecosystems. *Science* 277, 494–499.

# 11

# A FISHY BUSINESS

*Lisa Kemmerer and Bethany Dopp*

The ocean is a vast series of diverse yet interconnected ecosystems. Ocean waters cover the majority of the planet and are home to countless varieties of life. The volume of living space provided by the seas is 168 times greater than that of land-based habitats; 90 percent of earth's living biomass is in the sea (Clark et al. 2006). Despite this opulence, ocean life is in peril, but unlike other threatened ecosystems, the impending oceanic crisis is largely unseen and largely unmentioned even among environmentalists. While animal advocates and environmentalists rally on behalf of dolphins, seals, and whales, neither group seems too worried about Chinook salmon or Atlantic halibut—but they ought to be.

## Feeling Fish

Creatures of the sea lack the fuzzy bodies that tend to attract human empathy. Though they do not cry out in pain, or perhaps even grimace, they are sentient. It is "unthinkable that fish do not have pain receptors; they need them in order to survive" (Pittsburgh Independent Media 2005). Fish are vertebrates with a complex nervous system. Anatomical, pharmacological, and behavioral data, as well as evolutionary evidence and neurophysiological analogies, demonstrate that fish suffer (Chandroo et al. 2004; Rollin 1981). Fish have nerve endings designed to register pain just like other vertebrates, and fish produce the same brain chemicals that humans produce to counter pain: enkephalins and endorphins (Balcombe 2006).

Fish interact socially, remember, learn, suffer, fear, and struggle mightily to preserve their lives. They are intelligent: fish interact socially, learn, and remember what they learn. The brains of fish, like those of other sentient vertebrates, provide a means by which to avoid suffering, including a good memory and the ability to learn (Dionys de Leeuw 1996). Fish are intelligent—intelligent enough to use tools (Balcombe 2006). Fish use their long-term memories to survive in waters riddled with predators, and to navigate a complex social world (Balcombe 2006). We now know that fish are able to recognize "shoal mates," acknowledge hierarchy, track relationships, and eavesdrop on others in their community (Balcombe 2006).

> Gone (or at least obsolete) is the image of fish as drudging and dim-witted pea brains, driven largely by "instinct," with what little behavioral flexibility they possess being severely hampered by an infamous "three-second memory." … Now, fish are regarded as steeped in social intelligence, pursuing Machiavellian strategies of manipulation, punishment and reconciliation, exhibiting stable cultural traditions, and co-operating to inspect predators and catch food.
>
> (Laland et al. n.d.)

We now know that "fish have all the relevant characteristics attributable to animals requiring humane treatment" (Dionys de Leeuw 1996). The pain that fish experience on a hook is likely comparable to "dentistry without Novacain, drilling into exposed nerves" (Pittsburgh Independent Media 2005). Unfortunately, as with knowledge of farmed animals, this information does not seem to affect our dietary choices.

## Fishing Methods—Destroying Ecosystems and Individuals

Ecological impact and suffering associated with industrial fishing is largely determined by indiscriminate methods. Three common fishing methods—long lines, nets, and trawling—are central to industrialized fishing, and each has extremely negative effects on ecosystems and sea life.

### *Bycatch*

Bycatch is marine life that is *accidentally* fished from the sea. The very existence of bycatch demonstrates one of the fundamental problems of industrial fishing: Lines and nets are indiscriminate, catching whatever becomes entangled, including endangered and protected species. When sea gulls or cormorants are caught on a longline hook and pulled underwater, they drown. When seals, turtles, or dolphins are caught in a net, they drown. When deep sea fish are pulled to the surface, blood vessels and air bladders rupture. For obvious reasons, bycatch mortality is 90–100 percent (Clucas 1997).

Shrimp trawlers produce the most bycatch. Worldwide, the Food and Agriculture Organizations (FAO) of the United Nations estimate that 85 percent of any shrimp haul is bycatch (Clucas 1997). With every 15 pounds of shrimp consumed, those who buy shrimp pay for 85 pounds of bycatch—living, breathing creatures that need not have been caught, including sea turtles, seals, and whales. Shrimp trawling destroys more endangered turtles than all other human endeavors *combined* (Decline 1990).

Thanks to bycatch, at least five species of deepwater fish—only pursued since the 1970s—are now on the critically endangered list (MacKenzie 2006). "Roundnose and onion-eye grenadier were once commercially fished, but are now taken almost entirely as accidental bycatch alongside Greenland halibut, another deep-water fish that has also begun to decline" (MacKenzie 2006). Three other species—blue hake, spiny eel, and spinytail skate—*have never been taken except as bycatch*. Between 1978 and 1994 these five species "lost between 87 and 98 percent of their initial abundance," and between 1995 and 2004 grenadiers "declined still further—93.3 percent for the onion-eye and an astonishing 99.6 percent for the roundnose over 26 years. Their average size has also halved, showing that few fish are getting a chance to mature and breed" (MacKenzie 2006). Bycatch further depletes decimated fish populations and endangers otherwise healthy populations.

### Longlines

Longlines are one of the most common fishing methods worldwide. Longlines can stretch over sixty miles: "Bottom longline fishing uses hundreds or even thousands of baited hooks along miles of lines laid behind fishing vessels and stretching down to the reef and ...[sea] floor" (Gulf Sea Turtles 2009). Longliners set an estimated one billion razor-sharp hooks each year; about five million of these hooks are sent down into the ocean every day (Sea Turtle Restoration Project 2003). Though intended to catch large predator fish, like tuna and swordfish, longlines kill an estimated 4.4 million seabirds, marine mammals, sharks, billfish, and sea turtles annually (Ovetz 2004). Some 40,000 endangered sea turtles fall victim to longline hooks each year (Sea Turtle Restoration Project 2003; Bycatch Reduction Devices n.d.). Longlines incidentally hook an estimated 4.4 million seabirds, marine mammals, sharks, billfish, and sea turtles annually (Ovetz 2004). Most often turtles drown when caught by these hooks, but if they escape, hook injuries "affect a turtle's ability to feed, swim, avoid predators, and reproduce" so that even turtles who escape are rarely able to "recover from the extreme physiological stress of being caught" (Gulf Sea Turtles 2009).

Every year longline hooks also kill 300,000 seabirds, including twenty-two endangered species (Sea Turtle Restoration Project 2003). Fishing further threatens nearly half of the earth's endangered seabirds (Seabirds 2007). For example, nineteen of the twenty-one known albatross species are considered threatened with extinction worldwide and longline fisheries are the core problem (Black-Footed 2007). Cetaceans (whales, dolphins, and porpoises) such as the Pseudorca, a smaller cousin to the orca (living in and around the Hawaiian Islands) also fall victim to longlines. Many of these cetacean die as a predictable "accidental side effect" of longline fishing (False 2009).

### Nets

Nets—drift nets, gill nets, trawling nets, and seine nets—are the most common industrial fishing methods. The basics of net fishing are simple: Sea life swims or is swept into a net where they are trapped and pulled from the water. Netting, like every other method of fishing, is indiscriminate: Nets catch everything that comes into their web, including endangered species and an abundance of bycatch. Seals, whales, and dolphins often drown in nets because they are unable to surface for air.

Tuna nets have caused the deaths of "more than seven million dolphins since the introduction of purse seine tuna fishing in the late 1950s" (Palmer 2009). The vaquita, endemic to the northern part of the Gulf of California, is currently the world's most endangered cetacean, and biologists have determined that gill nets are primarily responsible. Even though reserves have been set aside to protect the vaquita and their essential habitat, these busy swimmers still turn up in fishing nets at alarming rates—a minimum of 20–30 each year—an alarming number for a population that has fallen to only 300 individuals (Dalton 2004; Vanishing 2008; Rodríguez-Quiroz 2012).

Fishing nets are invariably lost at sea, becoming "ghost nets" that ensnare and kill sea creatures for as long as their strong nylon fibers last—which is to say, a very long time. Making a questionable practice yet more shameful, mega-fleets lose hundreds of kilometers of drift nets each year, and they continue to ensnare ocean life for decades, if not centuries (Mission Blue 2013). At present, more than 10,000 plus miles (16,000 km) of ghost nets are estimated to be floating the seas, luring new victims with their floating glob of decaying matter, destroying millions of additional sea creatures each day (Watson 2012, p.643).

## *Trawling*

The Gulf of Thailand has suffered a 60 percent decline in large sharks, skates, and finfish with just five years of trawling (Myers and Worm 2003). Smaller fish are also threatened; the New England herring population collapsed in the 1970s:

> The herring industry is becoming increasingly dominated by high-volume industrial ships known as midwater trawlers—which drag massive small-mesh nets behind them, catching everything in their path. The trawlers sometimes work in pairs so they can drag even bigger nets between them. The practice can lead to localized depletion of herring and contribute to the overfishing and stalled recovery of severely depleted populations of cod, hake, haddock, and other fish that live near the ocean floor.
>
> (Under 2007)

Clear cutting does to landscapes what trawling does to the sea floor. Yet, while most environmentalists understand the negative effects of clear cutting, few seem aware of the harm caused by trawling. The massive weight of a trawl, as much as five tons, leaves only finely ground debris in its wake (Destructive 2004). An ocean-bottom environment forms over centuries; trawls destroy these ecosystems in one pass. Coral grows slowly, averaging only 0.2–1.1 inches per year (Coral 2006). Trawling thereby has a long term negative affect on fish populations, on sea life more generally, and on ocean ecosystems. Sponges, coral, and rock formations on the sea's bottom provide feeding and breeding grounds for marine life, and shelter from predators.

Bottom trawling also decimates fish populations, scooping up tons of fish in a single haul, ravaging habitat along the way, and making species recovery difficult, if not impossible. Because they are critical and fragile, requiring centuries to recover, the earth's sea-floor ecosystems ought to be protected.

## Solutions

Industrial fishing wreaks havoc on species, habitats, and on the food web, causing injury and death to countless living beings, many of whom are sentient. What can we do to ease the momentous problems caused by our taste for sea flesh?

### *Aquaculture—Factory Fishing*

Aquaculture, more commonly referred to as fish farming—and more accurately labeled factory fishing—is on the rise. But factory fishing comes with an additional series of worrisome problems, including disease, bio-contamination, pollution, and suffering. Just as grass fed beef and organic chicken do not mitigate the massive environmental problems inherent in animal agriculture, factory fishing does not mitigate the horrific environmental problems inherent in the consumption of fish. It simply creates a different set of environmental problems.

Factory fish are raised in netted-off areas of the open ocean. Consequently, there is risk of bio-contamination, of "domestic" fish escaping into and breeding with wild populations. As with factory farms, factory fish live in crowded conditions, where they suffer from parasite infestations and the spread of disease. While crowded factory farmed fish are treated with chemicals and antibiotics (like factory farmed cattle, hogs, and hens) to reduce the spread of disease and infestation wild salmon, for example, have no such protection, so they are more vulnerable to infection (Rosenberg 2008). The free flow of water from captive fish into the open ocean allows food, waste, parasites, chemicals, and antibiotics from fish farms to contaminate surrounding waters and wild fish populations (Cufone 2008). Factory fishing has put "several wild salmon populations at risk of extinction" (Salmon 2008). For example, there is "a direct connection between the aquaculture industry's rapid growth in the Broughton Archipelago off British Columbia and the sharp decline in its wild pink salmon" (Salmon 2008). Though older salmon can handle the parasite, young wild salmon migrating through these areas are much more vulnerable.

> In the natural system, the youngest salmon are not exposed to sea lice because the adult salmon that carry the parasite are offshore. … Fish farms cause a deadly collision between the vulnerable young salon and sea lice. They are not equipped to survive this, and they don't.
>
> (Salmon 2008)

While harming and endangering seas and sea life, factory fishing increases demand for wild-caught fish. As with the terrestrial flesh industry, raising salmon (and any other carnivorous fish) burns up more food than is ultimately produced. Three pounds of fish are required to produce one pound of farmed salmon (Pauly and Watson 2003). On factory farms, grain is inefficiently cycled through animals to create flesh, dairy, and eggs. On factory fish farms, wild-caught fish are inefficiently cycled through farmed fish to create fish flesh. The University of British Columbia recently reported that "90 percent of the global small fish catch—which includes anchovies, sardines, and mackerel—is processed into fish meal and fish oil and used in animal feed" (A Broken 2009). Factory farms and factory fisheries are both dependable places to sell bycatch.

## Technologies Reducing Bycatch

Fishing technologies have been developed to protect endangered species (such as sea turtles) and to reduce bycatch (especially birds and cetaceans). For example, net design modifications allow dolphins to escape from nets through trapdoors, sound-emitting devices keep dolphins away, and streamers that flap in the wind frighten birds away from longlines. Other methods include placing lines deeper, so that birds will not see baited hooks, and fishing at night, when most birds sleep (Humane Society of the United States 2007).

Turtle excluding devices (TED) are designed to prevent turtles from being caught by shrimp trawlers, and have had some measure of success (depending on the features of the ocean floor, such as debris, vegetation, and rock formations) (Decline 1990). TEDs have bars in a grid pattern with an opening at the top or bottom of the net. Small animals (such as shrimp) pass through the bars, but when larger animals (such as sea turtles) are caught "they strike the grid bars and are ejected through the opening" (National Oceanic and Atmospheric Administration Fisheries Service n.d.). Unfortunately, many fishing boat operators have resisted adopting TEDs, partly because they are expensive, and partly because TEDs reportedly reduce the volume of shrimp caught (Decline 1990). In some areas, such as the Gulf of Mexico, TEDs are a legal requirement for all shrimp trawlers, but any successful solution to declining sea-life populations must be international and cooperative, not local and individual.

While these technological advances are a step in the right direction, they address just a few industrial fishing problems, and they lack both international application and the strength of law enforcement. Protective technologies are only effective in conjunction with well-developed management programs, including both supervision and enforcement (Herring Alliance n.d.). New technologies are only a partially effective way of plugging some of the smaller holes in the environmental catastrophe of industrial fishing.

### *Enhanced Regulations, Limited Harvests*

One frequently voiced solution to the fisheries problem is to lower catch limits, and to reduce the number and size of fishing fleets, but "[h]arvesting long-lived species on a sustainable basis is much more difficult than we used to believe. Populations don't seem to bounce back just because fishing effort is reduced" (Schiermeier 2003). This is especially true in areas where trawling has damaged or destroyed the ocean floor. Coral and other deep-sea habitats will require centuries to heal (Destructive 2004). Enhanced regulation and stricter limits is simply too little, too late.

### *Marine Parks*

Wide-ranging networks of marine parks—closed to fishing—offer a measure of hope. Industrial fishers have fished themselves out of a job. Governments currently pay between $15 and $30 billion in fishing subsidies every year to fishers who are out of work due to government imposed fishing restrictions (Harder 2004). Vast marine parks will also provide jobs for displaced fishers, employing their knowledge of marine life and habitat to oversee and protect marine parks. Rather than subsidize a destructive, devastating, and dying industry, it would be more sensible—ecologically and financially—to pay fishers to manage marine parks.

### *Consumer Choices*

Fish change hands so many times between ocean and plate that it is impossible to dependably purchase "sustainable" sea flesh. Sea flesh is often renamed and/or mislabeled, foiling even the most conscientious consumer. DNA analysis on seventy-seven Pacific red snapper fillets determined that about 60 percent of the flesh came from some other species, in which case "there's no way you can know whether you are buying an over-fished species" (Fox 2008).

The only dependable way that consumers can do their part to prevent further decline of sea species, decimation of underwater habitat, degradation of ocean ecosystems, and the suffering and premature death of billions of sea creatures is to *stop eating sea flesh*. There is no "sustainable" source of fish flesh—especially when compared with other dietary options, such a vegan diet.

## Conclusion

Fish and oceanic ecosystems are in peril because we choose to eat animals who live in the sea—lots of them—frequently destroying their habitat in the process. Which is more important—our habit of consuming shrimp, crabs, salmon, tuna, and innumerable other disappearing sea creatures, or protecting and preserving the earth's oceans? Do we want dipping dolphins, ponderous sea turtles, and wheeling albatross in underwater seascapes rich with undulating kelp beds and colorful coral reefs? If we do not alter the items we put on our plate, we will continue to alter ocean ecosystems. The decision is ours, and we decide where we stand every time we sit down to eat.

## Discussion Questions

1 What are your levels of concern with regard to salmon or sharks as opposed to bears or wolverines? If there is a difference, how do you explain this difference? Are you willing to defend any interest in one over and above the other? If so, how?

2 Do you think fish and the ocean ecosystem have an equal footing in the world of wildlife conservation and research? If not, why do you suppose this is the case? Should they have an equal place in this discussion? Why or why not?

3 Do you think fish feel pain? If you don't, how do you dismiss the scientific evidence presented at the outset of this chapter? How do you know that dogs—or for that matter, other human beings—are sentient?

4 How do fishing and hunting compare environmentally? What similarities and/or differences seem most important environmentally?

5 If you eat fish, how many fish do you generally eat in the course of a week or month? As an environmentalist, how do you justify this choice?

## Essay Questions

1 Do an internet search for "aquarium" or "SeaWorld." From your reading and research do you feel that these institutions can be ethically maintained? Explain your point of view. Be sure to factor in sentience and intelligence and welfare laws designed to protect sentient nonhumans.

2 At your local grocery store, examine the "seafood" section. Which fish species are represented, how many species of fish are represented, how are they displayed, is there information on sourcing? How does it make you feel to look at these individual fish and pieces of dead fish knowing what you know about fish and sea ecosystems?

3 Write a detailed job description or job announcement for a conservation position that would put unemployed fisherman back to work. Write an essay detailing a handful of alternative livelihoods that might be available and suitable for those currently employed in factory or industrial fishing.

4 Search the internet for environmental organizations working to protect the seas. How many of these organizations advocate for a change of diet? How would you assess the importance of fish in their platform (as opposed to concerns with sea ecosystems)? Does it make sense to be concerned about sea ecosystems but not concerned about fish or dietary choice?

5 Watch *The Cove*. Would you risk your freedom—would you risk imprisonment—to fight industries that, by nature, damage/destroy ecosystems? What about industries that imprison wildlife? What might you do and how do you justify such acts? Is it possible to rationally justify such acts for some animals but not others?

6 Watch *Blackfish*. What is the government's role in controlling capitalist enterprises that are damaging to the environment and/or other animals? If you feel that wildlife should not be exploited by the entertainment industry, what actions can you take to bring change? Choose one of your actions and carry it forward.

7 Watch *Blackfish* and *The Cove*. One film highlights environmental destruction and animal suffering in the food industry, the other focuses on the entertainment industry. Which film do you find more compelling and why? In light of information in this chapter, in what specific ways are you implicated in either or both of these forms of exploitation, and what specific changes might you make to come clean as an environmentalist?

## Suggested Further Reading


Clover, C. (2006). *The end of the line: How Overfishing is Changing the World and What We Eat*. New York: New Press.

Lichatowich, J. (1999). *Salmon without rivers: A History of the Pacific Salmon Crisis*. Washington, DC: Island Press.

Pauly, D. (2010). *Five easy pieces: The Impact of Fisheries on Marine Ecosystems*. Washington, DC: Island Press.


## References


A Broken Food Chain. (2009). *AWI Quarterly*, 58(1), p.18.

Balcombe, J. (2006). *Pleasurable kingdom* London: Macmillan, pp.187-188.

Black-Footed Albatross Advances Toward Protection: Seabird threatened by Longline Fishing. (2007). *In Brief*, p.6.

Bycatch Reduction Devices. (n.d.) Australian Fisheries Management Authority. Australian Government. Accessed Dec. 20, 2013. http://www.fish.gov.au/fishing_methods/Pages/bycatch_reduction_devices.aspx.

Chandroo, K., Duncan, I. and Moccia, R. (2004). Can fish suffer?: perspectives on sentience, pain, fear and stress. *Applied Animal Behaviour Science*, 86(3), pp.225–250.

Clark, M., Tittensor, D. and Rogers, A. (2006). *Seamounts, deep-sea corals and fisheries*. Cambridge, United Kingdom: UNEP World Conservation Monitoring Centre.

Clucas, I. (1997). A study of the options for utilization of bycatch and discards from marine capture fisheries. *FAO Fisheries Circular*.

Coral. (2006). *Britannica Concise Encyclopedia*. [online] Available at: http://www.credoreference.com/entry/ebconcise/coral [Accessed 14 Jul. 2010].

Cufone, M. (2008). Ocean Fish Farms and Public-Resource Privatization". *The American Prospect*, 19(12), p.A17.

Dalton, R. (2004). Net Losses Pose Extinction Risk For Porpoise. *Nature*, 429(6992), p.590.

*Decline of the Sea Turtles: Causes and Prevention*. (1990). Washington, DC: National Acadamies Press, pp.74–117.

Destructive Power of Deep-Sea Bottom Trawling on the High Seas, The. (2004). *DSCC Policy Paper*. [online] Available at: http://www.savethehighseas.org/publicdocs/DSCC_Bottom_trawling.pdf [Accessed 10 Jan. 2010].

Dionys de Leeuw, A. (1996). Contemplating the Interests of Fish: The Angler's Challenge. *Environmental Ethics*, 18, pp.37–39.

False Killer Whales in Decline—Suit Seeks to Rescue Hawai'i Population. (2009). *In Brief*, p.5.

Fox, D. (2008). Imposter Fish. *Conservation*, 9(4), pp.14–19.

Gulf Sea Turtles Get a Breather: Government Orders Review of Long-Line Fishing. (2009). *In Brief*, p.11.

Harder, B. (2004). Cost of Protecting the Ocean. *Science News*, 165(26), p.414.

Herring Alliance (n.d.). *Bycatch and Monitoring*. [online] Available at: http://www.pewtrusts.org/~/media/Assets/2010/09/10/Bycatch_Monitoring.pdf [Accessed 23 May. 2013].

Laland, K., Brown, C. and Krause, J. eds, (n.d.). Learning in Fishes: An Introduction. In: *Fish and Fisheries*, pp.199–202.

MacKenzie, D. (2006). *Deep-sea Fish Species Decimated in a Generation*. [online] *New Scientist*. Available at: http://www.newscientist.com/article/dn8533-deepsea-fish-species-decimated-in-a-generation.html [Accessed 13 Aug. 2010].

Mission Blue, (2013). *Ghost Nets, among the greatest killers in our oceans…* [online] Available at: http://mission-blue.org/2013/05/ghost-nets-among-the-greatest-killers-in-our-oceans/ [Accessed 15 Jun. 2014].

Myers, R. and Worm, B. (2003). Rapid worldwide depletion of predatory fish communities. *Nature*, 423(6937), pp.280–283.

National Oceanic and Atmospheric Administration Fisheries Service (n.d.). *Turtle Excluder Devices: NOAA Fisheries*. [online] Available at: http://www.nmfs.noaa.gov/pr/species/turtles/teds.htm [Accessed 20 Jun. 2010].

Ovetz, R. (2004). *Pillaging the Pacific: Pelagic Longline Fishing Captures and Kills About 4.4 Million Sharks, Billfish, Seabirds, Sea Turtles, and Marine Mammals Each Year in the Pacific Ocean*. Forest Knolls: Sea Turtle Restoration Project.

Palmer, M. (2009). Mexico Tries to Crush "Dolphin Safe" Tuna Label Via WTO. *AWI Quarterly*, 58(1), p.22.

Pauly, D. and Watson, R. (2003). Counting the last fish. *Scientific American*, 289(1), pp.42–47.

Pittsburgh Independent Media, (2005). *Sports fishing: bad for fish and other living things*. [online] Available at: http://pittsburgh.indymedia.org/news/2005/07/19545.php [Accessed 4 Jun. 2011].

Rodríguez-Quiroz, G. (2012). *Fisheries and Biodiversity in the Upper Gulf of California, Mexico*. [ebook] Available at: http://cdn.intechopen.com/pdfs/33951/InTech-Fisheries_and_biodiversity_in_the_upper_gulf_of_california_mexico.pdf [Accessed 22 May. 2012].

Rollin, B. (1981). *Animal rights and human morality*. Buffalo, N.Y.: Prometheus Books, p. 31.

Rosenberg, A. (2008). Aquaculture: The price of lice. *Nature*, 451(7174), pp.23–24.

Salmon Farming Threatens Wild Populations. (2008). *AWI Quarterly*, p.18.

Schiermeier, Q. (2003). Europe dithers as Canada cuts cod fishing. *Nature*, 423(6937), p.212.

Seabirds Needn't Die in Vain. (2007). *New Scientist*, 195(261), p.6.

Sea Turtle Restoration Project, (2003). *Long Line Fact Sheet*. [online] Available at: https://seaturtles.org/wp-content/uploads/2014/06/CCF02062014_00001.pdf [Accessed 2 June. 2014].

Under Pressure, Government Moves to Protect Herring. (2007). *In Brief*, pp.6–7.

Vanishing Vaquita, The. (2008). *AWI Quarterly*, 57(4), p.21.

Watson, P. (2012) "Tora! Tora! Tora!" In D. Schmidtz and E. Willott (eds) *Environmental Ethics: What really Matters, What Really Works*, 2nd edition. Oxford: Oxford University Press, pp.639–643.

# 12

# FARM GONE FACTORY

## Industrial Animal Agriculture, Animal Welfare, and the Environment

*Chris Hunt*

When prompted to contemplate the origin of burgers, chicken wings, egg salad, or cheddar cheese, most Americans picture the idyllic family farms of yesteryear, complete with red barns, green pasture, and farmed animals grazing freely in open fields. Unfortunately, the vast majority of U.S. meat, eggs, and dairy products are produced by industrial animal operations. Also known as factory farms or CAFOs (concentrated animal feeding operations), these enormous facilities raise hundreds (and in some cases, hundreds of thousands) of animals in confined conditions without access to pasture. Lamentably, this system of production damages the environment, threatens human health, devastates local communities, and harms animals. Factory farms adversely affect farmed animals and the natural environment, ultimately underscoring the dual importance of transitioning to a more sustainable system of food production.

### Factory Farms and Animal Welfare

The fundamental problem with industrial animal operations is that pigs, chickens, cattle, and turkeys are simply confined too closely together. As a result, they are stressed, and are susceptible to illnesses caused by crowded, squalid living conditions and constant exposure to toxins released during the decomposition of manure. Extreme confinement also prevents animals from engaging in natural, instinctual behaviors, which further increases their stress. Furthermore, industrial farm operations utilize a variety of practices that cause these animals to suffer additional discomfort, pain, and stress (e.g., debeaking, tail docking, administration of rBGH, etc., described below). People generally feel that farmed animals should be given a minimum standard of welfare, including freedom from unnecessary suffering (Animal Welfare Institute n.d.), but industrial farms treat animals more like inanimate commodities than sentient beings, often ignoring even their most basic needs, and compromising their health and wellbeing in order to boost profits.

## *Cattle in Dairy and Beef Production*

Factory farmed cattle are often closely confined. They can suffer diseases of the feet and udders from standing in their own manure. Crowded into feedlots, exposed to extreme weather conditions during transport, and forced through gates to slaughter, the lives of factory farmed cattle are marked by physical and mental suffering.

Somewhat counter intuitively, dairy production entails some of the most egregious suffering. Calves born on industrial dairy farms are usually separated from their mothers within 12 hours of birth. Male calves are usually raised for veal, often in intensive confinement, and are slaughtered before six months of age. Female calves are either raised to produce calves and milk before they are sent to slaughter, or they are also slaughtered for veal after about six months. Calves kept for milk production are often subjected to tail docking, a procedure in which approximately two-thirds of the tail is removed. Tail docking prevents cows from using their tails to swat flies (American Veterinary Medical Association 2010). Although this painful procedure is ostensibly intended to prevent cows from developing infections caused by constant exposure to manure, studies have shown that it not only causes the animal to suffer significant pain and stress, but has little to no health benefit (Comis 2005).

Industrial dairy farms breed cows to produce increasingly large volumes of milk. In 1950, a cow produced about 5,300 pounds of milk per year; on today's dairy farms a cow is likely to produce more than 18,500 pounds annually (Hallberg 2003). This increased milk production is extremely stressful for cows, and can cause higher incidence of disease (Halverson 2001). Nonetheless, many industrial dairies boost milk production further by injecting cows with rBGH, an artificial hormone developed by the agribusiness corporation, Monsanto. Studies conducted by Health Canada indicate that administration of rBGH increases the risk of mastitis (an infection that causes painful inflammation of the udders) by 25 percent, interferes with reproductive functions, and increases risk of clinical lameness by 50 percent (Health Canada 1998). Use of rBGH is currently banned in the EU and Canada.

Cattle raised for beef production also suffer. When they reach market weight, they are sent to slaughter. Although slaughterhouses are required to meet USDA standards for stunning before slaughter, oversight is extremely limited, and operators are expected to maximize profits by killing animals as quickly as possible. As a result, mistakes are common. Reports indicate that slaughterhouses frequently fail to stun cattle before killing them, which means that many are conscious while slaughtered (Grandin 2001).

## *Pigs*

Factory farmed pigs are typically born in crates so small that the sow (mother pig) is unable to turn around, make a nest before birthing, or separate her piglets from feces, though these are natural behaviors for a sow. A factory farmed sow lies immobilized, unable to move about while her young suckle. Piglets are also unable to engage in the most basic piglet behaviors, such as running, jumping, or playing (Van Putten 2000).

Separated from their mothers at an unnaturally young age, pigs spend their lives confined in concrete pens without bedding or soil. These conditions prevent them from rooting and cause stress, which can lead pigs to bite one another (a phenomenon that has only been observed on factory farms) (Van Putten 2000). Rather than simply providing pigs with adequate space and straw, factory farms often cut off pigs' tails to prevent tail biting. As in the case of cows, tail docking has been shown to cause chronic pain (Halverson 2001).

By the time pigs are sent to slaughter, they often suffer from pneumonia caused by high ambient levels of ammonia released during the decomposition of manure, and/or leg deformities caused by standing on concrete and slatted floors (Horrigan et al. 2002). Because of stress, illnesses, and rough handling, many pigs die en route to slaughter (Grandin 2002).

### *Chickens*

Chickens raised in industrial poultry and egg farms are either crowded into small cages or packed in large warehouses. In both cases, birds spend their entire lives in a space smaller than a piece of writing paper (Grandin 2001). Because they are unable to forage for food—or engage in any normal behaviors—chickens tend to peck one another (Fölsch et al. 2002). In order to prevent this, factory farmed hens are usually debeaked. Debeaking is painful, cutting through sensitive tissues, and robs birds of a critical part of their physiology, which makes it difficult for them to feed normally (Halverson 2001). Chickens in factory farms exhibit high levels of stress (Grandin 2001). "Broilers" (chickens raised for meat production) are bred to grow extraordinarily quickly, reaching market weight in half the time required in the 1940s (Mench and Siegel 1997). Their unnaturally rapid body growth rate is often unmatched by bone growth, which can cause serious deformities and pain, leaving many hens unable to walk (Halverson 2001).

## Environmental Damage

Farmed animals aren't the only living beings adversely affected by industrial agriculture; factory farms also cause severe damage to the environment, destroying natural habitats, decimating wildlife populations, and impairing human health. In fact, irresponsible management practices, inadequate regulation, and insufficient oversight place factory farms among the worst polluters in the U.S. The two most significant environmental problems linked with factory farms are water and air pollution.

## Water Pollution

Factory farms keep too many animals in too little space. In addition to compromising animal welfare, this produces a tremendous amount of waste. For instance, a cow on an industrial dairy facility produces about the same amount of waste as 21 humans[1]—so a dairy CAFO with 15,000 cows generates as much animal waste as a city of 315,000 people! Yet factory farms do not have sewage treatment facilities. Instead, animal waste is stored in giant piles or lagoons, then spread or sprayed onto surrounding fields. This process spews toxins into the air, and can pollute surface water and groundwater, damaging the environment, killing aquatic organisms, and threatening human health.

Factory farm waste enters surface and groundwater through lagoon spills and leaks, intentional release of waste, and from the routine application of waste to farmland. Manure lagoon leakage is legal; for instance, in Iowa, lagoons can leak up to 1/16" (0.15875 cm) of liquid waste per day. Consequently, pollutants seep into the surrounding soil, which can eventually contaminate groundwater and/or surface water (Ham 2002). The EPA reports that waste generated by hogs, chickens, and cattle has polluted over 35,000 miles (56,300 km) of river (U.S. Environmental Protection Agency and U.S. Department of Agriculture 1998).

Lagoon spills sometimes involve millions of gallons of waste. These spills are caused by storms or heavy rainfall, collapse of lagoon walls, intentional releases (to drain waste), or mechanical failure in pumps, pipes, or other farm equipment (Hodne 2005). Large spills disrupt ecosystems, kill fish and other aquatic life, contaminate drinking water, and make waterways unsafe for human use—even recreation. In 1999, heavy rain from Hurricane Floyd caused manure lagoons throughout North Carolina to overflow, resulting in widespread contamination of the state's waterways.

Factory farm waste impairs water quality in several key ways:

***Nutrients:*** Animal manure is rich in nitrogen and phosphorus, which, in small quantities, can improve soil quality and foster plant growth. But factory farms generate enormous quantities of animal waste holding far more nutrients than land can absorb. When excessive manure is applied to farmland, excess nutrients can be carried into groundwater and surface water, causing serious damage to the natural environment.

***Eutrophication:*** Eutrophication is the process through which excess nutrients are added to bodies of water. The addition of excess nutrients causes algal growth, which creates two problems: First, algae cloud the water and block sunlight, which kills underwater plants. Since these plants ordinarily provide food and shelter for fish and other aquatic life, fish and other organisms die, reducing biodiversity. Furthermore, algae eventually die and are decomposed by bacteria that consume oxygen. This can create hypoxic (low oxygen) areas, causing additional fish kills, further reducing biodiversity. In extreme cases, this process wipes out all aquatic life, creating "dead zones."

***Dead zones***: Nutrient pollution can create extensive hypoxic areas in oceans and lakes. Hypoxic conditions kill any organism that is unable to move to waters with more oxygen, creating "dead zones." The world's second largest dead zone currently exists in the Gulf of Mexico; once teeming with living organisms, this 7,700-square-mile (12,400 $km^2$) section of water (an area approximately the size of New Jersey), is now devoid of aquatic life (Howarth et al. 2000). A 2008 study noted that more than 400 dead zones have been reported throughout the world (Diaz and Rosenberg 2008).

***Toxic algal blooms:*** Eutrophication can also cause rapid growth of highly toxic species of algae. One notorious toxic alga, *Pfiesteria piscicida*, emits toxins thought to break down the skin tissue of fish, causing bleeding sores or lesions, and resulting in massive fish kills (Silbergeld et al. 2000; U.S. Environmental Protection Agency 1998a). Exposure to *Pfiesteria* also harms humans; this alga's potent toxins are believed to cause skin irritation, memory loss, confusion, and cognitive impairment (U.S. Environmental Protection Agency 1998a). Numerous *Pfiesteria* outbreaks have occurred in the Chesapeake Bay and in the coastal waters of North Carolina, both of which are severely polluted with nutrients from factory farm manure.

***Nitrate:*** Factory farm pollution can also increase nitrate levels in drinking water. Infants exposed to nitrate-contaminated water can develop methemoglobinemia ("blue baby syndrome"), a potentially fatal condition that reduces the blood's oxygen-carrying capacity. In U.S. counties where factory farms are located, approximately 1.3 million households must now rely on well water in which nitrate levels exceed the Maximum Contaminant Level (U.S. Environmental Protection Agency 2002).

***Ammonia:*** In addition to poisoning aquatic life, ammonia reduces dissolved oxygen in water, which can decrease aquatic species diversity and lead to fish kills. Since ammonia is a nutrient (and can also be easily converted to the nutrient, nitrate), ammonia pollution can contribute to eutrophication, further reducing biodiversity and degrading aquatic ecosystems. Volatilization of ammonia can also contribute to acid rain.

***Pathogens:*** Pathogens are living microorganisms (e.g., bacteria, viruses, and parasites) that cause diseases. Because pathogens attach to sediment and are released when a streambed is disturbed, they pose a long-term threat to human health (U.S. Environmental Protection Agency 1998b). Pathogens cause a range of symptoms, including stomach cramps, vomiting, diarrhea, fever, kidney failure, and death. Animal manure can carry a variety of harmful pathogens, including *Campylobacter*, *E. coli*, and *Salmonella*. When factory farm waste enters ground or surface water, pathogen contamination can render water unsafe for human consumption, and in some cases, unsafe for swimming or other recreational uses. Manure from cattle operations was among the suspected causes of the 1993 *Cryptosporidium* contamination of Milwaukee's drinking water, which sickened an estimated 403,000 residents (MacKenzie et al. 1994).

***Antibiotics:*** Factory farms administer huge quantities of antibiotics in order to promote growth and prevent widespread disease among animals in crowded, unhealthy conditions. In the U.S, according to the FDA, 80 percent of all antibacterial drugs are currently administered to farmed animals (Slaughter 2011). Since 25 to 75 percent of antibiotics pass into manure unchanged, factory farm waste can carry antibiotics into the environment, contaminating surface water and groundwater (Chee-Sanford et al. 2001). An Ohio study found that 67 percent of the water samples taken near industrial poultry farms contained antibiotics (Campagnolo et al. 2002). Industrial agriculture's use of antibiotics promotes the development of antibiotic-resistant bacteria, making antibiotics less effective as a treatment for humans.

***Hormones:*** Industrial beef and dairy operations regularly administer hormones to cattle to boost growth and increase milk production. Some of these hormones pass into their waste and can contaminate surface and groundwater, disrupting the development and reproductive systems of aquatic organisms (Orlando et al. 2004; Soto et al. 2004). For instance, a study of fathead minnows exposed to feedlot effluent revealed demasculinization of male fish (demonstrated by lower testicular testosterone synthesis, altered head morphometrics, and smaller testis), and defeminization of female fish (demonstrated by a decreased estrogen/androgen ratio of *in vitro* steroid hormone synthesis) (Orlando et al. 2004).

***Salts and trace elements:*** Animal manure also contains salts (*e.g.*, sodium, calcium, magnesium, potassium, chloride, sulfate, bicarbonate, carbonate, and nitrate) and elements (e.g., arsenic, copper, selenium, and zinc). When huge quantities of manure are concentrated in one area, these substances can contaminate ground and surface water (U.S. Environmental Protection Agency 1998b). In addition to rendering water unsuitable for drinking, these contaminants damage soil, restrict plant growth, and harm wildlife (U.S. Environmental Protection Agency 1998b).

***Organic matter and solids:*** Waste from factory farms may contain various solid materials (e.g., spilled feed, bedding materials, litter, and hair), which can increase the turbidity of surface waters, harming populations of fish, crabs, and other aquatic organisms. Factory farm waste also contains biodegradable, carbon-based material that occurs naturally in animal manure. In surface waters, this organic matter is decomposed by bacteria and other microorganisms that deplete oxygen levels in the water. As noted above, decreased oxygen levels can kill aquatic organisms and reduce biodiversity.

## *Air Pollution*

Factory farms also damage the environment by releasing a host of harmful pollutants into the air. These substances are released from farmed animal waste stored in manure lagoons, and from the land on which manure is spread or sprayed. The most significant air pollutants are hydrogen sulfide, ammonia, particulate matter, odor, and greenhouse gases.

***Hydrogen sulfide ($H_2S$):*** This colorless, highly toxic gas is one of the most dangerous pollutants emitted by factory farms. Exposure to low concentrations of hydrogen sulfide—even levels well below the thresholds for irritation or toxicity—can be harmful; studies have linked low-level exposure to cough, throat irritation, eye symptoms, nasal symptoms, and headache (Merchant et al. 2002). Other reported symptoms include nausea, stomach distress, dizziness, and blistering of the lips (Doss et al. 2002). Exposure to high levels of hydrogen sulfide causes rapid loss of consciousness, shock, pulmonary edema, coma, and death (Merchant et al. 2002). Tragically, a significant number of factory farm workers have been killed by unexpected releases of hydrogen sulfide from manure lagoons. In many cases, multiple victims have died while attempting to rescue others, entering manure pits without realizing that they contained deadly levels of the gas. For instance, a 27-year-old hog farm worker died of hydrogen sulfide poisoning in Minnesota, after climbing into a manure pit to remove a pump. The 46-year-old farm co-owner also died after entering the pit to attempt rescue (National Institute for Occupational Safety and Health Fatality Assessment and Control Evaluation Program 1992).

***Ammonia ($NH_3$):*** Approximately 80 percent of all U.S. anthropogenic ammonia emissions are released by manure (Doorn et al. 2002). Ammonia eventually falls to the ground, either dissolved in precipitation (which creates "acid rain") or as dry particles. Ammonia raises nitrogen levels in water, exacerbating eutrophication. Acid rain can also increase soil acidity, which causes nutrients to be leached out of soil, harming plants and impairing ecosystem productivity (Ad Hoc Committee on Air Emissions 2003). In mild cases, acid rain can damage leaves, limit nutrients, and expose plants to toxic substances found in soil, slowing growth. In extreme cases (for instance, in New York's Adirondack Mountains), acid rain can cause severe damage to large forests.

Ammonia also clouds the air and threatens human health. Research indicates that exposure to moderate concentrations of ammonia can cause severe cough and increased mucous production. Higher concentrations cause scarring of the upper and lower airways, reactive airways dysfunction syndrome, persistent airway hypersensitiveness, lower lung inflammation, and pulmonary edema (Merchant et al. 2002).

***Particulate matter:*** Solids or liquid particles suspended in the air are called particulate matter. Factory farms emit particulates such as fecal matter, feed materials, skin cells, and bioaerosols (e.g., bacteria, fungi, spores, viruses, endotoxins, exotoxins, and products of microorganisms) (Thorne et al. 2002). Long-term exposure to factory farm dust can lead to persistent respiratory symptoms and a decline in lung function (Merchant et al. 2002). Exposure to bioaerosols may induce asthma, cough, chest tightness, and hypersensitivity pneumonitis ("farmer's lung") (Merchant et al. 2002). Particulate matter also reduces visibility and creates a cloud of haze.

***Odor:*** The most obvious air pollutant emitted by factory farms is the odor caused by a complex mixture of chemicals released by bacteria during the digestion of feed and the decomposition of manure. One air sample study identified 331 fixed gases and volatile organic compounds (VOCs) emitted by factory farms (Schiffman et al. 2001). Factory farm odors have been linked to headaches, eye irritation, nausea, nasal irritation, diarrhea, hoarseness, sore throat, cough, chest tightness, nasal congestion, palpitations, shortness of breath, stress, drowsiness, and altered moods (Schiffman et al. 2005).

***Greenhouse gases:*** Globally, the livestock sector generates roughly 18 percent of all anthropogenic greenhouse gas (GHG) emissions measured in carbon dioxide equivalent (Steinfeld et al. 2006). This is due primarily to the emission of potent and persistent GHGs, methane and nitrous oxide. Remarkably, the livestock sector produces 37 percent of the world's anthropogenic methane emissions and 65 percent of anthropogenic nitrous oxide emissions (Steinfeld et al. 2006).

Animal agriculture, now concentrated on factory farms instead of small, local family farms, transports products long distances to reach consumers. This transportation emits even more carbon dioxide, nitrous oxide, sulfur oxides, particulate matter, and other air pollutants. These elevated gases impair human health and exacerbate climate change, acid rain, and the acidification of ecosystems.

Factory farmed animals also consume tremendous amounts of animal feed, which is produced by industrial farms that use huge quantities of synthetic fertilizers and pesticides. Heavy use of fertilizers contributes to nutrient pollution, which causes severe damage to waterways and releases substantial quantities of nitrous oxide, which contributes to climate change.

## The Power of One

The industrialization of animal agriculture has enabled agribusiness to sell meat, eggs, and dairy products at historically low prices—but cheap food comes at great cost. Factory farms compromise the health and welfare of farmed animals and damage the environment, threatening human health and degrading surrounding communities. Though we prefer lower prices for food, few would argue that cost savings should be achieved by polluting waterways, contaminating the air, or causing farmed animals to suffer. Unfortunately, agribusiness's tremendous wealth and political power have enabled the industry to successfully thwart regulations. Moreover, corporations have begun to utilize industrial farm production techniques around the world, selectively choosing to construct factory farms in countries with fewer regulations and less public awareness of the problems the industry creates.

Fortunately, as awareness of the ills of industrial livestock production reaches the public conscience, informed consumers change their purchasing patterns. By rejecting factory farmed foods—by choosing an environmentally friendly, animal friendly diet—consumers are slowly reshaping the U.S. food system. Indeed, protecting animals is protecting the environment: The best way to avoid supporting the animal abuse and environmental damage caused by factory farms is to simply avoid buying their products.

## Discussion Questions

1 In what specific ways does our diet affect the environment?

2 Does the suffering of a chicken matter? What about a pig? 1 million pigs? Does it matter who suffers (species, race, age, class), or how many suffer (one, three, or ten billion)? Is suffering always morally relevant, or only in certain instances?

3 If we reduce suffering in animal agriculture, will it be morally acceptable to eat meat and dairy and eggs? How much must we reduce the suffering? What sorts of suffering will be considered unacceptable? Are these levels of suffering also acceptable when inflicted on human beings?

4 Abolitionists want no cages, animal welfarists want larger cages. Where do you stand and why?

5 For those concerned about animals, can larger cages be considered a short-term benefit for hens, or does this type of action simply make abolition more difficult to accomplish, and less likely?

6 How would you describe Monsanto to someone who knows nothing about the industry? Why do U.S. laws permit such corporate power? How would you describe the relationship between the U.S. government and Big Ag corporations such as Monsanto?

## Essay Questions

1 How is an understanding of Monsanto critical to dietary choice?

2 What is your responsibility now that you are aware of the environmental implications of dietary choice?

3 If you consume factory farmed animal products, how do you morally justify your support of industries that cause such suffering for cattle, pigs, turkeys, and chickens?

4 Is it possible to pose a moral justification for consuming other animals (or their nursing milk or their reproductive eggs) when we have no nutritional need to do so, when factory farming harms the environment, and when such a diet is harmful to other animals? If so, describe those particular animal products and the conditions under which they might be consumed as *you* present your arguments. Does your argument provide *you* with moral justification to eat animal products? If *you* think there are morally permissible ways to consume animal products even when we have no nutritional need for these foods, and even though we have other food options, describe these animal products and what makes them morally permissible.

## Suggested Further Reading

*Sustainable Table* (2014) Sustainable Food and Agriculture Issues, GRACE Communications Foundation. Available from: www.sustainabletable.org/issues.

Imhoff, D. (ed.). (2010) *CAFO: The Tragedy of Industrial Animal Factories*, Earth Aware Editions, San Raphael.

Kim, B. F., et al. (2013) *Industrial Food Animal Production in America: Examining the Impact of the Pew Commission's Priority Recommendations.* Johns Hopkins University Center for a Livable Future. Available from: http://www.jhsph.edu/research/centers-and-institutes/johns-hopkins-center-for-a-livable-future/research/clf_publications/pub_rep_desc/pew_report.html.

## Note

1 This figure was calculated using dairy and human waste characteristics reported in the USDA's *Agricultural Waste Management Field Handbook* (1999). The calculation assumes an average lactating cow weight of 1,400 lbs. (635 kg), and an average human weight of 175.8 lbs. (79.7 kg) (Average human weight was estimated using sex and age group weight data reported by the CDC (Ogden et al. 2004) and 2005 population estimates reported by the Population Division of the U.S. Census Bureau (U.S. Census Bureau: Population Division 2006).

*Waste Calculation:* Weight of waste excreted by a lactating dairy cow: 80.00 lbs./day/1,000 lbs. of live weight (36.3 kg/day/453.6 kg of live weight) (USDA, Natural Resources Conservation Service 1999). Weight of waste excreted by a 1,400 lb. (635 kg) lactating dairy cow: 112.0 lbs./day. (50.8 kg/day). Weight of waste excreted by humans: 30.00 lbs./day/1,000 lbs. of live weight (13.61 kg/day/453.6 kg of live weight). Weight of waste excreted by a 175.8 lb. (79.7 kg) human: 5.274 lbs./day (2.392 kg/day) (USDA, Natural Resources Conservation Service, 1999). Thus a 1,400 lb. (635 kg) lactating cow excretes 21.24 times as much waste per day as a 175.8 lb. (79.7 kg) human.

## References

Ad Hoc Committee on Air Emissions from Animal Feeding Operations, Committee on Animal Nutrition, National Research Council (2003) *Air Emissions from Animal Feeding Operations: Current Knowledge, Future Needs.* Washington, DC: National Academies Press.

American Veterinary Medical Association. (2010) *Welfare Implications of Dairy Cow Tail Docking.* Available from: https://www.avma.org/KB/Resources/LiteratureReviews/Pages/Welfare-Implications-of-Tail-Docking-of-Cattle.aspx. [13 October 2012].

Animal Welfare Institute (AWI). (n.d.) *Consumer Perceptions of Farm Animal Welfare.* Available from: http://www.awionline.org/ht/a/GetDocumentAction/i/25067. [29 July 2011].

Campagnolo, E. R., Johnson, K. R., Karpati, A., Rubin, C. S., Kolpin, D. W., Meyer, M. T., Esteban, J. E., Currier, R. W., Smith, K., Thu, K. M. and McGeehin, M. (2002) 'Antimicrobial Residues in Animal Waste and Water Resources Proximal to Large-Scale Swine and Poultry Feeding Operations.' *The Science of the Total Environment*, 299(1–3), pp. 89–95.

Chee-Sanford, J. C., Aminov, R. I., Krapac, I. J., Garrigues-Jeanjean, N. and Mackie, R. I. . (2001) 'Occurrence and Diversity of Tetracycline Resistance Genes in Lagoons and Groundwater Underlying Two Swine Production Facilities.' *Applied and Environmental Microbiology*, 67(4), pp.1494–1502.

Comis, D. (2005) 'Settling Doubts About Livestock Stress.' *Agricultural Research*, March, pp. 4–7. Available from: http://www.ars.usda.gov/is/AR/archive/mar05/stress0305.htm [1 August 2012].

Diaz, R. J., and Rosenberg, R. (2008) 'Spreading Dead Zones and Consequences for Marine Ecosystems.' *Science*, 321(5891), pp. 926–929.

Doorn, M. R. J., Natschke, D. F and Meeuwissen. P. C. (2002) *Review of Emissions Factors and Methodologies to Estimate Ammonia Emissions from Animal Waste Handling*. U.S. Environmental Protection Agency and State of North Carolina Department of Environment and Natural Resources. Available from: http://tammi.tamu.edu/epaemissions.pdf [29 July 2011].

Doss, H. J., McLeod, W. and Person, H. L. (2002) 'Beware of Manure Pit Hazards.' *Michigan State University Extension*, April.

Fölsch, D. W., Höfner, M. Staack, M and Trei, G. (2002) 'Comfortable Quarters for Chickens in Research Institutions.' In: Reinhardt, V. and Reinhardt, A. eds *Comfortable Quarters for Laboratory Animals*, 9th ed. Animal Welfare Institute. Available from: http://labanimals.awionline.org/pubs/cq02/Cq-chick.html [8 July 2011].

Grandin, T. (2001) 'Corporations Can be Agents of Great Improvements in Animal Welfare and Food Safety and the Need for Minimum Decent Standards.' Paper presented at *National Institute of Animal Agriculture*, 4 April 2001.

Grandin, T. (2002) *Welfare of Pigs During Transport* [Online]. Department of Animal Sciences, Colorado State University. Available from: http://www.grandin.com/welfare.pigs.during.transport.html [28 July 2011].

Hallberg, M. C. (2003) 'Historical Perspective on Adjustment in the Food and Agriculture Sector.' *Workshop on Agriculture Policy Reform and Adjustment*, Imperial College, Wye. Available from: http://ageconsearch.umn.edu/bitstream/15755/1/cp03ha01.pdf [28 July 2011].

Halverson, M. K. (2001) 'Supplementary Literature Summary and Technical Working Paper for the Minnesota Generic Environmental Impact Statement on Animal Agriculture: Farm Animal Health and Well-Being.' *Minnesota Generic Environmental Impact Statement on Animal Agriculture*. Available from: http://www.researchgate.net/publication/235954611_FARM_ANIMAL_HEALTH_AND_WELL-BEING_TECHNICAL_WORKING_PAPER_MINNESOTA_GENERIC_ENVIRONMENTAL_IMPACT_STATEMENT_ON_ANIMAL_AGRICULTURE [28 July 2011].

Ham, J. M. (2002) 'Seepage Losses from Animal Waste Lagoons: A Summary of a Four-Year Investigation in Kansas.' *Transactions of the American Society of Agricultural and Biological Engineers*, 45(4), pp. 983–992.

Health Canada (1998) *Report of the Canadian Veterinary Medical Association Expert Panel on rbST*. Available from: http://www.hc-sc.gc.ca/dhp-mps/vet/issues-enjeux/rbst-stbr/rep_cvma-rap_acdv_tc-tm-eng.php [4 June 2011].

Hodne, C. J. (2005) 'Concentrating on Clean Water: The Challenge of Concentrated Animal Feeding Operations.' For *The Iowa Policy Project*. Available from: http://www.iowapolicyproject.org/2005docs/050406-cafo-fullx.pdf [8 June 2011].

Horrigan, L., Lawrence, R. S. and Walker, P. (2002) 'How Sustainable Agriculture Can Address the Environmental and Human Health Harms of Industrial Agriculture.' *Environmental Health Perspectives*, 110(5), pp. 445–456.

Howarth, R., Anderson, D., Cloern, J., Elfring, C., Hopkinson, C., Lapointe, B., Malone, T., Marcus, N., McGlathery, K., Sharpley, A. and Walker, D. (2000) 'Nutrient Pollution of Coastal Rivers, Bays, and Seas.' *Issues in Ecology*, Fall, 7, pp. 1–17.

MacKenzie, W. R., Hoxie, N. J., Proctor, M. E., Gradus, M. S., Blair, K. A., Peterson, D. E., Kazmierczak, J. J., Addiss, D. G., Fox, K. R., Rose, J. B. and Davis, J. P. (1994) 'A Massive Outbreak in Milwaukee of Cryptosporidium Infection Transmitted through the Public Water Supply.' *New England Journal of Medicine*, 331, pp. 161–167.
Mench, J. A. and Siegel, P. (1997) 'Animal Welfare Issues Compendium: Poultry' in *Plant and Animal Production, Protection and Processing. Animal Welfare Issues Compendium: A Collection of 12 Discussion Papers*, U.S. Department of Agriculture Cooperative State Research, Education and Extension Service. Available from: http://www.nal.usda.gov/awic/pubs/97issues.htm#Issue10 [4 April 2011].
Merchant, J. A., Kline, J. . Donham, K. J., Bundy, D. S and Hodne, C. J. (2002) 'Chapter 6.3: Human Health Effects' in *Iowa Concentrated Animal Feeding Operation Air Quality Study*, eds Iowa State University and the University of Iowa Study Group, pp. 122–145.
National Institute for Occupational Safety and Health Fatality Assessment and Control Evaluation Program (1992) *Hog Farm Co-Owner and Employee Die of Hydrogen Sulfide Poisoning in Manure Pit—Minnesota*. FACE 9228. Available from: http://www.cdc.gov/niosh/face/in-house/full9228.html [5 August 2011].
Ogden, C. L., Fryar C. D., Carroll, M. D. and Flegel, K. M. (2004) 'Mean Body Weight, Height, and Body Mass Index, United States 1960–2002.' *Advance Data from Vital Health Statistics*, October, (347), pp. 1–20.
Orlando, E. F., Kolok, A., Binzcick, G, Gates, J and Horton, M. (2004) 'Endocrine-Disrupting Effects of Cattle Feedlot Effluent on an Aquatic Sentinel Species, the Fathead Minnow.' *Environmental Health Perspectives*. 112(3), pp. 353–358.
Schiffman, S. S., Bennett, J. L. and Raymer, J. H. (2001) 'Quantification of Odors and Odorants from Swine Operations in North Carolina.' *Agricultural and Forest Meteorology*, 108(3), pp. 213–240.
Schiffman, S. S., Studwell, C. E., Landerman, L. R., Berman, K. and Sundy, J. S. (2005) 'Symptomatic Effects of Exposure to Diluted Air Sampled from a Swine Confinement Atmosphere on Healthy Human Subjects.' *Environmental Health Perspectives*, 113(5), pp. 567–576.
Silbergeld, E. K., Grattan, L. Oldach, D. and Morris, J. G. (2000) '*Pfiesteria*: Harmful Algal Blooms as Indicators of Human: Ecosystem Interactions.' *Environmental Research*, 82, pp. 106–112.
Slaughter, L. (2011) *Confirmed: 80 Percent of all Antibacterial Drugs used on Animals, Endangering Human Health*. Available from: http://louise.house.gov/press-releases/confirmed-80-percent-of-all-antibacterial-drugs-used-on-animals-endangering-human-health/ [24 February 2011].
Soto, A. M., Calabro, J. M., Prechtl, N. V., Vau, A. V., Orlando, E. F., Daxenberger, A., Kolok, A. S., Guillette Jr., L. J., le Bizec, B., Lange, I. G. and Sonnenschein, C. (2004) 'Androgenic and Estrogenic Activity in Water Bodies Receiving Cattle Feedlot Effluent in Eastern Nebraska, USA.' *Environmental Health Perspectives*, 112(3), pp. 346–352.
Steinfeld, H., Gerber, P., Wassenaar, T., Castel, V., Rosales, M. and de Haan, C. (2006) *Livestock's Long Shadow: Environmental Issues and Options*. Rome: Food and Agriculture Organization of the United Nations. Available from: http://www.fao.org/docrep/010/a0701e/a0701e00.HTM [3 February 2011].
Thorne, P. S. (2002) 'Chapter 3.0: Air Quality Issues' in *Iowa Concentrated Animal Feeding Operation Air Quality Study*, eds Iowa State University and the University of Iowa Study Group, pp. 35–44. Available from: http://www.public-health.uiowa.edu/ehsrc/cafostudy.htm [8 June 2011].

U.S. Census Bureau: Population Division (2006) *Annual Estimates of the Population by Sex and Five-Year Age Groups for the United States: April 1, 2000 to July 1, 2005*. Available from: https://www.google.com/url?sa=t&rct=j&q=&esrc=s&source=web&cd=2&cad=rja&uact=8&ved=0CCQQFjAB&url=https%3A%2F%2Fwww.census.gov%2Fpopest%2Fdata%2Fnational%2Fasrh%2F2005%2Ftables%2FNC-EST2005-01.xls&ei=6VU9VMiCMI-AygSPvoKoBw&usg=AFQjCNFrmrkoAFqNr0So2sMflLk3ElesIA&bvm=bv.77412846,d.aWw. [3 March 2012].

U.S. Department of Agriculture, Natural Resources Conservation Service (1999) *Agricultural Waste Management Field Handbook*. Available from: http://www.nrcs.usda.gov/wps/portal/nrcs/detailfull/national/technical/ecoscience/mnm/?cid=stelprdb1045935. [3 March 2012].

U.S. Environmental Protection Agency. (1998a) *Pfiesteria piscicida. Fact Sheet: What You Should Know About Pfiesteria piscicida*. Pp. 1–6. Available from: http://nepis.epa.gov/Exe/ZyPURL.cgi?Dockey=200050HU.txt [4 June 2011].

U.S. Environmental Protection Agency. (1998b) *Environmental Impacts of Animal Feeding Operations*. Available from: http://water.epa.gov/scitech/wastetech/guide/cafo/upload/1999_02_22_guide_feedlots_envimpct.pdf [8 July 2011].

U.S. Environmental Protection Agency. (2002) *The Benefits of Reducing Nitrate Contamination in Private Domestic Wells Under CAFO Regulatory Options*. Available from: http://www.epa.gov/npdes/pubs/cafo_benefit_nitrate.pdf [3 March 2012].

U.S. Environmental Protection Agency and U.S. Department of Agriculture (1998) *Draft Unified National Strategy for Animal Feeding Operation*. Available from: http://agrienvarchive.ca/bioenergy/download/an_feed_operations.pdf [3 March 2012].

Van Putten, G. (2000) 'An Ethological Definition of Animal Welfare with Special Emphasis on Pig Behaviour.' *Proceedings of the 2nd NAHWOA Workshop*, Driebergen-Rijsenburg, Netherlands.

# 13

# EATING ECOSYSTEMS

*Lisa Kemmerer*

Cheap meat, dairy, and eggs are an illusion—we pay dearly for these food choices with depleted freshwater reserves, deforestation, soil degradation, and wildlife manipulation. "Livestock activities have significant impact on virtually all aspects of the environment, including air and climate change, land and soil, water and biodiversity . … The overall impact of livestock on the environment is enormous" (Food and Agriculture Organization of the United Nations, 2006, 5). Consequently, diet is the most critical decision we make with regard to our environmental footprint—and it is a decision that we make every day, several times a day.

## Freshwater Depletion

Critical sources of freshwater are drying up around the world. Life-sustaining waterways such as the Colorado, Nile, and Yellow River now run dry before they reach the sea (Brown, 1996). Important underwater aquifers such as the Ogallala Aquifer are also shrinking. The Ogallala Aquifer spans "800 miles [1,287 km] from north to south, and 400 miles [644 km] from east to west," supplies 30 percent of U.S. groundwater irrigation to 27 percent of the nation's irrigated land, and is critical to an area often described as "America's Breadbasket"—though this area would more correctly be termed "America's Meat Plate." The Ogallala provides drinking water for 82 percent of the people living in South Dakota, Nebraska, Colorado, Wyoming, Oklahoma, Texas, New Mexico, and Kansas (Worm, 2004). But the Ogallala Aquifer is more than half gone, and looks likely to run dry in the next 25 years.

Animal agriculture drains freshwater in several key ways. First, farmed animals require freshwater. And thanks to consumer choices, the earth currently supports at least:

- chickens, 550 billion gallons (2082 billion liters),
- hogs, 2190 billion gallons (8289 billion liters), and
- bovines, 18,907 billion gallons (71,563 billion liters).

Combined, on average, chickens, pigs, and bovines suck up at least 21,647 billion gallons (81,943 liters) of freshwater every year. To put this in perspective, an Olympic sized swimming pool holds 660,430 gallons (2,500,000 liters) of water, and there are more than 1,500 Olympic swimming pools in just one billion gallons of water. Every year the world's bovines, pigs, and chickens drink an average of about 33 million Olympic swimming pools of freshwater, and this does not even include the many goats, turkeys, and sheep who also turn to troughs for refreshment.

Animal agriculture's most excessive expenditure of freshwater is not the global herd's water requirement, but their food intake—irrigation for feed crops. Not only do we feed 70 percent of U.S. grain crops to farmed animals, but we waste "1000 tons of water to produce one ton of grain" (one U.S. ton = 907 kg) (Brown, 1996). This extravagance is aggravated by the fact that grains are generally grown on lands that require irrigation (United States Department of Agriculture: Animal and Plant Health Inspection Service: Wildlife Damage Management, n.d.). If we consumed plants directly, we would require only 37 percent of current croplands.

Slaughterhouses and dairies also require large quantities of freshwater. These facilities are cleaned several times each day—hosed down (literally blasted with water). "Water tables are falling on every continent" (Brown, 1996). Roughly 80 percent of the world's freshwater is expended for agriculture, including agricultural lands irrigated and harvested as feed for farmed animals. When cleaning facilities, processing, packaging, and all other aspects of flesh, dairy, and egg consumption are taken into account—as they ought to be—animal agriculture is responsible for 90 percent of freshwater depletion worldwide. Our dependence on animal agriculture causes "severe environmental degradation through water depletion" (Food and Agriculture Organization of the United Nations, 2006).

Because of the amount of freshwater required for animal agriculture, the average American omnivore "consumes" 4,200 gallons (15,900 liters) of freshwater per person per day, while an American vegan averages just 300 gallons (1,136 liters) per person, daily (Schwartz, 2001). Consider these comparisons:

- 5,200 gallons (19,700 liters) of water are necessary to produce one pound of beef, but only 23 gallons of water are necessary to produce one pound of tomatoes (Cassuto, 2010).
- 500–2,500 gallons (1,893 to 9,464 liters) of water are necessary to produce one pound (0.45 kg) of beef, but only one-hundredth as much water is necessary to produce a pound of wheat (Kauffman and Braun, 2004).
- 5,200 gallons (19,700 liters) of water are necessary to produce one pound of California beef, but only 25 gallons of water are necessary to produce one pound of California veggies (Schwartz, 2001).

- Roughly 100 times more water is necessary to produce one pound of "beef" than is necessary to produce one pound of potatoes (Schwartz, 2001).

Why bother to shorten your shower if you plan on poultry or pot roast for dinner?

## Deforestation

For the sake of grazing and raising feed for farmed animals—for the sake of such things as ham, cheese spreads, steak, egg sandwiches, and milkshakes—one fifth of the world's rainforests were destroyed between 1960 and 1990. Between 1985 and 1990, 210 million acres of forest were turned to pasture, "an area nearly the size of Texas and Oklahoma" (Kauffman and Braun, 2004). A section of rainforest roughly the size of 20 football fields is destroyed pretty much every minute of every day, and in "the Amazon, cattle ranching is now the primary reason for deforestation" (Food and Agriculture Organization of the United Nations, 2006). In just 50 years, 50 percent of Costa Rica's forests disappeared—60 percent were cleared for bovines (Wikipedia.org, n.d.a). Only 13 percent of Costa Rica's original rainforest remains, and what remains is now "highly fragmented and degraded" (Reynolds and Nierenberg, 2012).

Brazil continues to lead the way (by a considerable margin) in rainforest destruction, and agriculture is responsible for roughly 98 percent of Brazil's deforestation (Monga Bay, n.d.). Ranchers are responsible for 65–70 percent of Brazil's lost forests (Monga Bay, n.d.). There were about 10 million bovines in Brazil in 1980, and there are now upwards of 55 million (Global Warning Science, n.d.). The U.S. and the EU are implicated: The U.S. imports some 80 million pounds of Brazilian beef every year; 85 percent of EU beef originates in Brazil. South America is still at the top of the list for loss of forests. All this ecological devastation for a mere spot of flesh—55 feet (17 meters) of tropical forests yield just a quarter pound (120 grams) of hamburger. If we continue as we are, primary forests will be altogether gone by 2050 (Pimm and Raven, 2000).

Not only beef-eaters, but those who consume turkeys, pigs, chickens, eggs, or dairy products are also implicated. The primary reason for loss of forests is conversion of lands to agriculture—both for grazing *and for feed crops*. In Latin American, land is converted from forests to agriculture largely for feed crops, "notably soybeans and maize" (Food and Agriculture Organization of the United Nations, 2006). Brazil's soy crop grew more than 3,000 percent in the last 40 years, becoming the world's second larger soybean producer (Wikipedia.org, n.d.b). Worldwide, 80 percent of the soybean crop is fed to farmed animals (Reynolds and Nierenberg, 2012).

In the U.S. 98 percent of soy is turned into meal to feed poultry, bovines, hogs … and farmed catfish; more than fifty percent of the U.S. soy produced is fed to poultry. Those who cast an accusing eye at soy-eaters have missed a vital point:

*Worldwide, 80 percent of soybean crops are planted, tended, and harvested for farmed animals, implicating those who eat cheese and chicken—not those who eat tofu and tempeh* (Reynolds and Nierenberg, 2012).

For purely selfish reasons, most of us do not wish to see rainforests disappear. Rainforests hold a good deal more than half of the world's animal and plant species—if we lose most of the rainforests, we lose at least 50 percent of Earth's biodiversity (Pimm and Raven, 2000). Many of these species have not even been classified by Western scientists—plants and animals rich with medical and nutritional possibilities remain safely hidden from chemists and cooks. Rainforests harbor untapped recreation sites for flush travelers, and protect some of the most hidden and "untainted" human cultures. Additionally, rainforest soils are critical to land stability—devoid of this rich canopy and reaching tree roots, soils are prone to mudslides and desertification. Perhaps most important in light of global climate change, rainforest tree canopies are powerful converters, turning carbon dioxide into oxygen, helping to mitigate some of the effects of climate change.

Forests are home to 70 percent of the Earth's land animals and plants. Forests are also important to the Earth's water systems, holding water and generating rain for drought-plagued landscapes (Schwartz, 2001). Nonetheless, we have destroyed about 80 percent of Earth's natural forests, including more than half of the Earth's rainforests (Schwartz, 2001). Estimates of the rate of species extinction vary widely, but conservative estimates indicate that we are losing at least 137 species every day (Raintree.com, 2010) and the "leading cause of this deforestation is meat production" (Hawthorne, 2012). We are unwittingly chewing anteaters, armadillos, jaguars, and untold species—which we have not yet even marveled at—into oblivion.

## Soil Degradation

Worldwide, the leading causes of soil degradation are overgrazing (35 percent), deforestation (30 percent), and agriculture (28 percent)—all directly linked to our consumption of animal products (University of Michigan, 2010). Forty percent of the world's agricultural lands were degraded in the last century "by the hard hooves and heavy bodies" of millions of farmed animals (Wardle, n.d.). Dry areas have suffered the most—73 percent of dry rangeland worldwide is already degraded (Food and Agriculture Organization of the United Nations, 2006). Farmed animals are the primary cause of desertification, both through overgrazing and from the production of feed crops—especially monocultures (Western Watersheds Project, 2010).

Many lands are not suited for grazing, and so they can only be used for grazing for a handful of years before they are depleted of nutrients, at which point they become wastelands (Hawthorne, 2012). When grazing lands are depleted of nutrients, people clear and till neighboring plots of land in order to continue production. But these lands, too, can only sustain grazers for a short period of time. Consequently, "every year a new chunk of real estate the size of Rhode Island is being swallowed by sand" (Hawthorne, 2012). Depleted soils from these dusty graveyards blow across continents. Seoul, for example, often falls under a perpetual brown haze caused by China's newly created northern wastelands—the result of overgrazing in the hope of satisfying China's ever-growing demand for flesh.

## Land Use

"The livestock sector is the world's largest consumer of the Earth's land resources" (Reynolds and Nierenberg, 2012). Worldwide, feed crop production claims about 1,164 million acres (471 million hectares)—"33 percent of the total arable land" (Food and Agriculture Organization of the United Nations, 2006). About 8.6 billion acres (3.5 billion hectares) are devoted to grazing lands (The Sustainable Seed Project, 2003). The amount of land tilled and tended on behalf of animal agriculture worldwide (both grazing and crops) is roughly the size of Africa (Hawthorne, 2012, 39).

Animal agriculture requires monumentally more land than any other human enterprise (Food and Agriculture Organization of the United Nations, 2006). In the U.S., 73 million acres (30 million hectares) are devoted to growing corn, 80 percent of which is fed to farmed animals at home and abroad. An additional 73 million acres are devoted to soybeans, of which not quite half are fed to farmed animals. Another 53 million U.S. acres (22 million hectares) are put into wheat, some 22 percent of which is used for animal feed; 8 million acres (3.2 million hectares) are devoted to sorghum, almost all of which is fed to farmed animals, and 60 million acres (24 million hectares) are devoted to hay and alfalfa—all for farmed animals (Environmental Protection Agency Agricultural Center, 2009a). This means that 58 million U.S. acres (23.4 million hectares) of corn, more than 22 million acres (9 million hectares) of soybeans, some 12 million acres (5 million hectares) of wheat, and about 7 million acres (2.8 million hectares) of sorghum are planted, irrigated, doused with pesticides, herbicides, and chemical fertilizers, harvested, transported, packaged, and stored in order to feed animals—just shy of 100 million acres (40 million hectares) are dedicated to crops raised not for human consumption, but to be cycled inefficiently through farmed animals. U.S. animal agriculture claims an additional 613 million acres (248 million hectares) for grazing (Environmental Protection Agency Agricultural Center, 2009b).

Some 98 million bovines live in the U.S. When planted with soy and corn for human consumption, 2.5 acres (1 hectare) of land can produce 2,200 pounds (1,000 kg) of protein, but an unnerving 25 acres (10 hectares) are required to produce the same 2,200 pounds of protein from beef. A plant based diet "uses less than half as many hectares as grass-fed dairy and one-tenth as many hectares as grass-fed beef to deliver the same amount of protein" (Matheny, 2003). If we adopt a plant based diet, hundreds of millions of acres can be returned to wild lands—forests and grasslands and prairies—an environmentalist's dream come true.

## Wildlife

Aldo Leopold wrote, "Predatory animals are the common enemy of both the stockman and the conservationist" (Leopold, n.d.). On behalf of ranchers (and hunters) the U.S. government's Wildlife Service predator control program has been wiping out wildlife since the early twentieth century. The agriculture lobby, which views large predators as a "threat to livestock," asks government wildlife agencies to eliminate any and all animals deemed hazardous to their investments—and U.S. government's Wildlife Services willingly complies (Fox, 2009): Every year since 2004 somewhere between 100,000 and 125,000 mammalian carnivores have been killed by Wildlife Services. Every year at least since 2004, some 600 badgers, 400–500 black bears, 10,000–13,000 raccoons, and 70,000–90,000 coyotes have been killed by the U.S. government—at taxpayer expense—on behalf of agriculture interests (Wild Earth Guardians, n.d.a). Though "motivated to undercount" reported numbers, in 2011 Wildlife Services reported killing 11,061 birds—hawks, falcons, owls, and vultures (up 42 percent from 2010), and 116,093 land predators such as wolves, coyotes (83,695), bears, bobcats, fishers, cougars, weasels, skunks, raccoons, and foxes (up 3 percent) on behalf of predator control. Also in 2011, Wildlife Services poisoned 18,587 animals (up 31 percent, including an additional 2,300 coyotes), and no doubt at considerable expense, gunned down 48,811 animals from helicopters (a 15 percent increase from 2010) (Wild Earth Guardians, 2012). The most recent report indicates that the U.S. federal government killed a whopping 3.8 million animals in 2011.

Traps and poisons are indiscriminate, maiming and killing hundreds of thousands of non-target animals (nearly 200 owls, nearly 400 falcons, and more than 1,000 hawks in 2010). Every year predator control kills thousands of animals incidentally, including armadillos, bluebirds, bears, bobcats, turtles, and alligators (United States Department of Agriculture: Animal and Plant Health Inspection Service: Wildlife Damage Management, n.d.). Inevitably, among these casualties are threatened and endangered species, including nearly 2,000 gray wolves unintentionally killed by predator control since 2004 (Wild Earth Guardians, n.d.a). Predator control has killed "nearly 1,400 house cats, [and] more than 400 domestic dogs" (Wild Earth Guardians, 2012). It is simply not possible to maintain the integrity of ecosystems while killing thousands of wild animals from specific target species, yet the U.S. Federal Wildlife Service continues its war on wildlife—"at the request of ranchers," at the expense of taxpayers (Lange, 2012).

To make all this killing possible, the U.S. government maintains a tax-funded "livestock protection budget" of well over $10 million (O'Toole, n.d.); 53 percent of the Wildlife Service budget is spent "protecting livestock" (O'Toole, n.d.). Predator control is a waste of taxpayer monies because it offers only a short-term fix. Killing members of target species usually causes a population to *increase* in short order because births adjust for deaths. For example, even when more than half of a coyote pack is wiped out, the pack is likely to bounce back within a year

(Lange, 2012). Though predator control is a very short fix (perhaps even counter-productive), and though this program damages ecosystems and further threatens endangered wildlife, the U.S. government continues its multi-million dollar extermination program, decade after decade, on behalf of animal agriculture.

## U.S. Public Lands

Ranching "is the most widespread commercial use of public lands in the United States" (Western Watersheds Project, 2010)—an environmental and economic catastrophe. Bovines harm native wildlife, vegetation, and soils. They do not browse and wander like native mammals, such as elk or deer. Bovines tend to "remain in the same area until they have eaten all or most of the edible material," trashing waterways, soils, and delicate vegetation in the process, and "causing significant harm" to native species and their ecosystems (Brown, 1996; Center for Biological Diversity, n.d.).

Bovines are especially hard on wetlands and water systems. For example, they add manure and bacteria to waterfowl nesting habitat (increasing water temperatures), trample waterfowl nesting sites, and destroy water retention (Brown, 1996). Bovines strip "riparian, or streamside, habitat favored by many small mammals and nesting songbirds," destroying root systems that stabilize stream banks, leading to erosion and loss of pebble beds where fish spawn (Chadwick, 2012). Bovines

> destroy native vegetation, damage soils and stream banks, and contaminate waterways with fecal waste. After decades of livestock grazing, once-lush streams and riparian forests have been reduced to flat, dry wastelands; once-rich topsoil has been turned to dust, causing soil erosion, stream sedimentation and wholesale elimination of some aquatic habitats.
>
> (Center for Biological Diversity, n.d.)

Ranching is also a primary cause "of native species endangerment in the American West" (Center for Biological Diversity, n.d.). Across time, innumerable plant species have been eliminated by overgrazing. In the arid Southwest, farmed animal grazing "is the most widespread cause of species endangerment" (Western Watersheds Project, 2010). Despite this plethora of serious and well-documented environmental problems, no "report has ever fully analyzed the incredible environmental costs of livestock grazing on federal public lands" (Wild Earth Guardians, n.d.b).

The U.S. government's grazing program is also a financial disaster, losing "money just as rapidly and consistently as it destroys habitat" (Center for Biological Diversity, n.d.). Permitting ranchers to graze farmed animals on federal lands (BLM, Forest Service) costs taxpayers as much as $1 billion annually, while ranchers enjoy "$100 million annually in direct subsidy" (Center for Biological Diversity, n.d.). Indirect costs are likely about three times this amount (Center for Biological Diversity, n.d.). Administering public lands to benefit ranchers creates a deficit. The U.S. government spends at least 144 million dollars managing farmed animals on federal lands—while collecting only $21 million in grazing fees—a net loss of $123 million (Wild Earth Guardians, n.d.b). Grazing fees are so low on public lands "that they amount to a subsidy" (Brown, 1996)—currently $1.35 for one month for one cow and her calf across 16 Western states (on public lands administered by the Bureau of Land Management and the Forest Service) (Bureau of Land Management, 2001).[1] In contrast, "[p]rivate, unirrigated rangeland in the West rents out for an average of $11.90" per cow and calf—the federal grazing fee is "a *de facto* subsidy for cattle owners," along with artificially low fuel costs (Center for Biological Diversity, n.d.)

On average, "three-fourths of all federal livestock funds, and three-fifths of cooperative funds, are spent on public lands" (O'Toole, n.d.). As much as 96 percent of these government expenditures—which ought to enhance public lands—"enhance livestock production *in direct conflict with legal mandates to restore the health of public lands*" (Western Watersheds Project, 2010; emphasis added). For example, in 2004 and 2005 in Arizona's Tonto National Forest, ranchers were given $3.5 million federal dollars for "range improvement" (Center for Biological Diversity, n.d.). "Range improvements" alter landscapes , damage ecosystems, and are therefore not in the best interest of the tax-paying public. The costs of grazing farmed animals on U.S. public lands is greater than greenbacks—even 500 million greenbacks. Public lands are being managed—and damaged—for the sake of low-cost flesh. For those who hike many miles into a wilderness that is presumably preserved for the public, it is an outrage to come upon a herd of bovines standing in a trampled, manure-riddled wasteland. Bovines "reduce aesthetics with their fecal matter, with the trampling of vegetation, and with their mere presence" (Brown, 1996). Unfortunately, the vast majority of taxpayers are not aware of the environmental impact of ranching on public lands—or the financial losses associated with this environmentally disastrous government subsidy. Ecosystems, wildlife, and U.S. public lands *ought* to be protected—not on behalf of ranchers and ranching, but for their own value and on behalf of present and future generations of humans and other species.

Ranchers have a very strong lobby. Unfortunately, what ranchers prefer is not generally in anyone else's interest, including the interests of wildlife. Buying beef supports and encourages ranching, complete with grazing on public lands. Do environmentalists wish to support millions of nonnatives—bovines (and pigs and chickens)—at the expense of local ecosystems and native species? Kale salad and vegetable fried rice look pretty good compared with the environmental damage caused by animal agriculture.

## "Sustainable," "Organic," and "Local"

Even when animal products are labeled "sustainable," "humane," "local," "grass fed," "organic," "free-range," and/or "cruelty free," choosing to consume animal products (rather than grains and greens directly) greatly increases our environmental footprint. A diet that includes animal products is a diet laden with greenhouse gas emissions, water pollution, deforestation, soil damage, freshwater depletion, and the destruction of wildlife. Every farmed animal poops, polluting land and waters. Every farmed animal drinks water; when fed grains, farmed animals consume tremendous amounts of freshwater and contribute to freshwater depletion in a host of different ways. Every farmed animal requires land, altering and threatening landscapes and ecosystems. Every product stemming from animal agriculture contributes mightily to greenhouse gas emissions. No matter what "PC" consumer label is placed on the final product, purchasing meat, dairy, or eggs demonstrates reckless disregard for the environment and is unconscionable if vegan foods are available—especially given that bulk vegan staples are inevitably less expensive than animal products.

## Conclusion

Despite severe ecological consequences, environmentalists tend to be unaccountably silent on the topic of dietary choice. Rather than alter meal plans—and ask others to do the same—environmentalists prefer to urge that we use less water on lawns, choose gas-efficient cars and fluorescent bulbs, and recycle—each of which is important, but each is also environmentally irrelevant compared with dietary choice. Eating plants and grains without cycling them through farmed animals is the most important change we can make on behalf of the environment.

## Discussion Questions

1 How might you explain the fact that so many environmentalists shorten showers and ride bicycles rather than change their diet?

2 What is the government's rightful role with regard to helping corporations that are extremely harmful to the environment, nonhuman animals, and human health?

3 What is our rightful role in a democracy in which the government is obviously aligned with big businesses that harm the environment, nonhuman animals, and human health?

4 What are the main environmental concerns in your area? What do you think is the most serious environmental concern in your area? Is dietary choice one of the root causes of local environmental problems?

5 How many morally relevant distinctions can you find between humans eating animal products and nonhuman animals eating their prey?

## Essay Questions

1 Explain the connection between a diet rich in animal products and any two of the following: depleted freshwater reserves, deforestation, soil degradation, wildlife manipulation.

2 Is it justifiable to use public lands to reduce the price of animal products? Why or why not? Do you think the killing of wildlife by Wildlife Services is justified on behalf of animal industries? Compose a letter to the editor, or to your senators and representatives, expressing your point of view on this matter.

3 If you consume animal products, how do you maintain integrity as an environmentalist while supporting animal agriculture? If you are a vegetarian, how do you maintain integrity as an environmentalist while supporting the dairy and egg industries? If you are a vegan, try the next question.

4 What dietary changes might you commit to on behalf of the environment? (Vegans also need to ponder this question, considering specific changes that are environmentally beneficial.)

## Suggested Further Reading

Campbell, T. C., 2006. *The China Study: The Most Comprehensive Study of Nutrition Ever Conducted And the Startling Implications for Diet, Weight Loss, And Long-term Health.* Dallas: BenBella Books.

Food and Agriculture Organization of the United Nations, 2006. *Livestock's Long Shadow: Environmental Issues and Options,* Rome: Food and Agriculture Organization.

Kemmerer, L., 2014. *Eating Earth: Dietary Choice and Planetary Health.* Oxford and New York: Oxford University Press.

Moskowitz, I. C. and Romero, T. H., 2007. *Veganomicon: The Ultimate Vegan Cookbook.* Cambridge, MA: Da Capo Lifelong Books.

Oppenlander, R., 2011. *Comfortably Unaware: Global Depletion and Food Responsibility … What You Choose to Eat is Killing our Planet.* Minneapolis: Langton Street.

Scully, M., 2003. *Dominion: The Power of Man, the Suffering of Animals, and the Call to Mercy.* New York: St. Martin's Griffin.

Singer, P. and Mason, J., 2006. *Eating: What We Eat and Why it Matters.* London: Arrow Books.

## Notes

This chapter is an excerpt from *Eating Earth*, Lisa Kemmerer, Oxford University Press, 2014, reproduced with permission.

1 The BLM manages 245 million acres (99 million hectares) largely in 12 Western states—more than any other Federal agency—while the Forest Service manages 193 million acres (78 million hectares) in 44 states, Puerto Rico, and the Virgin Islands ("BLM").

## References

Brown, L., 1996. *Cattle Grazing on Public Lands: The Hard Fought Battle in the Southwestern United States.* [Online] Available at: http://www.thebeckoning.com/environment/cattle/grazing.html#3-1[Accessed 27 December 2011].

Bureau of Land Management, 2001. *BLM and Forest Service Announce 2011 Grazing Fee.* [Online] Available at: http://www.blm.gov/wo/st/en/info/newsroom/2011/january/NR_01_31_2011.html [Accessed 28 December 2011].

Cassuto, D. N., 2010. *The CAFO Hothouse: Climate Change, Industrial Agriculture, and the Law,* Ann Arbor: Animals and Society Institute.
Center for Biological Diversity, n.d. *Biological Diversity: Grazing.* [Online] Available at: http://www.biologicaldiversity.org/programs/public_lands/grazing/ [Accessed 27 Dec 2011].
Chadwick, D., 2012. Nose on the Range. *All Animals*, Volume January/February, pp. 22–27.
*The Economist* Online, 2011. *Counting Chickens.* [Online] Available at: http://www.economist.com/blogs/dailychart/2011/07/global-livestock-counts [Accessed 27 December 2011].
Environmental Protection Agency Agricultural Center, 2009a. *Major Crops grown in the United States.* [Online] Available at: http://www.epa.gov/agriculture/ag101/cropmajor.html [Accessed 26 December 2011].
Environmental Protection Agency Agricultural Center, 2009b. *Land Use Overview.* [Online] Available at: http://www.epa.gov/agriculture/ag101/landuse.html [Accessed 26 December 2011].
Food and Agriculture Organization of the United Nations, 2006. *Livestock's Long Shadow: Environmental Issues and Options,* Rome: Food and Agriculture Organization.
Fox, C., 2009. *Indiana Coyote Rescue Center: Predators in Peril: The Federal Government's War on Wildlife.* [Online] Available at: http://www.coyoterescue.org/newsletter/winter2009.html [Accessed 16 November 2012].
Global Warning Science, n.d. *Deforestation: The Leading Cause of CO2 Emissions.* [Online] Available at: http://www.appinsys.com/globalwarming/deforestation.htm[Accessed 13 April 2013].
Hawthorne, M., 2012. Planet in Peril. *Veg News,* Volume March/April, pp. 34–41.
Kauffman, S. and Braun, N., 2004. *Good News for All Creation: Vegetarianism as Christian Stewardship,* Cleveland: Vegetarian Advocates Press.
Lange, K., 2012. Coyotes Among Us. *All Animals,* Volume May/June, pp. 14–19.
Leopold, L., n.d. *The Aldo Leopold Archive Writing Reprints: Wanted—National Forest Game Refuges.* [Online] Available at: http://digicoll.library.wisc.edu/cgi-bin/AldoLeopold/AldoLeopold-idx?type=turn&entity=AldoLeopold.ALReprints.p0451&id=AldoLeopold.ALReprints&isize=text [Accessed 2 February 2013].
Matheny, G., 2003. *Least Harm: A Defense of Vegetarianism from Steven Davis's Omnivorous Proposal.* [Online] Available at: http://www.veganoutreach.org/enewsletter/matheny.html [Accessed 27 December 2011].
Monga Bay, n.d. *Causes of Deforestation in the Brazilian Amazon.* [Online] Available at: http://www.mongabay.com/brazil.html [Accessed 13 April 2013].
O'Toole, R., n.d. *Audit of the U.S.D.A. Animal Damage Control Program.* [Online] Available at: http://www.ti.org/adcreport.html [Accessed 16 November 2012].
Pimm, S. L. and Raven, P., 2000. Extinction by Numbers. *Nature,* Volume 403, pp. 843–845.
Raintree.com, 2010. *Rainforest Facts.* [Online] Available at: http://www.rain-tree.com/facts.htm [Accessed 28 December 2011].
Reynolds, L. and Nierenberg, D., 2012. *Worldwatch Report 188: Innovations in Sustainable Agriculture: Supporting Climate-Friendly Food Production,* Washington, DC: World Watch Institute.
Schwartz, R. H., 2001. *Judaism and Vegetarianism.* New York: Lantern.
The Sustainable Seed Project, 2003. *Ecological Footprint: What is the Ecological Footprint?.* [Online] Available at: http://www.sustainablescale.org/conceptualframework/understandingscale/measuringscale/ecologicalfootprint.aspx [Accessed 13 January 2013].

United States Department of Agriculture: Animal and Plant Health Inspection Service: Wildlife Damage Management, n.d. *Table G-2: Animals Euthanized or Killed by Wildlife Services-FY 2 0 11.* [Online] Available at: http://www.aphis.usda.gov/wildlife_damage/prog_data/2011_prog_data/PDR_G/Basic_Tables_PDR_G/Table_G-2_Euth-Killed.pdf [Accessed 16 November 2012].

University of Michigan, 2010. *Land Degradition.* [Online] Available at: http://www.globalchange.umich.edu/globalchange2/current/lectures/land_deg/land_deg.html [Accessed 26 December 2011].

Wardle, T., n.d. *Diet for Disaster.* [Online] Available at: https://veganfuture.files.wordpress.com/2013/03/diet-of-disaster-meat-dairy.pdf [Accessed 28 December 2011].

Western Watersheds Project, 2010. *Public Lands Ranching.* [Online] Available at: http://westernwatersheds.org/issues/public-lands-ranching?gclid=CIGqkcarua0CFQ5lhwodV2YG4w [Accessed 28 December 2011].

Wikipedia.org, n.d.a *Deforestation in Costa Rica.* [Online] Available at: http://en.wikipedia.org/wiki/Deforestation_in_Costa_Rica [Accessed 13 April 2013].

Wikipedia.org, n.d.b *Agriculture in Brazil.* [Online] Available at: http://en.wikipedia.org/wiki/Agriculture_in_Brazil [Accessed 28 November 2013].

Wild Earth Guardians, 2012. *Wildlife Services Under Fire, Releases Annual Kill Numbers: Overall Kills Declined, but Native Carnivores Endured Increased Mortality.* [Online] Available at: http://www.wildearthguardians.org/site/News2?page=NewsArticle&id=7639 [Accessed 16 November 2012].

Wild Earth Guardians, n.d.a *Wildlife Killed.* [Online] Available at: http://www.wildearthguardians.org/site/PageServer?pagename=priorities_wildlife_war_wildlife_killed_table [Accessed 2011 December 26].

Wild Earth Guardians, n.d.b *Fiscal Costs of Federal Public Lands Livestock Grazing.* [Online] Available at: http://www.sagebrushsea.org/pdf/factsheet_Grazing_Fiscal_Costs.pdf [Accessed 27 December 2011].

Worm, K., 2004. *Groundwater Drawdown.* [Online] Available at: http://academic.evergreen.edu/g/grossmaz/WORMKA/[Accessed 26 December 2011].

# SECTION IV

# COMMON GROUND

## Raising Questions, Pondering Connections

**Poem**: Cara Chamberlain

### Laridae

Many world populations of large gulls increased at unprecedented rates in the last decades (Pablo Almaraz and Daniel Oro, "Size-Mediated Non-Trophic Interactions and Stochastic Predation Drive Assembly and Dynamics in a Seabird Community," *Ecology*, October 2011)

The exotic part is the pastel iris, the pupil within
that beats as the heart beats, the voice at random,
and I can never decide if it's the laughter
of joy or schadenfreude. If I were to say
to those assembled on the dolosse under the foghorn,
I love you because you're common and your feet
are pink and webbed, your breath from satin
orange mouth linings thick with crabs,
sardines, congealed grease, catfood, whatever
the landfill has afforded, I don't guess they'd care.
They might survive the death of everything else,
acidifying oceans, radiation, collapsing fisheries.
Glaring, unmoved, they might live forever.

# 14

# THE MANY AND THE ONE

*Randall Gloege*

A Yellowstone Park photographer, while working at his craft, was eaten by a grizzly. On discovery of his fate, the photographer's wife pleaded with the Park Service to spare the bear. Why punish the grizzly, she argued, for being a grizzly?

Eric Toso, a Tuscon, Arizona, writing instructor, while walking home from a community swimming pool one evening, was struck by a rattlesnake. The snake injected venom into Toso's foot. About the rattler, Toso observed,

> No one can know how a snake perceives the world, what it sees from eyes that have never blinked, how it can wait coiled for days next to a mouse trail, alert as a samurai, still as a stone, for a rodent that may or may not pass within striking range.
>
> (Toso 2007, 19)

Toso never blamed the snake. Nearly dying as the venom did its work, he documented how his near-death experience and gradual recovery led to a deeper spiritual understanding of and affection for the desert and its creatures.

Howie Wolke, my friend and a radical wilderness warrior, while we were backpacking in the Escalante, warned me away from a patch of gray, popcorn-like dirt. I absolutely *must not* tread on what turned out to be soil in the process of formation.

Eldon Smith, one of my environmental mentors, years ago pointed me toward a short stretch of the Big Horn River, including river bed, stream banks, flora, and fauna. Because it was self-sustaining and self-perpetuating, he declared this tiny bit of biosphere to be beautiful.

In the process of encountering tropical and temperate rain forests, glaciated high country, new mountains, old mountains, temperate forests, fragmented vestiges of eastern hardwood forests, rolling prairie, arid badlands, deserts, high-mountain tundra, sculpted sandstone land forms, and dormant volcanoes, I became an ardent wilderness advocate. Inevitably, I faced criticism from animal advocates. For example, I was accused of preferring wild animals (that are and will remain wild) to domesticated species. I agree that I and other wilderness activists seem to be biased in favor of wild creatures. Wilderness activists tend to look down on domesticated animals because, although they should be cared for, they are ill equipped to survive without human assistance. This, of course, is not the fault of the animals. Animals born of selective breeding are often ill equipped for survival. Furthermore, animal "pets" are often spoiled to death by their owners. Wilderness advocates frequently are drawn to species capable of surviving with relatively little human interference. This first accusation is grounded in truth. However, preference for wild creatures does not prevent wilderness advocates from loving and respecting animals tamed by circumstance. Furthermore, many of us share with the general public a repugnance for the way domestic animals are bred, kept, and killed for profit.

A second accusation coming from animal activists is that wilderness advocates favor large mammalian species. This bias is more prevalent among hunters than among those who advocate for wild environments. Hunters tend to prefer larger species, especially favoring certain "big game" species. Bison, for example, are less attractive than elk to many hunters. Although hunters do not generally wish to eradicate bison as a species, neither are they particularly interested in preserving bison in their natural environment. Their eagerness to destroy bison is aided and abetted by many cattle ranchers who, without having any proof of the impact of cattle–bison encounters in the field, have convinced themselves that brucellosis-infected bison are a major danger to domestic cattle. There is no scientific baseline data supporting their view. Furthermore, elk are more common and widespread than bison. Nevertheless, brucellosis hysteria has been used to promote and justify a bison execution program on the periphery of Yellowstone National Park. Elk, regardless of whether or not they may be infected with brucellosis, are a more popular "game" species, partly because hunting bison is less a challenge than an outright slaughter. Adding force to this prejudice is the cattle rancher's unwillingness to share prairie grass with the bison. This in spite of the fact that bison are far better equipped than cattle to survive in western landscapes.

Because game management agencies are funded almost entirely by hunters, this bias in favor of elk is perpetuated, and even written into policy. Hunters who claim to be environmentalists must be considered separately from environmentalists who are not motivated by hunting interests. I once heard a representative of Montana Fish, Wildlife, and Parks defend hunting as a marvelous family recreation rooted in love for the "sport." Among the hunters I have known, too many are motivated by a love of killing—a kind of blood-driven climax. Hunters often rationalize their joyous killing by assuring critics that they eat the meat of the dead. Few hunters actually need this meat. In contrast with hunters, environmentalists are concerned for ecosystems in their entirety. Such environmentalists are no less concerned about the possible loss of the little snail darter than they are about the possible loss of the polar bear. Indeed, environmentalists favor endangered species, but not necessarily larger mammals.

A third allegation is that wilderness advocates selfishly relish the pleasures and benefits of wild places. Here evidence is mixed. On the one hand, the long-term *human* benefits of wilderness are often emphasized in the political process: We preserve wild country because we want our children to enjoy a quality wilderness experience. Wilderness can satisfy intangible human wants and needs such as the desire for solitude and the experience of a commonly-agreed-upon beauty. Wilderness provides a platform for outdoor exercise; wilderness is a source of medicinal substances; wilderness serves as a genetic bank for the future, supplying useful items we do not even yet know about. On the other hand, many wilderness advocates also defend wild places for their *intrinsic* value. For example, the Deep Ecology movement teaches that we should value wilderness for its own sake.

These three accusations are the primary objections that I have heard brought against wilderness defenders by animal advocates. I do not pretend that the understandings and commitments of wilderness advocates come automatically. Activist understandings and commitments require long and careful contemplation of critical and controversial questions. With regard to humans and wilderness, I would ask the following key questions:

- Are we hard-wired to use and abuse our environment, wild or otherwise?
- Is a love for wilderness and wild creatures innate, or must this be learned?
- Are we obliged only to love the living world and living beings, or must we further come to love the vast, ever-changing complex of things both living and lifeless?

In order to perceive wholes as well as parts, we must challenge our biases and commitments resulting from a life-long process of learning and unlearning.

There is no doubt that, in order to survive, human beings must use their surroundings. However, in the modern, technologically obsessed world, very few can encounter wilderness intuitively and spontaneously. We have lost connection with our natural surroundings as well as our recognition that we are part of a very complex mix of things living and dead. A great many modern humans value greed, selfishness, and rapacity. They delight in triumphing over those little able to cope with the harsh realities of survival. It is also true, however, that our remarkable success at technical improvisation, yes, even at warfare, is dependent on cooperating effectively with each other. If we are hard-wired, it would seem that we are prone not only to vicious competition, but also to effective teamwork. It may be our motives that should be called into question rather than our innate characteristics.

Cooperation should not be limited to the mere acquisition of power. Can there be any life-sustaining joy in triumphing over others? Are we nothing more than bullies? Are we entitled to treat others merely as "resources"? We need to adopt more compassionate attitudes: "It is difficult to speak of or practice love, friendship, generosity, understanding or solidarity within a system whose rules, goals, and information streams are geared for lesser human qualities" (Meadows et al. 1992, 234). In our American experiment, the elevation of the individual to a position of extreme importance has caused us to downplay the benefits of cooperative, compassionate behavior except insofar as such behaviors further individual interests. The doctrine of enlightened self-interest has become the doctrine of short-term self-preservation, acquisition, advancement, and domination. Learning to love one another, not to mention loving other inhabitants of Planet Earth, may well require a fundamental unlearning, learning, and relearning process. We must first unlearn our self-promoting contempt for one another. Then we must learn to take meaning and sustenance from one another. And we must relearn the necessity to cooperate with each other in a context where we abandon the illusion that we have the right and the ability to dominate all the things of the earth. If we cannot accomplish these basic first steps, there is little hope that we will extend our sense of community beyond selfish and confining human boundaries.

What we call "environment," whether wilderness or otherwise, is comprised of both living and non-living components, and we are part of a very complex mix of the living and the dead. These components are dynamic both in themselves and in relation to each other. Life/death is a dynamic *process*. Living things are constantly converted into non-living things, while non-living things are constantly converted to living ones. A healthy environment requires both the living and the dead. This realization does not seem to come naturally to us. Still, what may not seem natural is in fact natural: Constant evolution is fundamental to life itself. Coming gradually to this understanding requires experience, perception, and interaction. Engagement with wilderness is an engagement with a dynamic whole, the preservation of the parts of which can help foster and perpetuate the whole. Because we are part of this complex, dynamic whole, loving is a matter of survival: "Collapse [of living systems] cannot be avoided if people do not learn to view themselves and others with compassion" (Meadows et al. 1992, 234). Collapse of the world as we know it can only be averted if we foster compassion.

Compassion, therefore, guides environmentalists and animal advocates alike. Advocates of wilderness generally seem to worry more about the fate of species than that of individuals, whereas advocates for animals often appear to be engaged more with the fate of individuals than species. Ultimately, environmentalists are not concerned with the fate of either species or individuals, but rather, with the fate of both—the fate of the many and the one. Additionally, wilderness defenders are concerned with the capacity of the land itself to support and sustain species. Habitat, habitat, habitat. However, speculations about preferring the many to the one, or the one to the many, are unfair to both camps. Wilderness advocates are concerned with species and individuals, while animal advocates fight for both individuals and species. It is just a matter of emphasis—the chicken or the egg. We must foster and protect both.

I began with a cluster of anecdotal encounters stemming from a profound compassion and love for this world. These moments represent a profound love for the ever-changing here and now, almost overrun with things living and dead, permeated with odors and textures, teeming with visions and strident with voices enticing us into the manifold mysteries of being. Our perceptual selves need to be awakened. We must be attuned to *now*, willing to be connected, opening our perceptions to feel and become what is nothing more or less than

> our part of a vast, interpenetrating webwork of perceptions borne by countless other bodies—supported, that is, not just by ourselves, but by the icy streams tumbling down the granite slopes, by owl wings and lichens, and by the unseen and imperturbable wind.
>
> (Abram, 1996, 65)

Perception itself is fluid, always in motion, changing with the light, the time of day, our surroundings, our moods. We and the environment that sustains us are inseparable. We yearn to be wholly where we are when we are there. How else can we know the many and the one?

## Discussion Questions

1 What accusations might environmentalists level against animal advocates?

2 Are there any other likely complaints that animal advocates might level against environmentalists?

3 Does it make sense to kill wild animals that have harmed human beings? Is it just to do so? In what other ways might we respond when a bear or some other wild animal harms or kills a human being?

4 How might you respond to Gloege's question: "Are we hard-wired to use and abuse our environment, wild or otherwise?"

5 How might you respond to Gloege's question: "Is a love for wilderness and wild creatures innate, or must this be learned?"

6 How might you respond to Gloege's question: "Are we obliged only to love the living world and living beings, or must we further come to love the vast, ever-changing complex of things both living and lifeless?"

7 Are human beings bullies in relation to other living beings? Is it morally acceptable to treat others merely as "resources"?

8 Gloege encourages us to foster compassion, as well as love and connection. How can these help us to help animals and the earth? What role do human emotions play in social justice advocacy?

## Essay Questions

1 Respond to one of Gloege's bulleted questions (page 203).

2 Respond to Gloege's comment: "Wilderness advocates are concerned with species and individuals, while animal advocates fight for both individuals and species. It is just a matter of emphasis—the chicken or the egg."

3 On what points do you agree and disagree with Gloege's responses to common animal advocate criticisms against environmentalists?

4 Are you more of an animal advocate or more of an environmentalist? What characteristics in your lifestyle, values, advocacy, or outlook most clearly define you in this way? Given the close relationship between animal and earth advocacy, how can you bring your life more in line with both?

5 How might you foster and express compassion in your own heart and life?

## Suggested Further Reading

Beers, Diane L. (2006). *For the Prevention of Cruelty: The History and Legacy of Animal Rights Activism in the United States*. Swallow Press, Athens, Ohio.

Davies, Kate (2013). *The Rise of the U.S. Environmental Health Movement*. Rowman and Littlefield, New York.

Liddick, Donald R. (2006). *Eco-Terrorism: Radical Environmental and Animal Liberation Movements*. Praeger, New York.

Lytle, Mark Hamilton (2007). *The Gentle Subversive: Rachel Carson, Silent Spring, and the Rise of the Environmental Movement*. Oxford U.P., New York.

## References

Abram, David. *The Spell of the Sensuous: Perception and Language in a More-Than-Human World*. New York: Pantheon, 1996.

Meadows, Donella H., Dennis L. Meadows, and Jorgen Randers. *Beyond the Limits: Confronting Global Collapse, Envisioning a Sustainable Future*. Post Mills, Vermont: Chelsea Green, 1992.

Toso, Erec. *Zero at the Bone: Rewriting Life after a Snakebite*. Tuscon: University of Arizona Press, 2007.

# 15

# CONSERVATION RESEARCH AND ANIMAL ACTIVISM

## In Search of a Better Balance

*Jon E. Swenson*

I view animal activists as people who genuinely care for nonhuman animals, both wild and captive. These dedicated people play an important role in educating the public about animal abuse and the critical needs of nonhuman animals. Importantly, they also influence the political process. Nonetheless, the actions of animal activists sometimes seem to work against what is best for wildlife. In my view, animal activists sometimes work to protect individuals in a way that is contrary to the best interests of the population as a whole, at least when we define "best interests" via science.

This essay discusses the relationship between animal activism and conservation using brown bears as an example. I write from a researcher's point of view, arguing that animal activists would do well to consider all that researchers can contribute to the conservation of animal populations. Activists also contribute to conservation, of course, but animal activists sometimes oppose the capture of individuals for research, thereby undermining conservation projects. I contend that certain wildlife populations can sometimes benefit from capturing and radiocollaring random individuals for research purposes, and that for this reason both animal and earth activists who care about wild populations ought to support this research.

### Focusing on Bears

I currently lead the Scandinavian Brown Bear Research Project. Our research team studies a wide range of topics within the ecology of the brown bear in Scandinavia, such as population ecology, life history, social and mating behavior, habitat use, food habits and predation, and the effects of humans on bears (such as how bears avoid humans, the effects of hunting and of different hunting methods, why some bears visit human habitation, depredation on sheep and cattle, human injuries, etc.). Our research provides scientific information for management and conservation of brown bears in Sweden and Norway in the form of 506 publications, including 178 international scientific journal articles and 18 PhD dissertations. Our work also helps bears in other locations. I have helped research biologists and wildlife agencies to answer conservation questions regarding threatened and endangered bear populations in the USA, Norway, Spain, France, Austria, Italy, Poland, Bulgaria, Georgia, Pakistan, and Sri Lanka. I have also helped agencies deal with large, increasing bear populations in Sweden, Finland, Slovenia, Croatia, and Japan.

I do not lobby politicians on behalf of bears or other wildlife, because I feel that it is important for a scientist to remain as politically impartial as possible. Over the years, working as I have with wildlife conservation and management, I have often thought about the relationship between animal activists and conservationists.

## A Conservationist's Questions for Animal Activists to Ponder

### *Is it Ethical to Capture and Collar Bears for Research?*

In my experience, many animal activists answer this question with a firm "no," whereas researchers often answer "yes." To put radiocollars on bears, field workers with the Scandinavian Brown Bear Research Project immobilize bears from a helicopter. In the process, bears experience some stress and sometimes—though rarely—they are injured or killed. Injury and death are certainly negative outcomes, but from my perspective, this must be weighed carefully against scientific gains; i.e. information that can aid in the conservation of populations. Management agencies, politicians, and researchers often need scientific data to create or change conservation policies.

Because bears can be harmed or die, it is important for independent ethical committees to examine research methods, including the probable effects of research on individuals, and weigh possible harm against the probable value of the results. In truth, it is a comfort to many researchers, including me, that our methods and activities are examined independently, and rejected if found unacceptable. Of course this does not free us from our responsibility to seek the highest ethical standards independent from outside reviewers.

In the Scandinavian Brown Bear Research Project, we have worked systematically to reduce researcher impact on the bears we study—with positive results. As part of this process, a doctoral student studied the physiological effects of immobilizing drugs on bears. Based on her results, we now routinely give bears oxygen while they are immobilized (Fahlman et al. 2012). We have reduced deaths associated with capture from 3.8 percent (prior to 1992), to 0.3 percent by changing immobilization drugs, doses, and capture procedures (Arnemo et al. 2006). This means that the chance that a wild bear will die from immobilization and capture is now about the same as the chance that a dog will die from anesthetics in a veterinarian's office. And we continue to improve our methods, including the use of modern technology that allows us to obtain adequate data from fewer individuals.

We currently have several projects under way to evaluate whether or not our standard research methods have long-term effects. Furthermore, our research results continue to contribute to improvements in capturing methods throughout the world, and of course our research is designed to benefit wild bear populations.

But, this begs the question of whether or not we ought to capture bears in the first place. This is, of course, an ethical question, that often brings those working to protect individuals in direct conflict with those working to protect populations or species.

Research using radiocollars has benefitted bears in many instances that probably could not otherwise have been achieved. For example, the use of radiocollars has exposed a very high rate of poaching of large carnivores (brown bears, lynx, wolverines, and wolves) in certain areas, including national parks (Rauset 2013). It is difficult for me to think of any other method for documenting such illegal activities with such precision. Scientific information gathered by our researchers on poaching has shocked the general public, enraged animal conservationists and activists, and irritated poachers. This information moved authorities in both Sweden and Norway to prioritize efforts against poaching. Paradoxically, a ban (national and local) on radiomarking bears (and other large carnivores) has been proposed both by poachers, who hoped to protect their illegal activities, and by animal activists, who were concerned about harm that might come to collared bears.

Another example of how radiocollared bears have helped bear populations is through studies exploring bear aggression toward human beings. Surveys indicate that people in Sweden are becoming more afraid of bears as bear populations increase. (There are now about 2,800 brown bears in Sweden.) Some of those who favor reducing the bear population with higher hunting quotas stress how dangerous bears are, and how an increase in local bear populations prevents people from hiking, picking berries and mushrooms, and so on.

With radiocollars, we have documented how bears react when they encounter people in their habitat. In order to accurately document such bear reactions, researchers in the Scandinavian Brown Bear Research Project walked past radiocollared bears, upwind at a distance of 50 meters. We have conducted over 200 such experimental meetings without even one sign of bear aggression; bears either run away before we arrive, lie quietly in their daybed, or wait until we have passed, then leave. In fact, they were so demure that we only detected them 15 percent of the time, even though we knew exactly where they were (Moen et al. 2012). With this documentation in hand, we can inform the public that bears are not aggressive or dangerous to hikers, and hope this new information will help reduce human fears. Most recently we have turned our attention to the circumstances in which people are sometimes hurt by bears—mostly hunters who surprise bears at close range, then wound them, thus provoking a bear attack (Sahlén 2013). To help resolve these problems, we have provided educational information in hunter education courses for bear hunters (given by the Swedish Association for Hunting and Wildlife Management) about how people ought to behave if they meet a bear. With an increase in the number of hunters taking this course, the number of hunters injured or killed annually by bears has declined, even though bear populations have increased.

A third example involves dogs used to find wounded bears. Hunting with dogs is allowed in Sweden, but our research shows that one must train dogs differently for hunting bears than for tracking wounded bears. There are strict Scandinavian laws requiring hunters to have access to a trained tracking dog in case wildlife is wounded, so that the wounded animal can be found quickly and dispatched. Researchers in our project have tested tracking dogs on trails left by radiomarked bears. Thus, we knew the exact trail the bear had left, and could evaluate how well a dog tracked a specific bear. We found that most tracking dogs, even those who were certified, were unable to follow a specific bear's trail for any great distance. It is therefore likely that hunter-wounded bears have sometimes been left to suffer simply because most tracking dogs were not properly trained to follow a specific bear's scent. This information was a great surprise to hunters, hunting organizations, and wildlife management authorities alike, and instigated a completely new training and testing system for tracking dogs (with which we cooperate). We are hopeful that our work will reduce the sufferings of bears wounded by hunters.

Our work has led other researchers to investigate how effectively tracking dogs locate wounded ungulates. (Their results were similar.) Clearly, those responsible need to rethink and redesign training programs for dogs who track wounded wildlife. Radiocollars were critical to this research, and without this research the public, authorities, and hunters themselves, would continue to believe that certified dogs were capable of tracking large, wounded wildlife, which was not the case. I could offer other examples, but I think these examples stand on their own. As a researcher, I maintain that the stress—or even occasional death—caused by capturing individuals in order to radiocollar them for well-designed research purposes can be outweighed by the long-term value of research results in the areas of both conservation and animal welfare.

In each of the above examples, we had no alternative but to capture and radiocollar bears if we were going to conduct this research. Consequently, if we value this type of research, I believe that there are times when there is no better option than to capture and collar animals to obtain information. However, I also recognize that, beyond the question of animal welfare or rights, researchers must always be aware of the possibility that nonhuman animals who have been captured may act differently from those who have not—which would reduce the value of our research findings.

## *What about Noninvasive Research Options?*

Animal activists often promote noninvasive research methods, whereby information is gathered without capturing or collaring individuals. These methods can be very useful for population estimations and monitoring. But bears are notoriously difficult to monitor. They are shy of humans, hibernate during the winter (when we might track them in the snow), and generally live in thick forests. All of these factors make bears difficult to study with noninvasive methods.

Nonetheless, researchers and wildlife management personnel currently use several noninvasive methods thanks to modern genetics, which allow researchers to identify individual bears and determine sex from DNA from hair and scat. Consequently, several methods have been developed to estimate bear numbers in a given area by setting up "hair-trapping stations," where bears rub against barbed wire, often surrounding bait or wrapped around a tree. Researchers collect snagged hairs, extract and amplify the DNA, and determine a bear's identity and sex (but not age). Another noninvasive method is to cooperate with volunteers who frequent bear habitats, such as hunters or hikers, who are able to collect many bear scat samples, from which researchers extract DNA.

Currently, these are the least invasive methods to census elusive wildlife populations. All methods have pros and cons, and it is important for authorities to understand these pros and cons in order to determine the best methods to employ.

Public groups often try to influence policies, producing population estimates that are colored by personal agendas. For example, hunters often believe that there are more bears than there actually are and that populations can withstand more kills than do people who are opposed to hunting, who often believe that there are fewer bears and, consequently, that fewer can be sustainably harvested. Authorities want and need reliable, neutral data. Consequently, researchers capture and mark bears to test the efficacy of noninvasive census methods. In our project we have, for example, determined which proportion of radiomarked bears were located and counted through noninvasive methods, and whether or not this portion offers a representative sample, including any bias in which bears (age, sex, etc.) are counted (Bellemain et al. 2005). We also found that the number of bears seen by moose hunters (weighted for the number of hunters and time afield) during the first week of moose hunting season provides a good estimate of relative bear density and can be used to monitor population trends (Kindberg et al. 2009). Now that we know that both methods are reliable, bear population size and trends are estimated routinely using noninvasive methods in both Sweden and Norway and there is therefore no further need to capture and mark bears for this purpose. Without verification through radiocollaring, we would not have been able to document the dependability of various noninvasive methods. Noninvasive methods are now used to estimate and monitor bear populations in many countries, eliminating the need for radiocollars.

### *What is the Place of the Individual—in Relation to Population—in Wildlife Conservation?*

Animal activists tend to focus on the lives and welfare of individual animals. Wildlife biologists usually focus on populations. Individuals make up populations and any population consists of a number of individuals. In the case of endangered, small populations, concerned people—the public, scientists, animal activists, conservation authorities, and politicians—work to protect *individuals* in order to ensure the survival and viability of the *population*.

Nevertheless, the choice of whether to focus on individuals or on populations can be the source of misunderstanding and conflict. For example, conflicts frequently arise when scientific population goals have been met; the population is then considered viable and hunting is allowed. Consider the case in Sweden around 1930, when brown bear populations were estimated to be at around 130 individuals. Even with controlled hunting, these numbers grew to about 3,300 bears by 2008 (Kindberg et al. 2011). As the bear population grew, bears expanded into areas where many humans live, increasing human–bear encounters. Human beings are often afraid of bears, and thus some do not favor any further increase in bear numbers. Current birth control methods have not proven to be feasible for species that live in low densities across expansive areas. Therefore, hunting quotas have been set to curtail bear population growth in many areas of Sweden. Unchecked bear population growth could jeopardize the widespread public support for bear conservation that currently exists in Sweden, and which has allowed a rebound in bear numbers. In Sweden (as in most nations) humans now occupy most of the areas where bears (and other wildlife) might otherwise live, which means that 2,800 bears are now distributed across two-thirds of the country.

## *Do Ends Justify Means?*

As is obvious from this essay, I choose to risk the welfare and lives of a few bears for what I see as a benefit to the entire bear population—and to humanity. In my opinion wildlife conservation justifies planned and closely regulated capture and radiocollaring of individuals to allow for research that is needed for effective conservation—even if some wildlife casualties result.

I realize that my utilitarian point of view puts me in a delicate situation if I am to maintain consistency—would I be willing to sacrifice human individuals similarly for the benefit of the larger human population? Would I be willing to undertake research involving humans—without their consent—knowing that a very small number would die as a result, and that some would suffer at least somewhat throughout the rest of their lives, for the hoped-for benefit of humanity more generally? My answer is "no," which proves me to be biased in favor of human beings. Many animal activists would not sacrifice the lives of any nonhuman individual if they would not be willing to similarly sacrifice the life of a human being, considering it to be an unjustified prejudice—like sexism and racism. Indeed, I admit that my utilitarian equation is much more stringent with regard to the possible loss of human life.

As a utilitarian in my interactions with wildlife, I seek the greatest good for the greatest number, ultimately privileging humans. Radiocollars cause a measurable but small amount of harm to some bears on behalf of the species as a whole, and on behalf of humanity. I am willing to sacrifice 0.3 percent of the bears we work with as part of this utilitarian equation—a cost I would not accept among human beings.

## Working Together for Animals at Risk

In a functional democracy a well-informed majority guides legislative action. Democracy therefore requires an informed population, which requires ready access to information. As growing human populations increase pressure on wildlife and their habitats, we must have reliable information if we are to make necessary decisions in order to maintain viable wildlife populations in increasingly human-dominated landscapes.

We are currently in the middle of a global extinction caused, for the most part, by humans. My work is largely the work of a conservationist—aimed at maintaining populations and ecosystems. I recognize the importance of animal activists in this process: They influence public opinion and politicians, and work as "watchdogs," keeping an eye on researchers (as well as hunters and politicians) to be sure that our methods are as benign as possible. However, they may also try to influence politicians to conclude that invasive research is never necessary, which I believe can hinder a reasonable balance between protecting and respecting individual animals and gathering scientific information to ensure the survival of the species. I see the importance of both animal activists and conservation scientists in the larger process of protecting wildlife. We have the best chance of success if we work together.

## Acknowledgments

I would like to thank my colleagues in the Scandinavian Brown Bear Research Project for their commitment to the ethical treatment of bears and for their untiring work for the conservation of bears everywhere. I would also like to thank Dr Jonas Kindberg for his comments and the editor, Dr Lisa Kemmerer, who has helped me write my first essay on conservation and animal activism and, while doing so, has challenged me to explore the philosophical and ethical basis of my research. This has been a valuable exercise for me, as I believe it would be for any researcher or activist. It is my hope that animal and earth activists, as well as conservation biologists, will keep an open mind and allow themselves to be as challenged by this collection of essays, as I was in writing this essay.

My goal in writing this essay is to encourage animal activists, conservationists, and researchers to find commonality in our quest for common goals—helping wildlife. We will be much more effective if we can respect our differing philosophies and still find room to cooperate.

## Discussion Questions

1 Swenson mentions the importance of "independent ethical committees to examine research methods." What does the word "independent" refer to? What does it not refer to, for example, are these committees likely to be independent of human bias/speciesism? How might independent ethical committees better represent the interests of wildlife? Is it important to do so? Why or why not?

2 Like many environmentalists, Swenson uses the term "harvested" in reference to killing bears, the same term used for wheat and corn and carrots. Animal advocates tend to find this language offensive. Why do you suppose environmentalists use this term, and why do animal advocates find it offensive? Is "harvest" an acceptable way to refer to killing? Why do some people choose this term? Why or why not?

3 If animal activists object to research methods, what are some effective and creative ways they might respond?

4 What are the implications of referring to the death of an individual as a "negative outcome"? Why might scientists use this language and why might animal advocates object? Is a 0.3 percent research death rate acceptable for bears or humans?

5 Animal activists are not likely to accept any suggestion of "sustainable harvest" with regard to animals. Divide into two groups with opposing views on this matter. Through discussion, achieve the following goals: (1) Articulate the opposing point of view. (2) Articulate common ground/shared concerns. (3) Establish language that is mutually acceptable.

6 Is it possible to be "politically impartial"? Is it desirable? Why or why not? Is it likely that researchers have a personal agenda? If so, what might that agenda be? Do animal advocates have anything similar to gain from animal advocacy?

7 If radiocollars help us to see that bears show every indication of preferring to be left alone (running away whenever they smell human beings), is it acceptable to continue placing radiocollars on bears? Why or why not?

## Essay Questions

1 A reviewer accused Swenson of being an apologist for those who engage in invasive research with wildlife. Is this a fair accusation? Why or why not?

2 How is Swenson's bear research similar to poaching and hunting bears? How is it different?

3 Read up on James Elliot Sims's controversial medical research with African American women. What are the arguments for and against Sims's work? How does this help us to think about Swenson's research?

4 Swenson's research has helped hunters to track and kill injured bears with dogs. How might the bears respond to Swenson's assertion that this helps bears?

5 Under what conditions should animal activists consider cooperating with researchers who capture and mark animals and under what conditions should they not?

6 If a particular species is declining to almost certain extinction, and a research program based on capturing and marking animals may allow researchers to find the reason for the decline and therefore perhaps save the species, what are the

arguments for and against supporting such research? Under what conditions is it morally acceptable to risk human life in the hope of saving human life? Should human life be treated differently, and if so, on what non-self-interested basis?

7 Swenson argues that he is helping bears by doing work that is vital to preserving an endangered species. At least some animal activists view his work as morally problematic because he risks the lives of individual bears. Present Swenson's arguments on behalf of his research. Present the arguments of animal advocates who object to his research. Create a bear program that reflects and respects both points of view.

8 What sort of wildlife research is being conducted in your state? How can you find this information, and what can you do as an activist if you object to certain methods or intended outcomes? Carry one of your suggestions forward as a class project.

## Suggested Further Reading

Ferdowsian, H. R. and Beck, N. (2011) "Ethical and scientific considerations regarding animal testing and research", *PLoS One*, vol. 6, no. 9, paper e24059

Kreeger, T. J. and Arnemo, J. M. (2012) *Handbook of Wildlife Chemical Immobilization, fourth edition*, Terry J. Kreeger, Sybille, Wyoming, USA

Sikes, R. S., Gannon, W. L., and the Animal Care and Use Committee of the American Society of Mammalogists (2011) "Guidelines of the American Society of Mammalogists for the use of wild mammals in research," *Journal of Mammalogy*, vol. 92, pp. 235–253

## References

Arnemo, J. M., Ahlqvist, P., Andersen, R., Berntsen, F., Ericsson, G., Odden, J., Brunberg, S., Segerström, P., and Swenson, J. E. (2006) "Risk of capture-related mortality in large free-ranging mammals: experiences from Scandinavia," *Wildlife Biology* vol. 12, pp 109–113

Bellemain, E., Swenson, J. E., Tallmon, D., Brunberg, S., and Taberlet, P, (2005) "Estimating population size of elusive animals using DNA from hunter-collected feces: comparing four methods for brown bears," *Conservation Biology*, vol. 19, pp. 150–161

Fahlman, Å., Caulkett, N., Arnemo, J. M., Neuhaus, P., and Ruckstuhl, K. E. (2012) "Efficacy of a portable oxygen concentrator with pulsed delivery for treatment of hypoxemia during anesthesia of wildlife'" *Journal of Zoo and Wildlife Medicine*, vol. 43, pp. 67–76

Kindberg, J., Ericsson, G. and Swenson, J. E. (2009) "Monitoring rare or elusive large mammals using effort-corrected voluntary observers," *Biological Conservation*, vol. 142, pp. 159–165

Kindberg, J., Swenson, J. E., Ericsson, G., Bellemain, E., Miquel, C., and Taberlet, P. (2011) "Estimating population size and trends of the Swedish brown bear (*Ursus arctos*) population," *Wildlife Biology*, vol. 17, pp. 114–123

Moen, G. K., Støen, O.-G., Sahlén, V., and Swenson, J. E. (2012) "Behaviour of solitary adult Scandinavian brown bears (*Ursus arctos*) when approached by humans on foot," *PLoS One*, vol. 7, no 2, paper e31699

Rauset, G. R. (2013) "Life and death in wolverines, linking demography and habitat for conservation," PhD thesis, Swedish University of Agricultural Sciences, Uppsala, Sweden

Sahlén, V. (2013) "Encounters between brown bears and humans in Scandinavia—contributing factors, bear behavior and management perspectives," PhD thesis, Norwegian University of Life Sciences, Ås, Norway

# 16
# PLAGUE

*Bernard Quetchenbach*

Death is like the insect
Menacing the tree
Emily Dickinson ([1896] 2004)

We praise You, Lord, for all Your creatures
St. Francis of Assisi (*c.* 1225)

## In the Crazy Mountains

Its viability as a commercial center undermined by easy freeway access to cities—Billings to the east, Bozeman to the west—Big Timber, Montana, holds its own as Sweet Grass County seat and as an I-90 waystation for travelers bound for Yellowstone, the Boulder River, or the Crazy Mountains. Suspicious of public lands, the Big Timber City Council recently rejected a proposed riverside park for fear that the green space was part of a secret UN plan to eliminate private property. In truth, however, public land access brings a steady if modest trickle of dollars and a fighting chance against the windy extinction looming over settlements cast farther from mountains, out in the mostly privately owned or hard-to-access prairie. In much of Montana, public land is virtually synonymous with mountain forests. Campers and hikers who pass through Big Timber are likely on their way to national parks and forests where people of modest means share a legacy that would otherwise be the sole province of extractive industries, second-house millionaires, and the wealthiest ranchers.

An isolated "island range" shared by two national forests, the Crazy Mountains provide a rugged, photogenic backdrop for Big Timber's tiny retro downtown. For several years, my wife and I had wanted to explore those blue and white rock pyramids suspended serene and mysterious one hundred miles across the wheatfields from our Billings home. On a late July afternoon in 2009, after stopping for lunch at the Grand Hotel, a Big Timber landmark, we struck out through rolling rangeland, bound for an overnight camping adventure in the Crazies. At last, after forsaking the state highway for a jolting half hour on unpaved county and Forest Service roads, we slipped between the shaded walls of Big Timber Canyon, its dark coniferous woods broken by patches of orange-red needles flaring in afternoon sun.

Like anyone who has spent much time in Rocky Mountain forests during the last few decades, we knew what that red meant. Beetles of the genus *Dendroctonus*, Latin for "tree killer," burrow into the bark of large conifers and feed on the vital inner phloem layer. Mountain pine beetles are the most notorious of the group, but related bugs attack Douglas-firs, spruces, and other tree species. Generally present in western evergreens, bark beetle populations cycle between background levels and periodic irruption phases. The current outbreak, perhaps the longest, most complex, and most extensive regional insect epidemic on record, has left millions of acres of conifers in western North America "red and dead." During irruptions, beetles attack *en masse*, and mortality among the large, mature trees they favor can be close to 100%, especially if drought and a symbiont fungus accompanying the bugs have weakened conifer defenses. Around Big Timber Falls, just a short walk from our Gallatin National Forest campsite, we found an alarming number of trees reduced to red tatters.

Back in the campground, a mule deer grazed amid a scatter of healthy lodgepole pines, most of which were equipped with verbenone packets. According to a flyer taped to our site's picnic table, the jarringly unrustic plastic squares nailed to trunks were filled with "anti-aggregation hormones" formulated to confuse and discourage mountain pine beetles. Like hikers armed with bearspray, these trees stood ready to ward off attack. Verbenone makes an effective barrier for campgrounds and vista points, but outfitting an entire forest with packets would be both logistically demanding and prohibitively expensive. Though thousands of conifers in Colorado's high-profile Rocky Mountain National Park are being treated not with verbenone but with more heavy-handed chemical pesticides, landscape-level responses to beetle outbreaks are considered mostly impracticable. In the Crazies, as in most western forests, trees and beetles are basically left to work out their fates as they will.

The morning after our arrival, we set off on a longish day hike up to Twin Lakes, at the head of Big Timber Canyon. Confirming our apprehensions, the higher slopes, where whitebark and limber pines must have draped their greenery just a few years back, were as gray and devoid of foliage as a winter woodland in New England. Whitebark pine forests throughout the Rockies have been decimated by a combination of mountain pine beetles and the accidentally introduced blister rust; in the Crazies a localized needle blight has also affected these trees. Throughout the Northern Rockies, dedicated "whitebark warriors" have mobilized on the trees' behalf, searching for resistant trees (not very successfully in the case of the pine beetle) and petitioning for endangered species designation. But in Big Timber Canyon the bugs and blights plainly had the upper hand.

Not that the Crazies were completely deforested: Though the canyon's spruces and firs were beset by their own insect predators—especially spruce budworms and Douglas-fir beetles—these evergreens seemed to be holding their own. Nor were the mountains empty of animal life. Somewhere up in those still-snow-capped crags were mountain goats, a species introduced for big game hunters some decades back, now roaming the mountains with native deer, black bears, and smaller creatures. Nearer at hand, the sunny trailsides were blessed with an array of late season wildflowers and attendant butterflies. Of course, the beetles themselves are wildlife—teeming wildlife, in fact. That's easy to forget, especially in a fauna-rich place like Montana. Deer and elk, eagles and hummingbirds qualify as wildlife. Insects are, well, insects. Next time someone asks if you encountered any wild animals on a hike, try telling them you were surrounded by wildlife numbering in the millions—gnats, mosquitoes, budworms, engravers, mountain pine beetles. Odds are you will be met either with a look of sheer horror or a quizzical "I meant *animals*, not bugs."

## Agents of Change

It may be that the perceptual distinction between insects and other wild animals has eroded as our expectations and lifestyles have grown less compatible with wild creatures in general. When a kestrel flew into a New York City store a few years ago, customers cried out for someone to "get" even this lovely, common small falcon. In cities and suburban neighborhoods, lethal responses once limited to "vermin" are increasingly applied to merely inconvenient species like beavers and deer. Still, extreme remedies against such "nuisance" vertebrates are likely to be halting and controversial. Munching deer may be unwelcome backyard guests, but, dang, they're pretty, and we love to watch them—when they aren't crossing unlit city streets, that is. Insects are "creepy crawlies." We make some allowance for the colorful butterflies and, less enthusiastically, the necessary bees, but beetles and their ilk get routinely squashed in houses where residents would brandish towels and brooms to herd even a starling back outside. If our fellow animals are, as Henry Beston says, "other nations" (1928, 25), insects seem like *War of the Worlds* space aliens. Life cloaked in an exoskeleton strikes us as mechanical, a clicking dynamo, fascinating and beautiful if we stop to think about it, but not without a nightmare aspect. It *must* be wrong for such pests to destroy valued and beautiful forest trees.

Or is it? Though some tree-damaging bugs are introduced, tree killers like the mountain pine, Douglas-fir, and spruce beetles are native components of wooded western ecosystems, renewing landscapes by eliminating old growth and providing seedlings with open ground and access to sunlight. That's what they've done for millennia, sometimes on a grand scale. Analyses of environmental history indicate that insect outbreaks and wildfire have remade Northern Rocky Mountain forests every century or two, regardless, for the most part, of human forestry or fire-suppression policies.

Now, however, ample evidence shows that human-generated climate change has given the beetles a leg—or six—up on their hosts by perpetuating drought and by limiting early- and late-season cold snaps that can be fatal to overwintering beetles and larvae. Ever more sophisticated models show a warmer, drier climate emerging in the West, a long-term trend confirmed by enough observational data for anyone not deeply invested in climate change denial. Beetle destruction of timberline whitebark pine stands, almost completely dependent on frigid mountain weather to ward off the bugs, is an especially ominous signal that humans are fundamentally altering natural systems.

Even so, the beetles could be seen simply as adjusting the landscape to new circumstances, shaping more resistant mountain forests and clearing lower ridges for drought-tolerant grasses and sagebrush. At best, however, ecosystems take decades to rebound, longer where conditions are harsh. The high, steep, and thin-soiled Crazy Mountain slopes denuded by beetles may remain barren of both trees and tree killers for the rest of my life.

## Taking Sides

It's been a cold hard winter in Montana. Rumor has it that some beetles in some places may have been frozen out. Wishful thinking, perhaps. Tonight's lows in the Crazies will reach twenty below zero—probably not cold enough, almost certainly not for long enough. Overwintering beneath the bark, beetles can withstand temperatures down to minus forty, provided that the air warms up within a few days. They are more vulnerable in fall and spring, which is one reason why shorter winters are bad news for conifers. In the complicated, ever-shifting dynamic linking pine and beetle, extreme cold favors the trees. Heavy snowfall does, too, arming trees with ample summer sap to pitch the beetles out. Drought, on the other hand, drains the trees' resources. Moreover, longer summers have in some places accelerated beetle reproduction, enabling multiple broods per season.

Why is it that I can't help pulling for the trees? Plants are even more distant from us in life's evolutionary genealogy than insects—they don't have heads, after all, and they can't move on their own. Forest trees have a comforting anonymous dependability. We may identify with a backyard shade tree as an old friend, but when conifers close ranks in mountain woods, their communication silent and chemical, their *is-ness* ambient and pervasive, we see them collectively as a green blanket spread across the land. We speak of being *in* the forest, the same way that we say *in the mountains*. We fail to see the trees for the forest. But without them, there would be no forest.

Perhaps because we see plants as a different order of being altogether, we don't expect to connect with them anthropomorphically. A bark beetle, on the other hand, has legs, a head, eyes; in addition to flying, a beetle can walk and climb. Because they spend much of their lives inside trees, and because the dramatic red bloom doesn't mark their victims until after they've moved on, we rarely see them. Singly, they are unassuming creatures, diminutive in their distinctively curved shells, with legs and other appendages extending like the utility attachments of a Swiss army knife. A beetle in motion seems the perfect image of a clock-work device, a kind of miniature robot animal. Immersed as we are in a culture that subjects our mammalian modes of awareness to mind-numbing mechanized workplaces and hectic impersonal cities, it's hard not to reduce beetle life to an unpleasant caricature.

Vertebrates, especially mammals—*animals* in common parlance—inspire our highest degree of empathy; their facial features and evolutionary heritage make them appealingly familiar to our self-absorbed selves. Despite laying claim to a "higher" position, at a visceral level we respond to other mammals as beings essentially like us. Predators such as wolves are accused of "stealing," a tacit admission that we and they are close enough kin to be held to albeit a one-sided application of the same ethical codes; no one ever accuses beetles of stealing trees. Outlanders as opposed to outlaws, insects seem not like us at all. They have compound eyes, their "mouth parts" are sideways, they never change their minds. If we do see ourselves in them, it's generally in terms of an Orwellian dystopic image of industrial "anthills" and "hives," places where human identity has been all but lost. Insects, especially in communal "hatches" or "swarms," strike us as nature at its least conscious and most relentless. That sense of automatic pilot led nineteenth-century entomologist J. Henri Fabre to recoil in astonished horror when processionary caterpillars circled the rim of a vase for a week, unable to muster enough creativity or doubt to break the circuit (1916, 86). How can a living thing be so, well, determined?

Even more distressing, perhaps, is that the dignity of individual life seems so fleeting, perhaps unimportant, in the insect world. In "Six Significant Landscapes" humanist poet Wallace Stevens admits that "... I dislike / the way ants crawl / in and out of my shadow" (1990, 74). Seeing ourselves in insects or them in us rattles our sense of our own personal importance. While there's no real reason to suspect that beetle life must be empty of meaning, we can't read insects well enough to project *our* sense of meaning onto them.

## What's at Stake

Early next summer, overwintering bark beetles will emerge, looking for places that they haven't already used up. So will we, as mountain trails beckon. I'm fortunate to have recourse to a place that hasn't yet been exhausted by my kind. Everyone—two-legged, four-legged, six-legged—has a stake in the coniferous woods of the Northern Rockies. No one needs a vast, self-renewing forest more than the bark beetles. In Montana cities like Helena and Billings, beetles have invaded city parks and yards, but pickings are sparse in such environments, and park trees are likely to sprout verbenone packets as communities mobilize to protect green spaces. If only we could tell these tree killers how much they depend on intact, functioning forest habitat. Of course, dozens of beetle generations cross a single human lifespan;

if, like busy little industrialists, they eat their distant descendants out of house and home, humans are hardly in a position to call them on it.

The Crazy Mountains lie just across the Yellowstone River Valley from the Beartooth and Absaroka ranges, the northern ramparts of the Greater Yellowstone Ecosystem, a place where the Crazies' isolated significance is magnified to a planetary scale. Millions of people worldwide are as deeply invested in Yellowstone as Big Timber is in the Crazies, and not only in an economic sense. This fiercely defended ecosystem offers a tentative haven for endangered species, a rich trove of cultural history and national identity, cross-generational memories cherished by millions of families. Greater Yellowstone is one place that humans and a host of other creatures, beetles among them, cannot afford to lose.

Like the Crazies, the Greater Yellowstone Ecosystem is an island. The protected lands of the park and its buffer national forests constitute a big place, to be sure, but not compared to the network of towns, ranches, and highways that modern civilization has spun around them. An ocean of humanity chips away at Greater Yellowstone's edges as inexorably as a rising sea level encroaches on atoll shores. An environmentalist beetle would find ample justification to claim that we, not they, are the major destructive force: The wide-open spaces of the pre-settlement, pre-warming West could accommodate even major insect irruptions. The beetles are making their living as they have always done; if they have become a plague, it is the drier, more fragmented, environment—our environment—that has made them so.

But human culpability only bolsters the case for human intervention. Much of Greater Yellowstone, especially the park itself, is covered with nearly pure lodgepole pine forest, an inviting target for mountain pine beetles; area whitebarks, a key food source for grizzly bears and other animals, already face functional extinction due in part to pine beetle expansion into higher altitudes. Pine beetle congenator Douglas-fir and spruce beetles have also taken a toll on Greater Yellowstone forests. If human agency is shattering the equilibrium between beetles and trees, thus threatening to transform this World Heritage Site into a different place altogether, aren't we morally obligated to do something to stop the scourge we ourselves have freed from its natural restraints?

## At War in the Fields of the Lord

Along with rising sea levels, melting glaciers, increased drought, the nightmares associated with an open-ended warming trend include looming insect plagues—renegade tropical disease vectors, crop-destroying agricultural pests, forest-denuding bark beetles. We will have to decide what we should do, or try to do, about them. If we want to maintain any semblance of today's natural diversity, and if we want places like Yellowstone to retain as much as possible of their present character, a more interventionist approach than I, for one, prefer will probably be necessary. But, if intervene we must, we should do so judiciously, with an eye to the law of unintended consequences that got us into this fix to begin with. Even talking about indiscriminate "total war" against insect hordes should make us profoundly uncomfortable. We've been down that road before.

"The bugs were everywhere. Unseen. Unheard. Unbelievably universal. Beneath the ground, beneath the waters, on and in limbs and twigs and stalks, under rocks, inside trees and animals and other insects—and, yes, inside man" (quoted in Lear 1997, 431). Those terrifying words are from "The Desolate Year," a Monsanto parody of Rachel Carson's introduction to *Silent Spring*. In the early 1960s, industry shills tried to overwhelm Carson's eminently reasonable cautions regarding wholesale unrestricted pesticide use by ratcheting people's atavistic fear of insects into a desperation not far from sheer panic. The gendered ugliness that branded Carson an ignorant "spinster birdwatcher"—she was actually the most widely read and respected American science writer of her time—is ample evidence of the industry's ruthless willingness to sacrifice both honesty and basic decency in its single-minded pursuit of profits and power. Unmoved by their products' likely impacts on nature and human health, chemical firms parlayed American finickiness into a mania for dispersing minimally tested, potentially lethal, no-holds-barred "biocides" over towns, gardens, waterways, and agricultural fields. The results Carson uncovered included catastrophic fish kills, rapid decimation of raptor populations, and the persistence of pesticides in all of our bodies.

Immersed in the dazzle of scientific progress, post-World War II America proved fertile ground for anti-bug hysteria, insects serving as ubiquitous representatives of all things primitive, uncontrolled, and therefore dangerous. While horror movies like *Them*, *The Deadly Mantis*, and *The Beginning of the End* sicced nuclear-enhanced giant insects on unsuspecting towns, Green Revolution hubris spawned an arms race pitting "scientific" chemical agents against the evolutionary nimbleness of agricultural pests. Some of these same chemical insecticides made their way into American homes, as new highways, GI-bill college degrees, and Civil Rights-era backlash led the burgeoning middle class to flee "roach-infested" cities for squeaky-clean, racially homogeneous, insect-sterile suburbs. In cities, towns, and farms across the country, powerful chemicals were unleashed to keep the bugs at bay. And keep them at bay we must, the industry line went. The insect world—a soulless, mechanical horror only modern chemicals can stop—is chewing at the door. Not to worry, Raid Kills Bugs Dead!

Today's ongoing forest insect depredations are genuinely alarming, millions of acres in Canada and the western United States grayed by the mountain pine beetle alone. But the world we know is as much an insect world as the emerging one we fear. The pine beetle and its compatriots have helped shape today's mountain landscapes. Maybe the forest needs them: Lodgepole pine, the dominant component of mid-elevation montane forests in our region, is relatively short-lived and dependent on the occasional "stand-replacing" cataclysm. We should also keep in mind that nature is not without its own devices: Woodpeckers eat bark beetles, for example, and forest fires break up extensive swaths of even-age conifers, potentially inhibiting the forward sweep of beetle waves. Climate change is significantly disrupting ecosystems, but no one really knows how ecosystems will react. If we let our fears overwhelm our better judgment, we may actually make matters worse by unleashing forces more deadly than beetles or by succeeding too well and derailing nature's attempts to adapt to a changing climate. Either way, wholesale, expedient efforts to defeat the beetles may have consequences as far-reaching and unpredictable as the ones of which Carson warned.

At a fundamental level, the beetle epidemic challenges our understanding of the family of living things, the limits we place on the Ancient Mariner's dictum to "…loveth … / All things both great and small" (Coleridge 2009, 68). A guiding principle of post-Aldo Leopold wildlife ethics is that all wild creatures, not just obviously "useful" ones, have a right to survive in their natural habitats. Any definition of wildlife that doesn't include at least the native insects would be arbitrary and baseless. In extreme cases—smallpox comes to mind—we argue over the principle that all creatures have the right to survive, but as we face the complex decisions inherent in a future that seems both increasingly humanized and more likely to spin out of control, that principle seems as good a place to start as any.

## Down from the Mountains

Despite the brooding slopes shrouded with dead pines, those hours spent at Upper Twin Lake in a spectacular, still-green Crazy Mountain cirque centered our minds and, if you will, our souls on what it means to live on this most beautiful of planets. In a way, I suppose, the sense of impending threat brought us a little closer to understanding what life must be like for so many of today's wild creatures, huddled in corners mostly neglected, for now, by us. For that insight, we should perhaps be grateful to the tree killers. Like bugs searching for green in a sea of red and gray, we seek refuge in a damaged world even as we ourselves are the damage we fly from. Equally dependent on what we are inexorably chewing to tatters, perhaps beetles and humans aren't so different after all.

Reluctantly, we shouldered packs and turned toward camp. Big Timber Canyon blazed once more with late afternoon sun, the reddened walls recalling how the mountains must have looked in the months immediately after the beetles struck, before the needles fell. It will be longer than I have on this Earth—untold generations from a bark beetle's perspective—until the pines grow back; given the changing climate, there's no guarantee that they ever will. We can guess that the beetles are spared such an expansive view, that they go about their beetle business day by day and "… do not tax their lives with forethought / of grief" (Berry 1985, 69).

We like to think of ourselves, on the other hand, as beings gifted with something approaching cosmic awareness. In "Six Significant Landscapes," just before objecting to those shadow-crawling ants, Wallace Stevens ambitiously proclaims that "… I reach right up to the sun, / With my eye; / And I reach to the shore of the sea / With my ear" (1990, 74). Maybe so. But human routines, like those of other animals, are perhaps inevitably shaped by immediate concerns; if not exactly oblivious to the big picture, we are undeniably preoccupied.

Descending from the lakes, we could see ranchland lapsing, we knew, toward the quiet streets of Big Timber, where evening meals were no doubt being prepared. The here and now has its imperatives. Leopold would have us "think like a mountain" (1966, 140), adopting a perspective beyond the scope of our own lifetimes, beyond even those of the proverbial seventh generation. It's a noble idea, and we must do better by the future if our descendants are to have much of a present. Still, though we might learn to *think* like mountains, we aren't and can never *be* mountains. Held in a momentary, illusory eddy of time and space, our individual lives are small-scale affairs. That's one more thing we have in common—brother pine, sister beetle, and I.

## Discussion Questions

1 How do you respond to insects and other "creepy crawlie" animals? How does this compare with your response to vertebrate animals such as birds and mammals? If your responses are different, is this reasonable? Why or why not?

2 Quetchenbach writes, "we can't read insects well enough to project *our* sense of meaning onto them." Is it reasonable to use a collective "we" in this sentence? Why or why not? Do you feel you can "read" other animals? If so, what species and in what instances? How sure can individuals be about "reading" others correctly, whether other species or other people? What are the advantages and disadvantages of this way of relating to others? What is your "sense of meaning"? If you share space at home with another species, what do you suppose is their sense of meaning?

3 Quetchenbach writes, "Greater Yellowstone is one place that humans and a host of other creatures. … cannot afford to lose." Are there ecosystems that we can afford to lose? Is there some way humans might fairly assess which ecosystems are more valuable? Valuable to whom? Is the loss of ecosystems about us, or about something beyond us? Is there any ecosystem we can afford to lose?

4 Given the unprecedented and growing threat of climate change, what kind and how much intervention in natural systems is appropriate? If we intervene, what principles ought to guide intervention? Where do you stand?

5 Are beetles and pine trees a legitimate focus when we are the root cause of the problem? What should be done about humans, climate change, and our dangerous tendency to harm the environment?

6 How should we weigh the value of ecosystems as opposed to individual creatures who inhabit those ecosystems?

## Essay Questions

1 Some recent critics have branded Rachel Carson a dangerous extremist whose anti-pesticide activism resulted in the increased prevalence of malaria (and

other diseases). Is this a legitimate criticism? How does this view compare with criticisms directed toward her in 1962, when *Silent Spring* was published? What do you think of Rachel Carson and her work, and why?

2 What ecosystem do you know that has been affected by human beings, especially by climate change? What forces have altered this area, how has the ecosystem been affected/changed, and how do you think this area will be the same/different in 2050?

3 Quetchenbach suggests that "projecting our sense of meaning" onto animals may provide a measure of empathy, but are such judgments trustworthy, valid, or appropriate? Why or why not? How much can we know about how other animals understand and value their lives? Can we see their perspectives independent of our own values? Should we? Why or why not?

## Suggested Further Reading

Logan, Jesse A, William W. McFarlane, and Louisa Willcox. 2010. "Whitebark pine vulnerability to climate-driven mountain pine beetle disturbance in the Greater Yellowstone Ecosystem." *Ecological Applications* 20. 895–902.

Lear, Linda. 1997. *Rachel Carson: Witness for nature.* New York: Henry Holt and Company.

Nagi, Kelsi, and Phillip David Johnson, editors. 2013. *Trash animals: How we live with nature's filthy, feral, invasive, and unwanted species.* Minneapolis: University of Minnesota Press.

Nikiforuk, Andrew. 2011. *Empire of the beetle: How human folly and a tiny bug are killing North America's great forests.* Vancouver, British Columbia: Greystone Books.

Turner, Jack. 2008. *Travels in the Greater Yellowstone.* New York: Thomas Dunne Books.

Underhill, Linda. 2009. *The way of the woods.* Corvallis: Oregon State University Press.

## References

Berry, W. (1985) 'The peace of wild things', in *Collected Poems 1957–1982*, North Point Press, San Francisco, p. 69

Beston, H. (1928) *The Outermost House*, Viking Press, New York

Coleridge, S. T. (2009) 'The rime of the ancient mariner', in *The Major Works*. H. J. Jackson (ed.) Oxford University Press, Oxford, UK, pp. 48–68

Dickinson, E. (2004). Third Series (1896) Section IV: XIII. In *Project Gutenberg's Poems: Three Series, Complete*. W. Mabel Loomis Todd and T. W. Higginson (eds).

Fabre, J. H. (1916) *The Life of the Caterpillar*. Trans. A. Teixeira de Mattos. Dodd, Mead, and Company, New York, Google Books, Available from http://books.google.com/books?id=8r5RAAAAMAAJ&printsec=frontcover&source=gbs_ge_summary_r&cad=0#v=onepage&q&f=false, accessed 15 April 2014

Francis of Assisi (c1225). 'Canticle of Brother Sun.' *The Prayer Foundation*.

Lear, L. (1997) *Rachel Carson: Witness for Nature*. Henry Holt and Company, New York

Leopold, A. (1966) 'Thinking like a mountain,' in *A Sand County Almanac with Essays on Conservation from Round River*. Ballantine Books, New York, pp.137–141

Stevens, W. (1954) 'Six significant landscapes,' in *The Collected Poems of Wallace Stevens*. Alfred A, Knopf, New York, Reprint, Vintage Books, 1990, pp. 73–75

# 17

# THE MORNING OF THE WORLD

*Cara Chamberlain*

> Last night I was out with my buffalo brothers
> In songs and prayers,
> Where colors swirled and sparkled.
>
> Henry Real Bird, "Mass in Crow"

## If You Meet the Bison ...

Pay no attention to that bull striding the meadow above Lost Lake. Give him room. He's alone, happy, free—but he might, as the flyer you were given upon entering Yellowstone explains, instantly dash forward at more than thirty miles an hour to gore you. If he takes a notion to do so. For the most part he seems to be paying as much attention to you as you are to that green beetle on a blossoming hawkweed. Yes, he knows you're there because, as he walks by, he looks at you—skewers you, really—with one hefty eye beam. His snout is too broad to enable him to fix you with two. A good thing. A stare from just one bison eye will almost wither you to the size and shape of Uzzah after he touched the Ark of the Covenant. You will feel yourself in the presence of God nonetheless, a God who will flick his tail and bend to snatch another mouthful of grass before ambling up the hill. Normally he would not look at you. Today, you have almost—almost—made an impression.

So much has been written about the American bull bison that he resembles the long-suffering pachyderm in the story of the blind men and the elephant. One writer looks at the bison's personality, another at the satisfaction to be found in shooting him, another at the history of American Indian cultures that relied on his once-unfathomable numbers, and yet another at his surliness and inconvenience. A particularly ambitious writer might tackle the entire animal, one chapter at a time, lodging him safely within narratives, scientific facts, conservation mantras.

But what *is* a bison?

No one, apparently, can say. Or they say all too easily. He's an untamed member of the cattle clan, the raw material of all Angus and Hereford stock. If not their direct ancestor, he is at least a cautionary tale of what domestic cattle, left to their own devices, might become. Or he is what wildness *should* be, its grand embodiment. He is never a family man, and the wise matriarchal cow leading her sisters, cousins, and children to all that is good in their world is hardly ever the focus of Euro-American lore. Though prominent in Plains Indian religions, she is too gregarious and practical to suit white patriarchal symbolism and definitions that tangle in prejudice or predilection. But, male or female, bison resist human linguistic categorization. "Wild cattle" doesn't quite cut it. "Conservation icon" is too cold.

If you spend enough time with bison, you might end up like the biologist I met whose actions had come to resemble theirs, swinging her head when uncertain, moving otherwise deliberately and with purpose. Still, bison would not—and will not—give up their secrets to her or to any other researcher, even though they've been under scientific scrutiny for over a century. If it is so ordained, they will pack up their unspoken wisdom and dance—as they can when merry—into oblivion.

It's best, perhaps, to cajole bison with mythos, begging a sign. They are scapegoats, martyrs, emblems. If we cannot live with them in our midst, as it seems we cannot, then perhaps we are unworthy of them. Our way of life lacks the vision to bow to such deities, and, like so many others before, they have been shelved, a museum exhibit, holy relics that have lost their relevance.

## Bison and Brucellosis

During recent winters, as many as one-third of Yellowstone's bison have been killed when they stepped over the park boundary into Montana in search of food. Not even Wyoming, notoriously squeamish about wolves, has such a totalitarian reaction. Why the slaughter? The herd carries brucellosis and might transmit this bacterial disease to cattle. At least, that's the official rationale.

Brucellosis has largely been eradicated in the United States. The rare infected American, however, will experience those proverbial "flu-like" symptoms that most such diseases cause. According to an online CDC fact sheet, the mortality rate for humans is less than two percent. And how, the hypothetical fact sheet reader asks, may I avoid infection? The CDC sensibly recommends the following: "Do not consume unpasteurized milk, cheese, or ice cream while traveling. If you are not sure that the dairy product is pasteurized, don't eat it. Hunters and animal herdsmen should use rubber gloves when handling viscera of animals" ("Brucellosis" 2007).

I suppose I feel a bit flippant about this advice because I generally try to avoid "handling viscera of animals." I haven't been able to afford much travel out of the country, either. However, brucellosis is obviously not something to toy with considering such possible, if unlikely, sequelae as arthritis and inflammation of the heart muscle.

But never in my experience does talk of brucellosis revolve around human infection. The concern, really, is the possibility of a bison-to-cattle transfer. True, an infected cow will abort her fetus, which is bad enough. However, Montana's Yellowstone area ranchers who find an ailing cow in their herd must also render up any animals testing positive for brucellosis and, thus, potentially incur steep costs and lose much of their livelihood. It used to be that not just the affected rancher would suffer. The entire state would lose its brucellosis-free status and would exist in a kind of agribusiness purgatory until all danger of infection had passed. New federal rules loosen these stipulations. Still, discovery of brucellosis on a ranch can be financially disastrous.

This would all be very worrying and perhaps lead some politicians and rangeland managers to reason that bison are extremely dangerous and rightly deserve indefinite incarceration and, for the particularly obstreperous, capital punishment. However, more obvious carriers of brucellosis are rarely considered mere disease agents or thought to have overstepped their appropriate boundaries. On the contrary, these animals are encouraged to propagate just about everywhere. Hunters routinely handle their viscera. Unlike the homely, shiftless bison—welfare kings and queens of wild animals—the good, no nonsense, red-blooded American elk are treasured. There are elk protection societies, elk boosters, Elks Clubs. There are hunters who would kill a wolf that so much as sniffs at an elk track. A friend of mine once announced that bison are the ugliest, most unwholesome-looking animals in the West (excepting the wolf, perhaps) while elk are models of golden mean proportions.

I have nothing against elk. I rather like walking respectfully past them, the cows ushering young ones about or assiduously working their browse. I like the sound of bull elk bugling on a late September night. I like being in a meadow of elk thistle where recent elk hooves have tamped the elk bedding grounds. I like elk just fine. And, apparently, so do most people, excluding, perhaps, those few Eastern Kentuckians who are right now busy "controlling" the animals they helped reintroduce to their state because the ungrateful wretches have decided to eat zinnias and marigolds and pause dangerously on roadways.

But, if brucellosis is ruled out as the cause, why such ongoing and targeted persecution of bison? It can't be about economics, the competition for range, as has often been alleged. Or mostly not. In spring, mule deer, pronghorn, and elk feed on young grass shoots, including wheat. Yet no one speaks of confining these ungulates to a limited range. Ground squirrels and prairie dogs—though considered pests and damaging to grasslands and crops (and, therefore, subject to random violence delivered by shotgun, pistol, or poison)—hardly arouse the same kind of hatred. Perhaps it's not as romantic to imagine rodent pitted against alfalfa as it is to envision the clash of wild and tame cattle. From a mythic perspective, the vicious squirrel is about as frightening as the Monty Python rabbit, while the battle between Bison and Angus has the ring of Celtic legend. Perhaps workaday dryland farmers are less heroic figures than the storied cattlemen, so their rages are taken less seriously.

Two years ago, relief for bison finally seemed at hand. An article in *The Billings Gazette* announced what appeared to be good news. Former Governor Brian Schweitzer would prohibit the slaughter of bison in Montana, marking the end of that annual midwinter ritual in which bulls and cows who dared leave Yellowstone National Park, their last stronghold, in search of a less snow-covered square meal were rounded up, tested for brucellosis, and—often no matter the results of the test—sent away to be killed and butchered. Yet, reading more carefully below the headline "Schweitzer Blocks Yellowstone Bison Slaughter" (Brown 2011, B1), I found that the animals were so "disease-plagued" (B1) their mere presence—speeding through Montana in a livestock truck—was enough to spread brucellosis, perhaps by sorcery or ESP. The governor's action was a desperate gambit to force the federal government, lacking the stopgap of the slaughter, to do something, anything, to control a "diseased bison population that regularly spills out of the park and into Montana" (B1). No matter that, as the article sheepishly admitted in paragraph ten, "No bison-to-cattle transmissions have been recorded" (B1). Ever? Ever. Apparently. This tidbit after five rather serious-sounding references to the illness, which—not mentioned in the article—came to bison in the first place by way of domestic cattle.

So much for bison policy reform. The ban did not last, and last winter (2013–14), according to the Buffalo Field Campaign, almost 640 bison were killed by government agencies or by hunters. The worst slaughter in recent years, in the winter of 2007–8, was 1,631. To put these numbers in perspective, there are, according to the National Park Service's summer 2013 survey, approximately 4,600 bison in Yellowstone.

## An Environmentalist's Solution

A traditional environmental activist might argue that the annual slaughter of Yellowstone's bison is not only expensive but wasteful. Why just ignore these riches, this "treasure on the hoof"? Why not let bison "spill out," then, if they must, covering the prairies, reconditioning the grasslands, providing for the poor and downtrodden? Humans who consume flesh and use animal by-products could have a wealth of high-quality low-fat meat, good hides, fine horn. And all of us could have a restored prairie that would absorb carbon dioxide like no one's business. Ranchers could "harvest" animals who don't need growth hormones, coddling at calving time, or artificial feeding. Not to mention that some Plains Indian folks might want to rediscover their ancient relationship with bison.

Midwestern towns could build moats and ten-foot-high fences around their perimeters with ecotourism-ready towers from which Prius-driving bison fans could observe their favorites milling about—one herd taking two weeks to pass. Prairie Wal-Marts could sell free-trade bison binoculars and cuddly stuffed organic bison toys. Local McDonalds could offer McBison burgers and Happy Meals with bison products (all in recyclable packaging). Everyone would celebrate Bison Days with parades and festivals. It would be the kind of economic–ecological paradise environmentalists are often forced to promote. The cynical public, they know, is unmindful of billions of dollars in "services" nature provides every year and is always willing to toss even its own children to the Moloch of economic development. But if we'd just act sensibly, pragmatic environmentalists croon, not only would there be solid, long-lasting, well-paying jobs but also a healthy environment.

Only, when East Coast professors Frank and Deborah Popper suggested something similar a few years back, they were tarred and feathered and sent to bed without supper. Pretty much. There were death threats, to be sure, and their "Buffalo Commons" solution withered on the vine.

No doubt some environmentalists are guilty of treating bison in this manner, as a faceless "resource," but few midwesterners want to see these large animals return in any form. Those who do have a long acquaintance with beings they refer to as "benefactors." When the first bison came back to the Fort Peck Reservation in Montana after an absence of decades, "tribal elders were on hand to sing the old songs of welcome" (Chadwick 2011). Bruce Bauer, who "manages" the reservation herd, knows the animals by name. As Bauer says, "When I feel like taking a break from work, I just go sit with the buffalo. … They have a lot to teach us" (Chadwick 2011).

While the Assiniboine and Gros Ventre once viewed bison as honored instructors and kin (and many still do), whites apparently cannot tolerate the animals even as an amalgamated asset. Thus no matter how much the traditional environmentalist might advance an economically utopian outlook, she will be resisted. The activist will lose both her moral sense and her argument. The dominant culture is hostile to any conception—offensive or not—of free bison.

## Power Politics

What I really want to say is that hatred of bison runs deep in America. Maybe it's the history, the regret. White Americans, in particular, might feel guilty about how the destruction of wild bison herds was used as a genocidal tool. Guilty about the horror newcomers brought upon the indigenous peoples of this continent—and its nonhuman inhabitants—a reality that these decidedly uncuddly and uncooperative bison seem to remind us of at every turn. Some European-Americans don't want to remember this history. Some denigrate whoever pushes them toward awareness. Perhaps some are still racist enough to associate what they envision as "disease-plagued" human inhabitants of the prairie with the equally "disease-plagued" bison. Such emotional discomfort often creates bitterness and hatred.

Certainly, there have been many opportunities over the years to stop the bison's slide toward extinction. As the naturalist Archie Carr (1956) writes, their demise did not start in some remote corner of a dense rainforest or in the deeps of the ocean:

> The bison was in the public eye from the start. It cluttered land now Illinois real estate. It gave comfort to difficult red Indians [*sic*] and blocked the scant traffic on proud new railroads. The bison passed in a blaze, watched by

> everybody—not without lamentation here and there, but with little interference. It had to go, in the mind of the day, because it hindered progress.
>
> (238)

No, the 19th century death of the bison was a public spectacle often cheered by the weak and promulgated by the powerful. And the twenty-first century persecution is no different.

Perhaps like any big bully, the odd person here and there would like to avoid encountering an even bigger kid on the block, someone who outclasses and confounds him at the same time. A former seasonal worker at Yellowstone told me of one bison who hung around the Old Faithful Inn, enduring tourist taunts for hours until calmly walking right over an innocent bystander.

Maybe it's all a matter of power—politics in the rawest sense of the word. It's either us or them. As Montana State Senator John Brenden says in support of legislation to keep bison confined, "If we start spreading and disseminating buffalo all over the state of Montana, I think that creates a bigger problem … you're just opening up yourself for more liability" ("Bills" 2011, 3C). According to the *Billings Gazette,* Senator Brenden has thrown up his hands when it comes to the wild bovines, confiding "that there is no room for free-roaming bison in a modern society" (3C). And many residents of Gardiner, Montana, apparently feel the same way. In an experimental program, the federal government has responded to former Governor Schweitzer's challenge and "allowed," with decidedly mixed results, a few bison to migrate out of the park for the past few winters, pass through town, and dine on the richer forage to be found in Paradise Valley. Some folks complain of the nuisance, fear, and economic pain this loosening of the prison bars has caused. So far, however, only bison have died as a result.

Still, Montanans should rest assured. "Modern society" will undoubtedly trump all. I'm not sure what "modern society" is, though it sounds bad. It must have something to do with cars and freeways and agribusiness and hamburgers, and little to do with freedom and life. "Modern society" might be the pinnacle of human creation. Whatever it is, it lacks free-ranging bison. That much I understand. It certainly makes sense that the people of Gardiner are a bit alarmed about bison wandering past school bus stops, but I think a lot of less involved observers just get a lift knowing that such a mighty animal as the bison can be herded, corralled, buffeted, chafed, tortured, jailed, or slaughtered—all at the whim of a scrawny, naked primate. How far we've come! How modern we are!

On the other hand, it probably gives National Park rangers heartburn to know that certain scrawny primates insist on the flexing of modern society muscle. A ranger might try during the July rut to haze lovesick bison off the road with a truck that broadcasts noises meant to annoy them. How these particular scritchy sounds were decided upon, I don't know. Anyway, it doesn't seem to help. A few individuals skip out of the way and then, when the ranger drives on, cavort right back onto the pavement. It is a strange thing to witness—romantic bison losing all caution and gravity. It is also a wonderful sight … but it does back up traffic.

My husband and I once pulled off the road in Hayden Valley for a bison jam. I had plenty of time to photograph moths and beetles. I even found a boreal chorus frog—or, rather, she found me. When I looked up from the meadow, two hours had passed, and the traffic was still thick, barely moving.

I like it that love can stop 500 cars in their tracks. Watching a bison chase his sweetheart down a hill, through a meadow, across the Yellowstone River, and up the other side, I realize why Zeus appeared to Europa as a bull. I have also watched parted lovers reunite, racing—tails high with excitement—toward each other like the people in those old hair-coloring ads: "The closer he gets, the better you look." *Aww,* I wanted to gush before remembering that the loving bison could demolish my car.

How wonderful, then, it must make some feel to control and destroy such force.

## Taking the Bison God Out of the Museum

I love bison. Many people do. You don't have to be a buffalo-fancying billionaire like Ted Turner or a member of the richly endowed American Prairie Foundation to love bison. Ordinary folks love bison. Ordinary folks go out of their way to see bison. Even some ordinary folks in Gardiner, Montana, love bison. Hatred and exasperation aren't the only possible responses. After all, bison are bison. They're huge, noisy, regardless, dignified, self-absorbed, shaggy, humped, switch-tailed, rugged. They can survive temperatures between –50 and +120 degrees Fahrenheit. They once ruled the continent from Pennsylvania to Colorado, from boreal forest to Chihuahuan Desert. They moan and snort softly to each other as they wallow and graze. Their calves are golden and full of sudden mischievous notions. They are socially sensitive and diplomatic and responsible. Two adventurous bulls once reportedly spent a year on an island in Yellowstone Lake, and, when they didn't like that arrangement anymore, one chose to walk home over the ice and one chose to stay, voluntarily marooned, for another year (Captain Lindsey 2009). A group of cows once tried to spend winter outside the park near the North Fork of the Shoshone River in Wyoming but apparently vowed never to return (and never did) when some of their youngsters, finding the winter too hard, perished (Meagher 1993). Yes, a few unattached bulls still vacation there, but never a mom with young.

So imagine: You—a bison aficionado—decide that it is not your job to organize or define the world. You let each bison decide what she or he is going to do and who she or he is going to be. You avoid unpasteurized milk and cheese and the "handling of animal viscera." Your sternum relaxes. Your breath comes easy. You do not see malevolence or Senator Brenden's liabilities in wildness. It is powerful and might kill you, but it is a part of your life. Maybe you get smarter. Maybe you learn to watch creation with as much fussy discernment as the artists of Chauvet Cave, who, somewhere around 35,000 years ago, painted a few tiny mystical human figures and hundreds of expressive giant animals, including bison, in all possible poses. At any rate, you follow the Lost Lake trail through golden grasses, leaving the road far behind. It's been raining and the clouds are low. Up ahead is a bull bison, a garland of mist caught in his horns. You might be witnessing the morning of the world. You do not venture close, and this time he does not look up. He keeps grazing.

## Discussion Questions

1 Do you think people hate bison because they feel guilty about past (and present) indiscretions? Reflecting back on the previous essay on hunting in Chapter 6, what explanations might you offer as to why humans shoot bison?

2 Reflecting on Chamberlain's poem at the beginning of this section, "Laridae,"and on her bison essay, think about species that you favor. Why do you favor this particular species? Are there certain species that you strongly dislike? Do your likes and dislikes affect your actions toward individuals within these particular species? Would you accept a similar prejudice with regard to race?

3 Many authors assert that the Yellowstone bison herd is "genetically pure," that these animals, unlike most herds throughout America, have not been interbred with domestic cattle. How is this attitude toward bison speciesist? How is it harmful? Can you think of an example of how this same attitude, directed at domestic dogs or horses, harms these species? What about chickens and pigs? Can you think of any other examples of speciesism with regard to genetic purity?

4 Can racism be transferred to other species? How do essays in the first section of this anthology explain the forces behind this phenomenon?

5 Can sexism be transferred to other species? How is this evident in animal "husbandry"?

6 Must we love bison in order to leave them alone? What is the rightful role of human emotions in the protection of individual animals, species, and the environment?

7 Chamberlain refers to a cynical public. In what ways does the public tend to be cynical with regard to environmental problems? What can we do to change this tendency?

## Essay Questions

1 Greater and greater numbers of human beings compete with fewer and fewer other species for necessities such as land and freshwater. If you had absolute power, how would you solve this problem? Would you be happy if someone else had absolute power and enacted the same plan? Explain and justify your answers.

2 Propose a solution to the bison situation in and around Yellowstone National Park, and explain why you judge this to be the best possible solution.

3 Is it justifiable to be concerned with genetic purity in other species, but never in human beings? Why or why not?

4 Write a creative essay about an animal with whom you have interacted. How did this encounter affect or change you? How might you have affected the nonhuman animal that you interacted with? How would his or her life have been different (better and/or worse) with the complete absence of human beings?

## Suggested Further Reading

Carter, D. (2007) "Lord of the buffalo," from *Book of Sparrows*; Tracy Grammer (vocal soloist), Tracy Grammer Music

Dary, D. (1989) *The buffalo book: The full saga of the American animals*, Swallow Press, Athens, Ohio

Matthews, A. (1992) *Where the buffalo roam: The storm over the revolutionary plan to restore America's Great Plains,* Grove Press, New York

Rinella, S. (2009) *American buffalo: In search of a lost icon*, Spiegel and Grau, New York

Rosen, R.D. (2008) *A buffalo in the house: The extraordinary story of Charlie and his family*, Random House, New York

Sandburg, C. "Buffalo dusk," *Poetry Foundation* (2014) www.poetryfoundation.org/poem/238488 (includes sound recording), accessed 10 April 2014

## References

"Bills restricting bison move forward" (2011) *Billings Gazette*, 27 February, p. C3

Brown, M. (2011) "Schweitzer blocks Yellowstone buffalo slaughter," *Billings Gazette*, 16 February, p. B1, B4

"Brucellosis," Center for Disease Control and Prevention (2007), accessed 16 February 2011

Captain Lindsey (2009) Interpretive speech, Xanterra, *The Lake Queen II* tour on Yellowstone Lake, 14 August

Carr, A. (1956) *The windward road: Adventures of a naturalist on remote Caribbean shores*, Knopf, New York

Chadwick, D. (2011) "Where the buffalo now roam," *Intertribal Bison Council*, www.itbcbuffalo.com/node/47, accessed 18 April, 2014

Meagher, Mary (1993) *Buffalo Bill Historical Center Seminar*, Cody, Wyoming

Real Bird, Henry (2010) 'Mass in Crow', in *Horse Tracks*, Lost Horse Press, Sandpoint, Idaho

# PART II

# Politics, Organized Activism, and Personal Encounters

# SECTION I

# Foundations

## Community and Politics

**Poem:** Lisa Kemmerer

### Breathtaking Bit of Being

Wriggling reptile fast disappeared
under our green and going truck,
                rushing resolutely.

Quick to come back,
I hunted down those hastening hopes
                glimpsed in go-by
and beheld that bantam being
surging with such certainty
                on what must be a terminal trail.

She wore black lattice on dandelion yellow—
                colors of caution.
But there was no time to go for gloves—
                with tail-powered torque,
                skin wrinkling with each wiggly weave,
                her textured toes took to their heels,
                     paddling a pavement of ruin.

Kneeling, I pinched her pudgy belly gently,
lifting her between tentative tips,
                marveling at her frosty, fragile core.

With mini-migrant safely suspended,
I explored her vaulted tail, persistently pacing props,
and primordial pate with eyes bulging like new buds—
                suited for wary-watch on water
                with legs lax, tail trailing.

Why had this precious bit of being
passed under our rapidly rolling truck?
                    What did those little looking lenses know?

I toted my pinched yet paddling prisoner to a picturesque pond
to tuck that tiny traveler beneath a leaf
for careful keeping

and only later realized—
                    she had not thrust tacky toes over that rough roadway
                    to bring *me* much-needed knowledge—
she was quite taken with her own time.

It was I who happened by for her.

# 18

# ENFORCING HUMAN RIGHTS FOR PEOPLE, ANIMALS, AND THE PLANET

*Debra Erenberg*

In April 2008, I had the opportunity to participate in a fact-finding mission to Sarawak, Malaysia's largest state, on the island of Borneo. At the time, I worked for Rainforest Action Network, and I was excited to finally visit a real rainforest and, in particular, to see free-living orangutans (for whom I had developed a strong affinity as a board member for the Great Ape Project). Seeing my excitement, a colleague laughed at my naiveté, explaining that most of the rainforest in Sarawak had been logged or burned by the pulp and paper industry or palm oil producers. The only orangutans I was likely to see, she added, now relied on a wildlife rehabilitation center for food and medical care.

Though disappointed, I shouldn't have been surprised. After all, I knew that logging in Malaysia and Indonesia continued at an astonishing pace. Rainforest destruction in these two countries remains a major driver of climate change—not only because rainforests function as the lungs of the planet (breathing in carbon dioxide and exhaling oxygen), but also because burning centuries-old forests releases tons of carbon that is stored in trees and in peat (partially decayed vegetation matter) on the forest floor. A 2010 report suggests that 85 percent of Indonesia's greenhouse gas (GHG) emissions stem from land use activities; with 37 percent due to deforestation and 27 percent due to peat fires (National Council 2010).

As I learned on my trip, the devastation of Malaysian forests is just one affront among many in which profit-seeking corporations, often with government complicity, act against the best interests of every living being and the planet. Again and again, people and other animals find themselves brushed aside without a second thought, as the land they rely on is laid waste to profit the wealthy.

One reason we find ourselves in this situation is that contemporary societies have created a corporate structure designed to ignore long-term public interests.[1] The task of corporate executives is to deliver a profit to shareholders; anything that diverts profits is a breach of a CEO's fiduciary duty. This arrangement creates incentives to destroy long-term collective assets for short-term private gains. In short, it leads corporations to choose profit over the well-being of present and future generations and of the planet we rely on. This perspective—and the desire of most governments to create a "business-friendly environment"—allows industries to exploit workers, nonhuman animals, and the planet's resources and to endanger community health with near impunity.

In this situation, the exploitation of humans, other animals, and the environment has become so intertwined that successfully addressing any one of these issues necessarily leads to significant improvements in the other areas. For example, if nation-states took their obligations to protect and enforce internationally recognized human rights seriously, not only would humans be protected from mistreatment by corporations, but as a direct result, other animals and the environment would also benefit. This calculation is well illustrated by factory farming. Whether their "product" is meat, dairy, or eggs, factory farms are built on exploitation and cause a tremendous amount of suffering and devastation, including—but extending far beyond—the animals who live and die on their premises. Animal agriculture brings in profits at the expense of people, farmed animals, and the planet.

## Going Whole Hog for Exploitation: Dismembering Pigs, Harming People

To keep the cost of meat, dairy, and eggs low, animal agriculture industries pressure suppliers to drive down labor costs while increasing productivity. This pressure can be seen in the quality of the lives of exploited animals, in slaughterhouses, and in processing plants where their bodies are disassembled. After World War II, most meat-packing jobs were unionized, paying a decent, middle class wage. But over the past few decades, the industry has consolidated into a smaller number of major producers that have relocated to so-called "Right to Work" states in order to avoid unions. Most workers in the meat-packing industry are now low-paid immigrants (many undocumented) who, because of their economic desperation, lack of English language skills and insecure legal status, accept dangerous work conditions for low wages with few benefits. A 2012 study by the Midwest Coalition for Human Rights noting these issues found that undocumented immigrants were frequently insulted by supervisors and singled out for the worst jobs because they were unlikely to complain (Midwest Coalition 2012).

For example, between 1989 and 2006, Quality Pork Processors in Minnesota sped up its kill line by 50 percent, while increasing the workforce only minimally. Subsequently, workers at the "head table" (who blast the brains and meat from pigs' skulls), developed maladies beyond the usual repetitive stress injuries: They reported severe flu-like symptoms, numbness, and intense skin pain. After much study, doctors theorized that exposure to pig brains caused an autoimmune disorder—progressive inflammatory neuropathy (PIN). Quality Pork responded by firing six PIN-afflicted workers who had filed worker's compensation claims, ostensibly for using forged or stolen identities (which the company likely helped them obtain). A dozen other workers with this debilitating ailment accepted a settlement of a mere $12,500 each (Genoways 2011). In exchange, they gave up their jobs, their medical coverage, and their right to sue the company. Apparently, the company made it clear that if these employees didn't accept the settlement offered, they would receive nothing at all. When the USDA Office of the Inspector General inspected food safety practices related to line speed at the plant, it left out any mention of these health concerns, determining that they were "not food related" and therefore out of the USDA's purview (Genoways 2013).

## A Pig is a Plant is a Person: Environmental Racism

Today, four giant companies—Tyson, Cargill, JBS (makers of Swift Premium brand pork products), and Smithfield—process more than half of the beef, chicken, and pork consumed in the U.S. These four corporations control more than 90 percent of the beef industry and two-thirds of the pork industry (*Brattleboro Reformer* 2014). The few remaining small farmers have basically become serfs: They sell their produce to these giant companies at whatever price is offered. Consequently, in 2012 the average farm worker earned just $18,910 (U.S. Bureau of Labor Statistics 2014).

There would be no meat-packing industry to speak of if not for factory farms, which provide a constant influx of bodies. Among the most insidious factory farms are the massive hog operations that have proliferated throughout rural North Carolina. Factory farmed hogs live in metal confinement barns, similar to a solitary confinement cell. Standing in their cells, the state's 10 million hogs produce 40 million gallons (more than 151 million liters) of waste *every day*, which "farmers" store in giant lagoons up to 14 feet (4 meters) deep to use as fertilizer (Butzer and Hildyard 2014). These pits are an ecological nightmare and a plague to the health of the neighborhood. Because of the stench, nearby residents have increased rates of asthma and other respiratory problems. Studies have also found higher rates of psychological stress and significantly more tension, depression, anger and fatigue among people who live near hog factory farms, called hog factories (Nicole 2013). In communities where most people get drinking water from shallow wells, hog manure lagoons sometimes contaminate groundwater. Spills from lagoons run into waterways, causing algae blooms that consume oxygen in the water, suffocating fish and other marine life (Food and Water Watch 2010).

Not surprisingly, hog factories proliferate in communities where people lack money and political power. In North Carolina, many of these factories are located in an area where African Americans make up 90–98 percent of the population, where a quarter of the people live in poverty, and where almost half have not completed high school (Statter 2004). There's a name for this: Environmental Racism.

To add insult to injury, after almost 600 people filed lawsuits against North Carolina hog farms, the North Carolina General Assembly passed a bill (last year) changing the way farm nuisance complaints are handled, requiring complainants who lose in court to pay the hog farmers' legal defense costs (Neff 2013). The bill closely resembles model legislation from the American Legislative Exchange Council (ALEC), a group that connects state lawmakers and business interests (Institute for Southern Studies 2014).

## Back to the Rainforest: Palm Oil

The Pesticide Action Network fact-finding mission that I joined in 2008 regarding palm oil industries in Malaysian Borneo was co-sponsored by social justice organizations devoted to indigenous people's rights, food sovereignty, and the environment, indicating just how many rights were being violated by these palm oil plantations. Most U.S. palm oil comes from Malaysia, but the U.S. comprises only a small fraction of the overall global market for palm oil. The largest markets for Malaysian palm oil are India, China, the European Union, Pakistan, India and the United States (May 2012). However, since Malaysia supplies 90 percent of U.S. palm oil, and since palm oil is present in 50 percent of supermarket products in the U.S. (from soaps to cosmetics to junk food), I was about to witness the social and environmental cost of our insatiable appetite as consumers, and of increased demand for palm oil biofuel around the globe—especially in Europe.

Industrial-scale biofuels derived from food crops are distinct from community-based biofuel initiatives, like bio-diesel, in which fuel is derived from the conversion of waste products, such as used cooking oil. While the former contributes to deforestation, environmental degradation, and increased world hunger; biofuels created from waste do not. Resulting food shortages and competition for scarce land and water disproportionately affect the world's most vulnerable people.

What I witnessed in Sarawak is representative of agribusiness expansion across Southeast Asia, as well as in South America and, increasingly, Africa. Of Sarawak's 2.2 million inhabitants, roughly half belong to indigenous groups collectively known as the Dayak. The Dayak have lived in the area for centuries, but in order to gain legal rights to their land, Dayak communities must prove they have used a particular area continuously since 1958. This is difficult to do because traditional, sustainable farming practices require that farmers periodically leave fields fallow, and by definition, fallow fields are not continuously in use. Rubber and fruit trees, in contrast, are proof of continuous use, but the existence of rubber and fruit trees can be difficult to establish. In one Dayak community, for example, the state surveyed land only by helicopter. Since rubber and fruit trees are hidden beneath a taller rainforest canopy, the state concluded from these photos that the community had not continuously used the land, and therefore did not have a legal right to the area they had occupied for generations. Villagers in other communities told me that agribusinesses received a "Provisional Lease" from the state to use their lands, and that the newcomers bulldozed rubber and fruit trees, thereby eliminating any proof of continuous indigenous land use. Sometimes the villages themselves were bulldozed off the land: In one case, six houses were bulldozed while villagers were out working fields.

Near Bintulu, in Malaysia, an oil palm company, Golden Hope, offered villagers 30 percent of the company's profits if the community allowed Golden Hope to bulldoze their land and plant oil palms. The community had no legal representation and could not read the English-language contract. Trusting the company's assurances, community representatives provided their thumbprints to sign the written agreement, which actually contained very different terms than the company described. In fact, the contract contained an "admission" that the villagers were squatting on company land and stipulated that they would therefore dismantle their homes and leave the area. Needless to say, they haven't received the promised profits.

In the Kampung Wawasan area of Sarawak, an oil palm company repeatedly sent thugs to pressure villagers into selling their land. Although the villagers filed more than 20 complaints with law enforcement, the police never came to protect the indigenous community. Instead, police arrested a village leader for possessing a pistol without a license—a charge community members say is false (but which carries a heavy penalty in Malaysia). After the arrest, police asked other villagers to come in for questioning, then arrested them on undisclosed charges.

Such human rights abuses are common in Malaysia, as well as in other countries subject to similar land use pressures. The state of Sarawak has put all of its eggs into one monoculture basket, betting that consumer hunger for junk food and biofuels will bring wealth (at least for government officials), while disregarding the rights of those within their own country who stand to lose everything in the process. As 173 land rights cases drag on in Sarawak courts, year after year, corporations continue to clear-cut rainforests and bulldoze homes, unjustly displacing indigenous people along with orangutans and tigers, replacing forest homes with oil palm plantations that are constantly doused with pesticides that pollute water and kill aquatic animals (Aiken and Leigh 2011). Surviving fish deliver pesticide poisons to birds, humans, and others along the rainforest food chain.

As with U.S. hog farmers who exploit and mistreat hogs, employees, and the planet in search of the greenback, Malaysian palm oil operators trample anyone in their way to maximize profits. Rather than offer a living wage to the Dayak people, whose land they have stolen, many palm plantations import workers from Indonesia, who will accept even lower wages and harsher working conditions. As with U.S. agribusiness, many of these imported laborers lack legal documentation to work in Malaysia, and they are therefore unable to complain about unsafe working conditions (such as the prevalence of dangerous pesticides), wage reductions, or even the complete absence of promised wages.

Because much of Malaysia's palm oil travels to China and elsewhere across Asia, Western forms of consumer activism (such as boycotts and petitions) are not sufficient to protect the Malaysian rainforests or the many individuals who live therein. Nonetheless, international pressure is starting to affect industry practices. The largest single consumer of Malaysian palm oil, Anglo–Dutch Unilever, has announced efforts to "clean up." Unilever now tracks palm oil sources in order to avoid buying from companies linked with deforestation and human rights abuses.

Despite this much-praised commitment, Unilever continues to source palm oil from Wilmar International, even though this company was accused of illegally clearing forest and violently driving people off their land in Indonesia in 2011 (Oxfam 2013).

## Using Human Rights to Protect Animal Rights and the Environment

If we are to stop corporations from exploiting the vulnerable—whether humans, animals, or the earth, we must enforce internationally recognized human rights. These agreements, and institutions developed for their enforcement, make a clear case for industry-wide change to protect ecosystems, species, and communities (both human and nonhuman).

In 1948, the General Assembly of the United Nations adopted the Universal Declaration of Human Rights (UDHR). Under the UDHR (and subsequent human rights agreements), states have an obligation to protect individuals and groups within their borders who suffer from human rights abuses. Where domestic legal proceedings fail to address human rights abuses, individuals and groups can bring regional and international complaints to help ensure that international human rights are respected, implemented, and enforced.

What most people don't realize is that each state's obligation to protect human rights reaches beyond basic civil rights (such as the right to equal treatment under law) and beyond simply refraining from overt harms such as torture and arbitrary imprisonment. The UDHR protects economic, social, and cultural rights that are routinely violated by industrial agribusinesses—including animal agriculture and palm oil plantations. Note these five articles from the Universal Declaration of Human Rights:

> *Article 12.* No one shall be subjected to arbitrary interference with his privacy, family, home or correspondence, nor to attacks upon his honour and reputation. Everyone has the right to the protection of the law against such interference or attacks.
>
> *Article 17.* (1) Everyone has the right to own property alone as well as in association with others. (2) No one shall be arbitrarily deprived of his property.
>
> *Article 22.* Everyone, as a member of society, has the right to … the economic, social and cultural rights indispensable for his dignity and the free development of his personality.
>
> *Article 23.* (3) Everyone who works has the right to just and favourable remuneration ensuring for himself and his family an existence worthy of human dignity, and supplemented, if necessary, by other means of social protection. (4) Everyone has the right to form and to join trade unions for the protection of his interests.

> *Article 25.* Everyone has the right to a standard of living adequate for the health and well-being of himself and of his family, including food, clothing, housing and medical care and necessary social services. …
>
> (UDHR 1948)

As demonstrated by cases described in this essay—undocumented meat packers, rural North Carolina community members near hog factories, and the Dayak people of Sarawak—corporate interests routinely and flagrantly violate international human rights. States are obligated to hold corporations that operate within their boundaries (or are chartered therein) accountable for UDHR violations, as well as violations to International Labour Organization conventions, conventions on the rights of indigenous peoples, and other international agreements. They are further obligated to ensure that violators provide just compensation to victims.

And states aren't the only entities obligated to honor international human rights laws. In 2011, the United Nations endorsed a detailed set of principles designed to assure that corporations respect human rights, and to compel them to take steps to address and mitigate any adverse effects of their "business or relationships," extending down the supply chain. While these important new measures are not legally binding, such UN Guiding Principles merely elaborate what is already required by existing standards and practices for states and businesses in international and domestic law.

Under this international framework, factory farming industries must be held accountable for harming the health and well-being of surrounding communities. Meat packers must respect a worker's right to unionize, to receive a fair wage, and to safe working conditions. International agribusinesses must be called to answer—in both their home countries and in the countries in which they operate—for harassing, intimidating, and deceiving villagers to drive them from their ancestral lands, not to mention destroying water supplies, homes, endangered species, and food crops. By upholding these human rights, we will create immediate and substantial gains for other animals and the planet.

## Conclusion

During my fact-finding mission in Malaysia, I felt privileged that communities entrusted their stories to me, but I also felt powerless to bring change. My experiences as an activist at home simply did not apply. Unlike the U.S. government, the Malaysian government is not likely to be shamed into establishing reforms. Despite outward appearances to the contrary, Malaysia does not have a functional democracy or a free press, so pressure tactics rarely have any effect. What role do domestic consumer protests play in a market that is largely driven by overseas buyers, such as China and India?

But we *do* have the power to bring change for the earth and all living beings against an international market, though not necessarily as consumers. Any one of us can bring a case to regional and international courts against those who violate *human* rights. During periodic reviews of a nation's human rights record, individuals and groups can call attention to reported abuses that have been not yet been rectified by filing "shadow reports." If we—all of us who know of these abuses and of what the UDHR requires—hold states and corporations accountable, and collectively apply pressure as needed, human rights will be honored. Humans won't be the only beneficiaries. Factory farms, palm oil, and other exploitive industries rely on governments and citizens looking the other way while they violate human rights, destroy the lives of animals, and despoil the environment. Human rights provide a powerful tool to hold these industries accountable and to protect people, other animals, and the planet.

## Discussion Questions

1 What is environmental racism? What sort of laws need to be put into place against environmental racism?

2 What are the main industries located near your university? How do these compare with industries in your home town?

3 Where does American beef come from and why does this matter?

4 What is the impact of agribusiness in local communities in Southeast Asia, South America, and Africa?

5 Who are the Dayak and why are they important to our diet?

6 Who are the most disadvantaged people in your community? Why are they disadvantaged (racism, speciesism, classism, etc.)?

7 Is Erenberg simply refocusing on human beings, or does she offer a legitimate means of helping nonhuman animals and the environment through human rights?

8 Is it acceptable to use human rights on behalf of earth and animals? Why or why not?

## Essay Questions

1 What is your responsibility to the Dayak people, if any? What commitments must you make to honor this responsibility, or why do you feel you have no responsibility to the Dayak people?

2 Read the last chapter of *Slaughterhouse* (Gail Eisnitz), and "A Social History of the Slaughterhouse: From inception to contemporary implications" from the suggested further reading list. How would you explain the connection between human rights and animal rights? What examples would you use from this anthology more generally?

3 Read *Eternal Treblinka* (Charles Patterson) and *The Dreaded Comparison* (Marjorie Spiegel). Explain and explore the connection between racism and speciesism.

4 Research, locate, and report on contemporary examples of environmental racism in your region.

## Suggested Further Reading

Eisnitz, G.A. (1997). *Slaughterhouse*. New York: Prometheus.

Fitzgerald, A.J. (2010). A Social History of the Slaughterhouse: From inception to contemporary implications. *Human Ecology Review*, 17(1): 58–69.

Leonard, C. (2014). *The Meat Racket: The Secret Takeover of America's Food Business*. Simon and Schuster.

Longest, R. and Dove, R. (2014). The prolonged stench of inaction on NC hog farms. *Raleigh News and Observer*. February 24.

National Human Rights Commission of Malaysia (Suhakam) (2013). *Report of the National Inquiry into the Land Rights of Indigenous Peoples.*

Nebraska Appleseed (2014). Meatpacking. http://neappleseed.org/blog/tag/meatpacking. Last accessed April 24, 2014.

*North Carolina in the Global Economy: Hog Farming* (2007). http://www.soc.duke.edu/NC_GlobalEconomy/hog/overview.shtml. Accessed August 23, 2011.

Parker, D. (2013). Palm oil companies ignoring community rights, new study shows. *Mongabay*, November 7.

Philpott, T. (2011). How the meat industry turned abuse into a business model. *Grist*. June 29.

Rainforest Action Network (2014). Report: *Conflict Palm Oil in Practice*. April 2.

Taylor, Dorceta (2014). *Toxic Communities: Environmental Racism, Industrial Pollution, and Residential Mobility*. NYU Press, New York.

United Nations Office of the High Commissioner on Human Rights (2012). *The Corporate Responsibility to Protect Human Rights: An Interpretive Guide.*

## Notes

Views expressed in this article do not necessarily reflect those of Justice at Stake, its board, or partners.

1 Contrast this with the growing movement towards "B Corporations" that combine elements of traditional corporations with a commitment to broader social goals (Sharp 2014).

## References

Aiken, S.R. and Leigh, C.H (2011). In the Way of Development: Indigenous Land – Rights Issues in Malaysia. *The Geographical Review*. October.

*Brattleboro Reformer* (2014). Our Opinion: "The Meat Racket," March 13.

Butzer, S. and Hildyard, C (2014). Murky Waters: North Carolina Wrestles to Manage Hog Waste Disposal. *The Pendulum*. April 8.

Food and Water Watch (2010). *Factory Farmed Hogs in North Carolina* (Factsheet). August.

Genoways, T. (2011). The Spam Factory's Dirty Secret. *Mother Jones*. July/August.

Genoways, T. (2013). The Truth About Pork and How America Feeds Itself. *Bloomberg Business Week*. December 5.

Institute for Southern Studies (2014). *Institute Index: A Looming Legal Battle Over Hog Farm Pollution in North Carolina (*Accessed April 13).

May, C.Y. (2012). Malaysia: Economic Transformation Advances Oil Palm Industry. *Inform*. September.

Midwest Coalition for Human Rights (2012). *Always Working Beyond the Capacity of Our Bodies: Meat and Poultry Processing Work Conditions and Human Rights in the Midwest.* October.

National Council on Climate Change (Indonesia) (2010). *Setting a Course for Indonesia's Green Growth.*

Neff, J. (2013). Hundreds File Complaints Over Hog-Farm Waste. *Raleigh News and Observer.* July 7.

Nicole, W. (2013) *CAFOs and Environmental Justice: The Case of North Carolina. Environmental Health Perspectives.*

Oxfam (2013). *Behind the Brands: Food Justice and the "Big 10" Food and Beverage Companies*, Feb 26.

Sharp, K. (2014). Millennials' Bold New Business Plan: Corporations with a Conscience. *Salon.* Feb 9.

Statter, H. (2004). Hog Farming in North Carolina. *Environmental Justice Case Studies.* University of Michigan.

UDHR (1948). UN General Assembly, Universal Declaration of Human Rights, 10 December 1948, 217 A (III).

U.S. Bureau of Labor Statistics (2014) *Occupational Outlook Handbook, Agricultural Workers*, Mch 10.

# 19

# RECIPE FOR COOPERATION

## Omniocracy and the Definitional Good

*Charlotte Laws*

Chopped iceberg lettuce and chives may appear homogeneous from a distance—they are both green, both members of "team vegetable," and both are conducive to a healthy lifestyle—yet, they are fundamentally different: taste, vitamin and mineral composition, medicinal qualities, and climactic requirements, for example, represent a few areas of divergence. The same is true for environmentalism and animal activism. They appear the same because they are both social movements driven by dedicated activists who entreat the public to alter gluttonous, consumer-directed lifestyles in order to improve the world—but they are fundamentally different, and their inherent ideological differences have too often led to conflict. This conflict can and must be resolved. Just as lettuce and chives can come together to make a luscious salad, in much the same way the environmental and animal movements can be ordered into a mutually beneficial partnership.

This partnership can succeed, for example, in the area of policy. The current political landscape is outmoded and fundamentally exclusive. For example, democracy—like every other form of government—is anthropocentric and ignores the interests of nonhuman animals and Earth. These critical individuals and elements are invisible in democracy's lofty maxims, guidelines, and rules; every carefully crafted word reflects a system that is of, by, and for *Homo sapiens*, without even a footnote concerning those who are not human or the Earth itself. Environmentalists and animal activists must abandon their support of democracy and turn to a new and improved form of government—which I call omniocracy—in order to institute fundamental protections for nonhuman animals and the environment

### Defining the Problem

Environmentalists tend to focus on the whole, on species, on ecosystems, on natural processes, on diversity, and on the sustainable management of resources. They may promote overturning postindustrial society, returning to a hunter/gatherer type of existence, and decentralizing power (Devall 1994, 133–135). While there are many shades of thought, there are (arguably) two primary branches of the environmental movement. The first is "radical environmentalism" or "deep ecology." Deep ecologists perceive their ultimate goal as protection of the environs apart from its utility to *Homo sapiens*; theirs is a nature-centered perspective. The second and larger group, mainstream environmentalists (or conservationists), tends toward an anthropocentric perspective: These environmentalists protect the environment for the benefit of human beings—particularly future generations.

Animal activists focus on each sentient being (rather than the whole), because it is the individual who writhes in pain, not the species, ecosystem, forest, or stream. Although thousands—even millions—of sentient beings may be included within an aggregate such as "species" or "ecosystem," to ignore their status as individuals means that each being is ultimately expendable, a conclusion that is wholly unacceptable to animal advocates.

Within the movement to protect individual animals—as with any cause—there are a multitude of philosophical nuances, activist frameworks, and gradations of opinion. Some animal advocates rely on deontological (rights and rules) arguments; others prefer utilitarianism. Some help dogs and cats as an achievable first step; others target factory farms due to the number of animals involved. Some engage in illegal, underground operations to rescue frightened creatures from vivisection labs; and others—such as philosopher Tom Regan—encourage only nonviolent, upfront activism. Despite such diverging methods, animal activists agree that nonhuman animals have inherent value—and many animal activists argue that nonhuman animals are of equal value with human animals, and that they are therefore worthy of equal moral consideration. Animal activists rail against speciesism (a prejudice against nonhuman animals, much like racism, sexism, or anti-Semitism) and our tendency toward a human-centered worldview.

Welfarists are animal advocates who do not fight for the abolition of animal exploitation; they are primarily concerned with bettering conditions for exploited individual nonhumans. They mirror mainstream environmentalists in their support for the status quo's anthropocentric worldview: People are the priority and nonhuman animals may be used for our benefit, but they are to be treated with a modicum of decency. Welfarists accept animal exploitation (to varying degrees) in order to satisfy human needs and desires.

Conflicts between more stringent animal activists and welfarists, like conflicts between radical and mainstream environmentalists, parallel disagreements in social movements throughout history. There were abolitionists, for example, who wished to eradicate slavery in the South, and activists who merely sought compromise and incremental change to improve conditions for those forced into slavery. Purists—whether working for animal, environment, or other social justice causes—assert that pushing for incremental or minimal improvements is not only misguided, but hinders the overall struggle. Proponents of compromise say that extremism alienates possible allies with differing ideologies.

A theoretical disconnect also divides Earth and animal advocates into two groups: those who focus on the integrity of macro systems (environmentalists), and those who agonize over the suffering of individual beings (animal activists). The writing of Aldo Leopold, a highly revered, early member of the ecology movement, highlights this schism. His theory, "the land ethic," does not address the "management and use of ... 'resources '" (Leopold 1994, 39). For Leopold (and subsequently for the environmental movement more broadly), the term "resources" is a code word for "individual animals." Leopold's land ethic and the environmental movement in general green-light human exploitation of nonhuman individuals, whether in uninhabited areas or crowded cities. In addition, for Leopold (as for most environmentalists), domestic animals—whether chickens, cows, dogs, cats or laboratory victims—are irrelevant: Environmental ethics do not address the sufferings or needs of "domestic" animals. Instead, environmentalists tend to idealize "natural" terrain and harbor a special respect for individuals and species who are wild and free. Domesticated animals, submissive and confined, are considered unworthy of an environmentalist's time and energy, and they almost always slip under the wire with environmental philosophers.

Leopold writes that "a thing is right when it tends to preserve the integrity, stability, and beauty of the biotic community. It is wrong when it tends otherwise" (Leopold 1989, 146). This statement elevates the whole over the parts—as long as species numbers are stable the agony and deaths of specific members of that species—or their quality of life—are irrelevant, overlooked, not taken into consideration. For example, philosopher J. Baird Callicott notes that "to hunt and kill a white-tailed deer... may not only be ethically permissible [to Leopold], it might actually be a moral requirement, necessary to protect the local environment, taken as a whole" (Callicott 1992, 43). Environmental ethics tend to mandate that certain animals be killed in order to preserve rare plants or wild animals, a sentiment echoed by environmental organizations, such as the Sierra Club, whose members often hunt and/or fish (Absher 2010).

## Theory in Action: Conflicts

Disagreement between environmentalists and animal rights advocates is not merely theoretical. For example, an author in *The Australian* argues that fur is green. The author insists that animal skins are an ecologically and politically viable option because synthetic furs are comprised of "19 litres of petroleum, a non-renewable resource" (Albrechtsen 2008). The author does not suggest that fur ought to be avoided altogether, whether faux or real. One would assume that there were no clothing choices *except* fur. Such writings are seized by fur proponents, who claim that their product is environmental, drawing renewed public interest, causing immense suffering for furry animals, and inflaming animal activists.

Also in Australia (Canberra), animal liberationists protested while 400 kangaroos were killed in an "ecological" effort to protect a grassland ecosystem, complete with many rare plants and animals (Rao 2008). Ecologists have also advocated slaughtering and eating kangaroos to "help save the world from global warming" caused by cattle (Collier 2007). Other environmentalists have suggested consuming chickens rather than cattle (beef) in order to preserve the land and biosphere. For animal advocates, killing animals to gain other benefits is unacceptable, and swapping the lives of chickens for the lives of cattle causes innumerably more individual sufferings and premature deaths ("Table" 2007)—it takes many dead hens to acquire the amount of flesh in just one cow. Furthermore, their suffering is perhaps more egregious. For example, 280 million male chicks are tossed into garbage bags each year, or ground up alive, because only females are raised for flesh—males are too feisty when kept in such tight confinement (Montavalli 2002).

## Earth, Animals, and Democracy

Many authors note major differences between democracies, republics, fascist systems, monarchies, communist regimes, aristocracies, and other forms of government, but the differences are minor in comparison with one huge and inescapable commonality: All are anthropocentric. Political systems routinely overlook the interests of other animals and nature. Dogs, lions, fireflies, streams, and so on, would resist *all* political systems, past and present.

When the root word "demos," meaning "people" or "populace," is used to describe a country's political system, it places those who are not human in an inferior status, and in a compromised position. Demo-cracy is no more fair and equal than would be a white-ocracy or a rich-ocracy. Omni-ocracy ("omni" means "all') is one possible political system that offers an alternative vision. Democracy is routinely seen as "an ideal, an aspiration … intimately connected to and dependent upon a picture of what it is to be human" (Kompridis 2009, 31). Democracy is often touted as a superior political system, largely because it is said to give the weak and powerless an equal voice. There is an error in this thinking: People—whether the poor, elderly, physically disadvantaged, women, or people of color—are not the weak and powerless in comparison with nonhuman animals. All humans have clout, influence, and a voice when compared with our nonhuman brothers and sisters. Similarly, international laws and foreign nations may intervene with political pressure, sanctions, and sometimes war, on behalf of human victims, but never for other species. There is no such thing as a war crime against a cat, cow, or oak tree, nor has any government imposed trade sanctions or gone to battle to end bullfighting, deforestation, dog-eating, toxic waste, or seal hunts.

Democracy is nothing more than a totalitarian regime in which powerful humans use, abuse, murder, and manipulate the powerless (nonhumans and the natural world) for their own perceived gains. Democracy ignores the truly voiceless, marginalized, forgotten, and oppressed: animals and Earth. Even in the United States, a comparatively wealthy country, nonhumans have no standing in court, no legal rights of their own, and no protections under the Constitution. On rare occasions, when a law is passed to protect the interests of other species, that law is often trumped via an appeal to human desire or commercial profits … or by astutely applied political pressure. Unlike oppressed human groups, "others" (nonhumans and nature) cannot turn to civil disobedience to protest cruelty, exploitation, or their lack of rights. They cannot refuse to sit in the back of the bus, picket, go on hunger strikes, form underground railroads, throw tea overboard, or refuse to pay taxes. They must rely on concerned, compassionate people—environmentalists and animal activists—to achieve even the slightest socio-political improvements. Each microscopic gain is a colossal hurdle because today's world is of the human, for the human, and by the human.

Luckily, some counties are moving towards what I call "omniocratic results." Germany, Ecuador, and Switzerland, for example, have inserted provisions protecting nonhumans, to varying degrees, into their constitutions. Turkey is considering this option (Senett 2011). Unfortunately, the changes have largely been "all talk and no action," except in the case of Switzerland, which seems to be slowly creating and implementing pro-animal laws (Evans 2010, 231–250). The European Union attempted to draft a constitution with a provision for animal welfare, but the effort was abandoned in 2005, when France and Holland refused to sign (regardless of the animal-related provisions).

Although some nations are well ahead of the U.S. on legal protections for nonhumans, they each have a political system that generally holds animals and nature as means to human ends. In 2002, meat-eating averaged 180 pounds per person in Germany, 160 in Switzerland, and 99 in Ecuador ("Meat" 2009). (In the U.S., the number was a whopping 275 pounds per person.) The constitutional provisions that ostensibly protect nonhumans in Germany, Switzerland, and Ecuador do not mandate, or even recommend, an end to human omnivory, even though meat-eating causes more animal suffering and deaths, and more environmental destruction, than any other human activity.

## The Definitional Good

What would a new system of government entail and what rules might be devised? Speciesist, anthropocentric perspectives are problematic, but what might be a new and improved basis for legislation?

Omniocracy is a new and improved form of government with representation for *all* living entities. This form of government requires elected representatives to consider the interests of all living entities as part of the decision-making process. Constituents are not simply the elite, or humans more broadly; constituents are rabbits, trees, flowers, coyotes, pigs and so on. These nonhumans are of *equal* value to human beings, and must be granted equal consideration. In an omniocracy, the concerns and agendas of both environmental and animal activists are addressed; this is the fundamental principle of an omniocratic state.

Nonhumans are constituents in an omniocracy much like human infants in a democracy, and they are thereby granted political representation despite their inability to vote, speak out, or run for office. Despite their vulnerability, more powerful individuals or groups are not permitted to override the interests of the weaker. The decision-making process in an omniocracy mirrors philosophical utilitarianism in relevant ways. Omniocracy moves society, including humans, dogs, bears, bumble bees, palm trees, and so on, towards what I call the "definitional good," a move that both environmentalists and animal activists ought to joyfully embrace.

The definitional good (DG) constitutes an optimal state of affairs for society and is fully inclusive. Inclusive political theories consider the needs and interests of all instead of picking and choosing favorites like a corrupt king. Just as most societies have determined that it is erroneous and biased to overlook the interests of women and people of color, an inclusive political theory must embrace nonhuman living beings, even though they are unable to fight for their rights in human societies.

There are differing opinions as to the relationship between policy and legislation. Sociologist Max Weber writes that the law possesses its own rationality, apart from morality, and when we attempt to bring these two together, the legitimacy of law is threatened (Weber 1964, 2). Other scholars indicate that "all laws, whether prescriptive or prohibitive, legislate morality" (Bauman 1998). One cannot dispute that rules sprung from deeply held moral convictions. On the other hand, Weber seems correct in suggesting that legislation is better based on rationality rather than on conventional morality. The DG requires utilitarianism. (I use the term "utilitarianism" without reference to ethics, despite its parallel with philosophical utilitarianism in other ways.) In an effort to achieve the DG, decision-makers add up hedons (positives) and dolors (negatives) to conclude whether a particular course of action is likely to maximize "happiness," most appropriately envisioned as the satisfaction of interests. Decision-makers then create rules based on that which will best satisfy the interests of constituents. Of course, the members of society in an omniocratic state include all living entities—plants and animals—that have interests, desires, needs, or wants, and thus prefer a particular outcome. A deer, for example, has an interest in living as evidenced by her attempt to flee from a hunter. A tree strives to survive, nurture itself, flourish, and reproduce; this can be determined by watching branches grow towards the sunlight, and roots push down to seek underground sources of water.

There are two essential components to the definitional good. First, individual living entities have subjective perspectives that matter very much. A politician must attempt to become an impartial observer and engage in empathic understanding for nonhuman animals and plants in order to assess how best to satisfy the interest of "others." This is an exercise with which political leaders are already well-acquainted. Their job today arguably requires them to stand behind a veil of ignorance (Rawls 1971, 136–142) or become impartial spectators (John Stuart Mill's form of utilitarianism) (Singer 1993, 11) and try to understand how their constituents feel, a technique that is paralleled in the courtroom when public defenders argue on behalf of their clients. Lawyers need not agree with (or even like) their clients; their job requires them to fight for the clients' rights, perhaps even empathizing with their clients, or trying to comprehend what it feels like to be in their shoes. Politicians make decisions that require them to assess the subjective desires of their constituents with a stated goal of bettering the community as a whole. The process in an omniocracy is the same, only the constituents change—they grow exponentially. Decision-makers must expand their scope to cover the interests of all living entities.

Although those in power are affected by their sex, culture, and religion, this does not mean they should legislate from that perspective. Ideally, politicians attempt to set aside biases and make decisions from behind a hypothetical "veil of ignorance," as if they did not know their own race, sex, age, or abilities when making decisions (Rawls 1971). Although an omniocracy is based on utilitarian calculations, this is not to say that the majority will hold sway, or that decisions will involve only "numbers games." Politicians today may decide to support gay marriage or wheelchair ramps for the disabled—despite protests from a preponderance of dissenters—because such legislation leads to a greater satisfaction of interests for the community as a whole. In an omniocracy, the same reasoning applies for cattle, dandelions, and beetles.

As stated above, it is arbitrary and prejudiced to exclude nonhuman entities from our political equation. There have, of course, been other similar tendencies throughout history—many successful—to deny rights to the comparatively weak in order to benefit the powerful. Ethnic minorities throughout the world have struggled for equality, as have women. However, if the political goal is to satisfy interests in a given community—a utilitarian goal that is routinely promulgated by politicians—then reason dictates that all living entities be introduced into the DG calculation.

The second essential component to the definitional good is that subjective perspectives (of all living entities) are to be viewed as creating an objective reality. A happy world is one in which subjective perceptions are maximized to produce a situation of overall happiness. Individuals—whose interests are met—establish a world in which interests are, overall, satisfied. When our interests are satisfied, we tend to be much happier. Therefore, satisfying the maximum number of individual interests leads to a happier world, and a happier world is to be preferred to a world in which individuals are frustrated and dissatisfied.

Whether the tossing, swirling universe cares about the satisfaction of interests or accumulative happiness is irrelevant. Whether there is a God, goddess, or afterlife is also irrelevant. All that matters from a political context are the subjective perspectives of living entities—who are tied to the state of the Earth.

To describe the DG in simple terms, let's assume that there are only eight living entities in the world, all frogs. The DG is accomplished when the interests of all eight frogs have been satisfied; a definitionally bad situation emerges when each of their interests are unfulfilled. Of course, there are innumerable gradations between these extremes, but the point is that each of these frogs has interests—each wants to be alive, free to swim or mate, and so on. This better or worse state of affairs for each frog—as for all individuals—is dependent on the satisfaction of their interests, and has nothing to do with morality. It is not wrong or immoral for frogs to writhe in pain in their natural environment. From the perspective of the universe, it makes no difference what happens to frogs (or humans). But to frogs themselves, this matters very much. The legislator's task in this "frog world" is to strive to design laws that maximize the satisfaction of interests for the only inhabitants of this particular community—frogs. When the interests of each frog are satisfied as well as possible in that time and place, the DG is accomplished.

There is no such thing as perfection; decision-makers can only *strive towards* the definitional good in much the same way that law enforcement works towards a crime-free country. (I hope that no one is under the illusion that the police can eradicate crime.) Legislation is an evolving process, and one of the benefits of establishing laws—as opposed to working with moral absolutes—is that laws are malleable: They can be reconsidered, adjusted, revamped, or improved; politicians can make modifications based on new information. Moral absolutes, on the other hand, emanate from an authoritative source, such as natural rights, natural law, divine commands, or religious texts, and by their very nature, are set deep in stone. They have no place in politics because politics is about change, about negotiating conflicts, and about continually tinkering with legislation in an effort to improve society as a whole.

As noted, the satisfaction of individual interests is rooted in a healthy Earth, and includes endangered rhinoceroses and dogs and turkeys. Under omniocracy, working for the definitional good, all the various environmentalists and animal advocates can find common ground.

## Conclusion: Sharing the Salad

Revamping the political system and enacting innovative policy solutions can unite environmentalists and animal activists, and it can offer never-before-seen protections for "others," whether Earth or animals. To move in this direction, the first step is verbal: We must stop supporting democracy—with its venomous humanocentric sting—and advocate for omniocracy, making mention of constitutional strides in Switzerland and other nations. "Omniocracy" needs to become a household word—ever-present like a close friend; it should appear in dictionaries, in newspaper articles, and in activist meetings. Words are persuasive; they are the first line of offense when moving towards a new worldview.

Animal and earth advocates should also encourage political leaders to widen their lens, and to embrace nonhumans as constituents. In addition, they should run for political office themselves, even if only on a local level. Political officers hold power; nonhumans and the Earth need that power.

Policy-related changes tend to be sluggish and incremental— so now is the best time to begin shifting away from a world in which other animals are viewed as means to human ends. The U.S. is one of the wealthiest nations, and on many levels, one of the most progressive, yet we are failing abysmally in protections and respect for "others." Together, Earth and animal advocates can reverse that trend. We can create an ecologically friendly human community, and end the senseless suffering of sentient victims. We can sit at a common table and toss a bountiful salad. Let us prepare the feast!

## Discussion Questions

1 Where does dissent fit into Laws's vision of the definitional good? For example, in the scenario with eight frogs, what if one frog has a dissenting view of what is desirable? What if four frogs want something that is mutually incompatible with what the other four frogs prefer? How would such instances affect the idea of a definitional good?

2 What if the interests of each frog conflict with the interests of every other frog? In what ways is this situation currently created by capitalism?

3 Other animals cannot reasonably participate in human government. How does this place them in jeopardy under our current political system?

4 Is it legitimate to call a political system created and run by human beings an omniocracy in the sense that Laws intends?

5 We have a long history of envisioning extant "democracies" as offering all citizens a say in their governance, starting with Greece, which excluded both women and slaves. Ideally, who/what is included in the power of a democracy, and if things like streams and pigs should be included, how are they to be given power?

## Essay Questions

1 What is a "happy world"? Is this a meaningful concept?

2 What sorts of political changes would you implement to better protect the interests of those not fully protected under our political system? How might these best be brought into play?

3 Given our tendency to favor humans and ignore the needs of nonhumans, how can a system in which humans hold all power, and are elected by other humans to serve *their* interests, possibly be expected to represent other animals?

4 How does the poem "Breathtaking Bit of Being" (page 237) express both a vision of omniocracy and our tendency to be locked into an anthropocentric view?

## Suggested Further Reading

Hargrove, Eugene C. (1992) *The Animal Rights / Environmental Ethics Debate*, State University of New York, New York.

Spinoza, Baruch (1982) *The Ethics and Selected Letters, Hacket Publishing Company*, Inc., USA.

Zimmerman, Michael E. (1993) *Environmental Philosophy: From Animal Rights to Radical Ecology*, Prentice-Hall, Inc., New Jersey

## References

Absher, J.R. (2010) "Sierra Club: We Support Hunting But…" (25 March) http://www.outdoorlife.com/blogs/newshound/2010/03/sierra-club-we-support-hunting, accessed 8 April 2014.

Albrechtsen, Janet. (2008) "Climate Change Cure is Warm and Fuzzy." In *The Australian*, 9 January. http://www.freerepublic.com/focus/f-news/1950796/posts, accessed 8 April 2014.

Bauman, Michael. (1998) "Dispelling False Notions of the First Amendment: The Falsity, Futility, and Folly of Separating Morality from Law," *Christian Research Journal*, 21: 3. http://www.equip.org/articles/law-and-morality, accessed 19 August 2011.

Callicott, J. Baird. (1992) "Animal Liberation: A Triangular Affair" in Eugene C. Hargrove (ed.) *The Animal Rights / Environmental Ethics Debate*. State University of New York Press, Albany.

Collier, Karen. (2007) "Greenpeace Urges Kangaroo Consumption to Fight Global Warming." *Herald Sun*. 10 October http://www.heraldsun.com.au/news/national/eat-a-roo-save-the-world/story-e6frf7l6-1111114612144, accessed 8 April 2014.

Devall, Bill. (1994) "The Deep Ecology Movement" In Carolyn Merchant (ed.) *Ecology*. Humanities Press, New Jersey.

Evans, Erin. (2010) "Constitutional Inclusion of Animal Rights in Germany and Switzerland: How Did Animal Protection Become an Issue of National Importance?" *Society and Animals*. http://www.animalsandsociety.org/assets/443_evansnutshell.pdf, accessed 8 April 2014.

Kompridis, Nikolas. (2009) "Technology's Challenge to Democracy." *Parrhesia*. 8, 2009. http://www.parrhesiajournal.org/parrhesia08/parrhesia08_kompridis.pdf, accessed 8 April 2014.

Leopold, Aldo. (1989) *A Sand County Almanac: And Sketches Here and There*. Oxford University Press, Oxford.

Leopold, Aldo. (1994) "The Land Ethic" in D. VanDeVeer and C. Pierce (eds) *Environmental Ethics and Policy Book*. Wadsworth, Boston.

"Meat Consumption Per Capita." (2009) *The Guardian*. 2 September. http://www.guardian.co.uk/environment/datablog/2009/sep/02/meat-consumption-per-capita-climate-change, accessed 8 April 2014.

Montavalli, Jim. (2002) "So You're an Environmentalist; Why Are You Still Eating Meat?" *E Magazine* and *Alternet*. 3 January. http://www.alternet.org/story/12162, accessed 8 April 2014.

Rao, Shoba. (2008) "Kangaroos to be killed in Canberra under Culling Program." *The Telegraph*. 19 May http://www.dailytelegraph.com.au/news/national/kangaroos-to-be-killed-in-canberra-under-culling-program/story-e6freuzr-1111116381352, accessed 8 April 2014.

Rawls, John (1971) *A Theory of Justice*, Harvard University Press, Cambridge, Mass.

Senett, Keph. (2011) "Following Examples Set by Ecuador and Bolivia, Turkey is Considering Ecological Constitution." 23 May http://www.pvpulse.com/en/news/world-news/following-examples-set-by-ecuador-and-bolivia-turkey-is-considering-ecological-constitution , accessed 8 April 2014.

Singer, Peter. (1993) *Practical Ethics*. Cambridge University Press, Cambridge.

"Table of Estimates on Animal Suffering on Factory Farms Broken down by Species." (2007) http://www.utilitarian-essays.com/suffering-per-kg.html, accessed 8 April 2014.

Weber, Max. (1964) *Wirtschaft und Gesellschaft*. Kiepenheuer and Witsch, Cologne.

# 20

# DEEPER THAN NUMBERS

## Consumers, Condoms, Cows

*Lisa Kemmerer, Daniel Kirjner, Jennifer Gross, and Nathan Baillet*

Humans were not always so abundant. Our population increased only slowly with the discovery of critical technologies—stone tools, wheels, plows, horticulture, and irrigation. Human numbers reached one billion over a couple hundred thousand years. Then, with improved hygiene, antibiotics, and other key discoveries, human populations exploded in the twentieth century, moving from 1.65 billion to 6 billion ('The World,' n.d.). Just 11 years later, another billion human beings were added to the planet.

Halloween eve, 2011, the 7 billionth human touched down. Danica May Camacho was one of many newborns chosen (by the United Nations) from around the world to "symbolically represent" this "global population milestone" ('World's' 2011). "Milestone" doesn't quite capture the moment—the terror that ought to have accompanied the arrival of our 7-billionth baby given that human beings are now so abundant that we threaten the health of the planet, even life as we know it on this spinning globe.

On the surface, the connection between human populations and earth/animals seems obvious—despite its conspicuous absence from the platform of pretty much every earth/animal organization. Looking beyond numbers, which merely record human births, several critical connections emerge.

### Population and Consumption

> [T]here is little reason to doubt that over-consumption of resources is a problem and that it depends both on lifestyle choices and population sizes.
>
> (Sarkar 2012)

The United Nations predicts that the world will be burdened with somewhere between 8 and 11 billion human beings by the year 2050 (Brown 2008). If only a dozen humans lived on the planet, they could consume (even petroleum) to their hearts' content. Even if all 12 Earthlings chose to be carnivores (eating *only* flesh), the earth could easily sustain human gluttony. But with 7 billion humans on the planet, we can neither consume nor breed at will.

Demographers use a number of methods to make population projections. For example, total fertility rate (TFR) determines the average number of children a woman might be expected to birth in her lifetime. However, focusing exclusively on TFR can lead to the false conclusion that the problem of human fertility is merely a matter of numbers, and that women and less developed countries are the primary problem.

Distinguishing between more developed countries (MDCs—North America, most countries in Western Europe, and a few Asian countries—most notably Japan) and less developed countries (LDCs—most nations in South America, Africa, and Eastern Europe) is too simplistic. Such stark borderlines are problematic when it comes to assessing populations and environmental footprints. For example, the International Monetary Fund's list of the most advanced economies excludes a number of critical nations, such as Russia, China, India, and Brazil. Most listings of developed countries include Malta and Singapore (but not China), even though consumption in these nations is considerably lower than those of "undeveloped" countries such as Brazil and Argentina. Among nations listed as LDCs, there are even more discrepancies: those with higher rates of consumption, growing economies, and increasing political stability (India and Brazil, for example) are often lumped together with nations devastated by ongoing internal conflict (such as Algeria and the Democratic Republic of Congo) (Adejumobi 2001). In each case, these attempts at categorizing countries clearly miss something vital.

Additionally, nations change, while labels tend to be static. For example, one of the environmentally problematic tendencies of nations labeled as LDC is to adopt the bad habits of powerful, comparatively wealthy countries. Traditional diets in China typically contained very small amounts of flesh and no dairy products, but in recent years "consumption of meat—pork, beef, poultry, and mutton—has climbed several fold [since 1978], pushing China's total meat consumption far above that of the United States" (Brown 2008). Dairy products, traditionally identified as the nursing milk of *another* species (therefore not appropriate for human consumption), are increasingly popular in China ('India' 2011). It is ominously clear that the consumption patterns of the greediest and most materialistic nations are spreading: In nations labeled "developing," "meat production increased more than 450 percent between 1980 and 2010" (Reynolds and Nierenberg 2012). At what point do these nations join MDCs? Are they more likely to be considered "developed" with increased consumption—including the consumption of animal products—despite this being contrary to every indication of what is necessary for sustainability?

Before categorizing any groups with regard to environmental impact, lifestyle patterns must be taken into account. Most citizens of MDCs drive cars, take hot showers daily, and live in heated and/or air-conditioned homes, while a large percentage of people in LDCs do without these luxuries (International 2013). MDCs, which hold just 16 percent of humanity, consume 80 percent of the world's resources ('World's Wealthiest' 1999). In the United States, each individual (including children) consumes about 22,680 kilograms (25 tons) of raw material each year ('World's Wealthiest' 1999). If every nation burned through resources this fast, it would require four earths to support the current human population (Semon 2012). It is difficult to mitigate these consumer affects with *any* cycling or recycling plan.

Dietary habits are also critical to environmental footprint, but again, lumping nations together can be problematic. Humans in MDCs consume approximately 80 kilograms (180 pounds) of flesh per person per year, while people in LDCs consume roughly 30 kilograms (70 pounds) of flesh in the same time period ('Meat' 2011). Some of these MDCs consume excessive amounts of animal products, like the United States (120 kg/264 lbs), Australia (111 kg/244.2 lbs) and New Zealand (106 kg/233.2 lbs), and these extreme cases distort meat consumption statistics for MDCs. Meanwhile, some nations pushed into the LDC category, including Argentina (98 kg) and Kuwait (119 kg), have a higher per capita meat intake than most countries labeled as "developed" ('Agricultural' 2013).

The environmental problems of consumption are aggravated by waste. More than 40 percent of the foods that are produced in the United States are wasted (Bloom 2007). This comes to about 0.5 kg (1 lb) of food per person per day, creating 136 million kg (300 million lbs) of garbage *every 24 hours*—50 billion kg (55 million tons) every year (Martin 2008). People in the U.S. devour some 815 billion calories every day—about 200 billion more than the U.S. population requires—much of which is wasted ('Consumption' n.d.). But waste is not restricted to MDCs. Latin Americans throw away enough food every year to feed 300 million people—one hundred million more than can be fed with the food waste of Europe ('Agricultural' 2013).

Consumption patterns, though more complicated than counting heads, are just as important for assessing a nation's environmental impact. For example, TFRs in the U.S. are one quarter those of Niger, but the environmental impact of a U.S. citizen is, overall, *much* greater than the environmental impact of a citizen in Niger *because of U.S. consumption habits.* To demonstrate this point, consider freshwater depletion and deforestation.

## *Interconnections: Freshwater Depletion*

Freshwater reserves around the world are drying up. Critical waterways such as the Colorado, Nile, and Yellow River now run dry before they reach the sea (Brown 2008). Large water bodies like Lake Chad and the Aral Sea are vanishing along with underwater aquifers. The Ogallala Aquifer, which spans 800 miles (1,287 km) from north to south and 400 miles (644 km) from east to west, supplies 30 percent of U.S. groundwater irrigation to 27 percent of the nation's irrigated land,

and provides drinking water to 82 percent of the residents in eight states, is more than half gone and looks likely to run dry in the next 25 years ('High' n.d.). In 1995, Ismail Serageldin (then vice president of the World Bank) commented: "If the wars of this century were fought over oil, the wars of the next century will be over water."

Excessive human populations certainly threaten freshwater reserves, but so do consumption patterns. Because of the amount of freshwater required for animal agriculture, American omnivores consume 15,900 liters (4,200 gallons) of freshwater every day, while American vegans consume just 1,136 liters (300 gallons) (Schwartz 2001). Grains are often consumed directly in countries like India and Sri Lanka, but in nations where people have diets rich in animal products, like Argentina, Australia, and the United States, grains are cycled through pigs, chickens, and bovines, which requires many more tons of grain: While each person in the U.S. consumes roughly 800 kilograms (1,763 pounds) of grain every year, only 100 kilograms (220 lbs) are eaten directly (i.e. in breads, pasta, rice, etc.) (Brown 2008). Cycling grains through animals requires more grain and water to provide the same number of calories. In fact, consumption of animal products is the biggest threat to freshwater reserves: 500–5,200 gallons (1,893 to 19,700 liters) of water are necessary to produce one pound of beef, but only one-hundredth as much water is necessary to produce one pound of wheat (Kaufman and Braun 2004). In the U.S., irrigation is responsible for 70 percent of freshwater expenditure, primarily to grow grains (wheat, corn, and soy—70 percent of which are used to feed farmed animals). Roughly 80 percent of the world's freshwater is expended for agriculture; animal agriculture is responsible for 90 percent of freshwater depletion worldwide ('Livestock's' 2006).

Dietary choice is the *single largest factor* with regard to human water consumption. Consumption is therefore the primary indicator of how much water an individual will consume in his or her lifetime. While a couple in Mali might be likely to have six children compared with one or two per couple in England, the water consumption of two omnivorous Brits is almost certain to exceed that of six grain-eating Africans. In other words, the greedy consumption patterns of flesh-eating nations carry more environmental weight than high fertility rates in nations that maintain a plant based diet.

## Interconnections: Deforestation

One fifth of the world's rainforests were destroyed between 1960 and 1990. Between 1985 and 1990, 210 million acres of forest were turned to pasture, "an area nearly the size of Texas and Oklahoma" (Kaufman and Braun 2004). A section of rainforest roughly equivalent to 20 football fields (22 soccer/football fields) is destroyed pretty much every minute of every day, and in "the Amazon, cattle ranching is now the primary reason for deforestation" ('Livestock's' 2006). Only 13 percent of Costa Rica's original rainforest remains, and what remains is now in a "highly fragmented and degraded state" (Reynolds and Nierenberg 2012). In just 50 years, 50 percent of Costa Rica's forests disappeared—60 percent were cleared for bovines—for beef ('Cattle Ranching's Impact' 2012).

South America is still at the top of the list for loss of forests, and Brazil continues to lead the way (by a considerable margin) in rainforest destruction. Agriculture is responsible for roughly 98 percent of Brazil's deforestation (Butler 2012): Ranchers are responsible for 65–70 percent of Brazil's lost forests (Butler 2012). There were about 10 million bovines in Brazil in 1980, and there are now upwards of 55 million ('Deforestation' n.d.). The U.S. imports some 80 million pounds of Brazilian beef every year; 85 percent of EU beef originates in Brazil.

All this ecological devastation for a mere spot of flesh—55 feet (17 meters) of tropical forests yield just a quarter pound (120 grams) of hamburger. If we continue as we are, primary forests will be altogether gone by 2050 (Hawthorne 2012; Pimm and Raven 2000). Those who consume turkeys, pigs, chickens, eggs, and dairy products are also blameworthy: The primary reason for loss of forests is conversion of lands to agriculture—*both for grazing and for feedcrops.* In Latin America, land is converted from forests to agriculture largely for feed crops, "notably soybeans and maize" ('Livestock's' 2006). Brazil's soy crop grew more than 3,000 percent in the last 40 years, becoming the world's second largest soybean producer worldwide—80 percent of Brazil's soybean crop is fed to farmed animals (Reynolds and Nierenberg 2012). In the U.S. 98 percent of soy is turned into meal to feed poultry, bovines, hogs … and farmed catfish; more than 50 percent of U.S. soy is fed to poultry. *Worldwide, 80 percent of soybean crops are planted, tended, and harvested for farmed animals, implicating those who eat cheese and chicken—not those who eat tofu and tempeh* (Reynolds and Nierenberg 2012).

Again, consumption patterns prove critical to our environmental footprint. Nations comprised of individuals who habitually cycle grains through chickens and pigs and dairy cattle, are responsible for the destruction of South American forests, leveled for grazing cattle and feeding farmed animals. Those convinced that condoms are the key need to consider cattle versus kale.

## Deeper than Numbers

A fuller picture of environmental impact must take into consideration capitalism, materialism, and consumerism, as well as attitudes, values, and beliefs that undergird and foster excessive human fertility (more than two children birthed/fathered by any one individual), including sexism.

### *Sex Discrimination*

A woman's expected TFR—the number of children she is likely to birth—often correlates with her social and economic status. With access to education and gainful employment, birth tends to decline. The expected TFR for uneducated women in Botswana is nearly six; the expected TFR for those with four to six years of schooling is just three. In Senegal, the expected TFR for women with no education is seven, compared with 3.6 for women with just ten years of education (Knox and Marston 2007). In the United States, both boys and girls are expected to stay in school until they are 16 years old. Additionally, affirmative action in the U.S. reserves a portion of coveted jobs for females. Women in the U.S. have an expected TFR of just two—replacement level (Beauchamp 1998).[1] Less educated women in the U.S. suffer higher incidences of unintended pregnancies—17 percent among females with at least an undergraduate degree, compared with 41 percent among women with less education ('Intended' 2012). Education and opportunity are both critical for curbing and controlling human fertility.

Overall, one's sex is important in determining how much an individual's life will be affected by environmental degradation; the natural environment is "inextricably connected to rural and household economies," which tend to be governed by women (Warren 2000). Development and destruction of the natural environment therefore creates "particular burdens for women," especially in underdeveloped countries (Des Jardins 2001). For instance, women constitute 80 percent of Africa's farmers, and are therefore directly affected by deforestation, soil erosion, and water shortages (Riley 2003). Women are primarily responsible for gathering fuel and water, and Women in Africa, Asia, and South America now "spend up to forty-three hours per week collecting and carrying water," traveling many miles to find aspects of the natural world that used to be readily available (Warren 2000). In this way, consumption (especially of animal products) in wealthier, avaricious, capitalistic countries blocks opportunities for women and girls, preventing empowerment and the possibility of independence for many individuals. How can women and girls study, work outside the home, or have hope for a brighter future if they spend much of each day toting large bundles of wood long distances to prepare food because we choose to eat animal products? It is unreasonable for someone chowing down cheese to complain about those who have "too many children."

## *Larger Families*

Patriarchal attitudes, values, beliefs, and laws tend to encourage and lead to larger families in most nations. For example, property-owning men prefer to pass accumulated wealth to sons, and to do this they must have male offspring (Lerner, n.d.). Traveling in Kenya, a few questions can quickly establish the common belief that a man demonstrates his virility by siring many children (Chalcraft 2015). In Zambia, boys learn that masculinity requires dominance over women, including sexual domination, and "demonstrations of sexual potency" require vaginal penetration; Zambian men are therefore more likely to forego condoms in order "to prove their virility" (Simpson 2005). As a result, schoolchildren father children. Men and boys seek "multiple partners" yet refrain "from using condoms," fearful that condoms will "impair their sexual performance," especially if they have been drinking (Simpson 2005). In the U.S., norms also continue to encourage childbirth and large families, though not generally as a matter of proving a man's virility. Veneration of motherhood is perhaps more at fault—as if bearing (and raising) children were somehow exceptional ('Why' n.d.). This, too, is a function of patriarchy, whereby women are identified with their biological functions, and praised for creating babies rather than bridges or space craft, leaving the latter work to be dominated by men (Spelman 1982). Additionally decades of popular U.S. television shows, including *Eight is Enough*, *Sister Wives*, *19 Kids and Counting*, *The Brady Bunch*, and *Jon and Kate Plus 8*, glorify irresponsibly large families. In the U.S., the infamous "Octomom" used fertility treatments to more than double her brood—despite already having six kids and no practical means of support … and consequently attained celebrity status (Suze n.d.).

Additionally, tax breaks are offered for *every* child born in the United States: U.S. tax burdens amount to a whopping $1000 per child per year without even accessing a host of additional child deductions ('Ten' 2011). Some would argue that, on a planet overburdened by humanity, it is unconscionable and irrational for the U.S. government to offer tax reductions for *every child born*. It makes more sense to provide aid and service to *anyone in need*—whether or not they have a dependent child at home. There are thousands of parents who have no need of a tax break to raise and tend progeny. Meanwhile, there are many child-free individuals who are in great need of government aid, which they may or may not be able to access.

How might nations offer incentives for citizens to *not* bring more humans onto the planet? Why not provide free vasectomies on request? Why is it difficult for young women in the U.S. (and many other nations) to find doctors willing to perform tubal ligations on request? Why don't we have a law protecting a woman's right to decide not to be able to be impregnated—without having to ask multiple doctors and, if married, the permission of her husband? Married or unmarried, a woman ought to be able to decide for herself whether or not to alter her body so as to prevent pregnancy. Why don't wealthier nations supply birth control on request without limits both at home and abroad? This would seem our best return on small investments to help stabilize failing states and help nations that are struggling with water, land, and food shortages. Why are individuals and organizations so quick to criticize nations taking a proactive stance to help stabilize population growth while failing to offer viable alternative solutions? Fertility rights and freedoms must not be juxtaposed against critical environmental concerns that threaten all of us.

Archaic religious attitudes, values, and beliefs—sealed into sacred texts—also tend to encourage large families, assuring the continuation of religious institutions while demonizing family planning, including access to birth control. In the Americas, Mormons, Catholics, and several strains of Protestantism (such as the Straight Arrows) are known for excessively large families. These same households are more likely to have stay-at-home mothers who are busy raising kids, and who therefore are less likely to find gainful employment—let alone positions of power and influence.

This, in turn, discourages communities from prioritizing education for girls , especially higher education during a young womans' reproductive years.

While the world's many patriarchal nations encourage fathering children, childbirth, motherhood, and larger families, the environmental impact of larger families varies according to consumption patterns. Population growth in Kenya is not environmentally problematic in the same way as population growth in North America or Europe. For example, Kenyan water use in 2006 was 72.44 m3 (2,558.2 $ft^3$) per inhabitant per year ('Agricultural' 2013). In the same period, U.S. water use was 1,550 m3 (54,738 $ft^3$) per inhabitant per year—more than 21 times greater. A Kenyan mother could bear more than 20 children before her brood matched the water expenditure of just one child in the United States. Similarly, a Kenyan mother would have to bear more than 14 children to match the likely energy consumption of just one U.S. child (World Bank 2014).

Excessive fertility on a limited planet in a sexist world is problematic. Nonetheless, for those living in Europe, North America, Australia, and New Zealand—for those likely to be reading this book—the environmental problem of greatest concern is consumption—an issue that people reading this essay are likely empowered to change. Rather than point the finger at women in Latin America or Africa, *you* decide whether or not to have bacon and eggs for breakfast. With regard to environmental problems, as with most problems, changing our own behavior is much more effective than blaming others. Family planning is important, but meal planning is even more important for those in "developed" nations—minimizing our environmental footprint requires that we quit consuming animal products.

## Refocusing Activist Platforms

Depriving girls and women of education and opportunity increases the likelihood that they will birth a greater number of children. Bringing offspring into the world always increases our environmental footprint—much more so in "developed" nations that have higher rates of consumption (and waste)—especially higher rates of consumption for animal products. Environmental organizations should encourage people to grab a condom and leave cows alone, but a sampling from ten different environmental organizations[2] reveals that these groups fail to advocate for women's equality, do not foster a child-free lifestyle, and make no mention of the importance of adopting a plant based diet. Environmental organizations are surprisingly soft—most often just plain silent—when it comes to deep societal change. Yet their unwillingness to take a stand on these matters undermines their effort to protect the planet. It is only sensible that environmental organizations take a strong stand against sex discrimination and the consumption of animal products, while noting the benefits of a child-free lifestyle.

Lower fertility also benefits nonhuman animals: As humans expand they snatch habitat, deplete water, destroy forests, and so on—fewer humans are clearly a benefit to other living beings. Nonetheless, a sampling of websites from ten different animal organizations[3] indicates that these groups, while focusing a great deal of energy on dietary choice and change, fail to connect animal concerns with sex discrimination and reproduction. This lack of engagement is particularly surprising given the tendency among animal organizations to employ numbers to encourage change: People for the Ethical Treatment of Animals calculates that the average person consumes roughly 200 farmed animals each year, including 130 shellfish, 40 fish, 26 chickens, 1 turkey, nearly half a pig, and a little more than a tenth of a cow ('Vegans Save' 2010); the Farm Animal Rights Movement notes that some 10 billion animals are slaughtered each year in the U.S. ('10 Billion' 2012). In spite of this mathematical focus, these organizations fail to suggest that there is any connection between numbers of animals consumed and numbers of human mouths consuming—let alone that there is any need for sex equality or to offer women options beyond motherhood. The vast majority of animal organizations are surprisingly soft (most often just plain silent) when it comes to sex equality and human fertility. The unwillingness of animal advocacy organizations to take a stand on these critical issues undermines their efforts to protect both farmed and free ranging/wild animals.

Fighting for the earth and nonhuman animals requires that we also fight for universal sex equality, especially for education for girls and women and for employment opportunities for women. At a minimum, every environmentalist and animal activist ought to maintain memberships with Population Connection (or a similar group working on behalf of women and the earth) and an organization dedicated exclusively to educating girls and women. (Visit *Educating Girls Matters*—http://www.educatinggirlsmatters.org/howtohelp.html—to select your favorite organization.)

## Conclusion

It is generally easier to focus on surface matters than it is to ferret out and target underlying, core causes, but *addressing root causes is essential if we are to bring lasting change*. Every newborn is necessarily a consumer, and is likely to subsequently produce his or her own offspring, continuing the cycle of human reproduction and consumption that has caused severe environmental degradation, depletion of species and the reduction of biodiversity, and the exploitation of billions of nonhuman animals. In nations such as England, the United States, Australia, Luxemburg, Kuwait, and Argentina, the environmental footprint of any one individual is as much about consumption as it is about childbirth—most importantly, the consumption of animal products. Animal activists and environmentalists alike should advocate for a vegan diet and sex equality, while extolling the benefits of a child-free lifestyle.

## Discussion Questions

1 Explain the subtitle, "Consumers, Condoms, Cows."

2 Why is the essay called "Deeper than Numbers?"

3 Is it racist for environmentalists in Luxembourg, Germany, or the United States to complain about fertility rates in poorer countries? Why or why not? Under what conditions and in what ways would it be acceptable for such activists to focus on population control in poorer nations?

4 While it is ill advised to expect other nations to be interested in our particular ideas for positive change, it is admirable to help other nations bring about positive changes. How can we best help other nations provide education and opportunities for girls?

## Essay Questions

1 Visit *Educating Girls Matters* (http://www.educatinggirlsmatters.org/how-tohelp.html). Select your favorite organization and explain why you would support this organization over all the others that are mentioned on the site.

2 If you were in charge of an environmental or animal advocacy organization, how would you incorporate the connected concerns of fertility, girls' education and job opportunities, and consumption patterns into your platform?

3 Search on the internet for organizations that support women, such as those providing seed-funds to develop small businesses or those that provide skills training. Choose several and, with regard to what they are doing, explain what seems most effective and least effective, and why.

4 If you were in charge of teaching the class you are in, what interactive exercises would you add to help teach students about the importance of reducing fertility and consumption, while fostering girls' education and opportunities both at home and overseas? (The editor would like to hear about any interesting responses to this or any other question in the text. Instructor's quality-control is appreciated.)

## Suggested Further Reading

Cafaro, Philip, 2012, *Life on the Brink: Environmentalists Confront Overpopulation*, University of Georgia Press, Athens, GA.

Diamond, Jared, 2005, *Collapse: How Societies Choose to Fail or Succeed*, Penguin Group, New York, NY.

Dorling, Danny, 2013, *Population 10 Billion: The Coming Demographic Crisis and How to Avoid It*, Constable and Robinson Ltd, London.

Hartmann, Betsy, 1999, *Reproductive Rights and Wrongs: The Global Politics of Population Control*. South End Press, Cambridge, MA.

Kirby, David, 2010, *Animal Factory: The Looming Threat of Industrial Pig, Dairy, and Poultry Farms to Humans and the Environment*. St. Martin's Press, New York.

Mikell, Gwendolyn, 1997, *African Feminism: The Politics of Survival in Sub-Saharan Africa*. University of Pennsylvania Press.

*Population Education*, 2014, Supplemental Materials, viewed 14 June 2014, http://www.populationeducation.org/content/supplemental-materials.

## Notes

1 This is not to suggest that women in the U.S. have gained equal social or economic status. Sexism remains less blatant in the U.S. than in some other countries, but remains a serious problem. In the U.S., for example:

- Women's earnings are roughly 77 percent of men's ('Wage' 2012);
- In the 113th Congress, a record 81 women served in the House of Representatives, while 20 women served in the Senate (just 19 percent of seats in Congress);
- Only 12 of the Fortune 500 are run by women ('Women' n.d.).

2 We visited the websites of Nature Conservancy, Earth First, Environmental Defense Fund, Greenpeace, Wilderness Society, EarthJustice, Intergovernmental Panel on Climate Change, United Nations Environment Programme, Earth System Governance Project, and Global Environment Facility.

3 We visited the websites of PETA, Farm Sanctuary, Mercy for Animals, Compassion Over Killing, United Poultry Concerns, Farm Animal Welfare, Humane Society, Animal Defense League, International Primate Protection League, and Last Chance For Animals.

## References

Adejumobi, S., 2001, 'Citizenship, Rights and the Problem of Conflicts and Civil Wars in Africa,' *Human Rights Quarterly*, vol. 21, no. 1.

'Agricultural Outlook 2013–2022,' 2013, OECD – FAO, Available from: http://www.oecd.org/site/oecd-faoagriculturaloutlook/highlights-2013-EN.pdf. [1 May 2014].

Beauchamp, T. L. 1998, 'In Defense of Affirmative Action,' *The Journal of Ethics*, pp. 143–158.

Bloom, J., 2007, 'Food Waste: The Food Not Eaten: Out of Sight, Out of Mind,' *Culinate*, 19 November. Available from: http://www.culinate.com/articles/features/wasted_food. [27 January 2013].

Brown, L., 2008, *Plan B: 4.0*, Earth Island Institute, New York.

Butler, R. A., 2012, 'Causes of Deforestation in the Brazilian Amazon,' *MongaBay.com*, 20 May. Available from: http://www.mongabay.com/brazil.html. [13 April 2013].

'Cattle Ranching's Impact on the Rainforest,' 2012, *Mongabay*, 22 July. Available from: http://rainforests.mongabay.com/0812.htm. [11 October 2014].

Chalcraft, V. J., 2015, 'Essential Elements for Elephants: Problems and Solutions,' in *Animals and the Environment: Advocacy, Activism, and the Quest for Common Ground,* ed. L Kemmerer, Routledge, New York.

'Consumption by the United States,' n.d. Available from: http://www.mindfully.org/Sustainability/Americans-Consume-24percent.htm. [26 January 2013].

'Deforestation: The Leading Cause of CO2 Emissions,' n.d., *Global Warming Science*. Available from: http://www.appinsys.com/globalwarming/deforestation.htm. [13 April 2013].

Des Jardins, J., 2001, *Environmental Ethics: An Introduction to Environmental Philosophy,* 3rd edn. Wadsworth, Belmont, California.

Hawthorne, M., 2012, 'Planet in Peril,' *VegNews*, March–April, pp. 34–41.

'High Plains Regional Ground Water Study,' *USGS*, n.d., Available from: http://co.water.usgs.gov/nawqa/hpgw/HPGW_home.html. [29 March 2013].

'India, China to Boost Global Milk Production: Tetra Pak,' 2011, *The Economic Times*, 11 July. Available from: http://articles.economictimes.indiatimes.com/2011-07-11/news/29761312_1_flavoured-milk-liquid-dairy-products-tetra-pak. [12 December 2011].

'Intended and Unintended Births in the U.S.: 1982–2010,' 2012, *The Reporter*, vol. 44, no. 3, October 2012, pp. 45.

'International Energy Outlook,' 2013, *Energy Information Administration*, U.S. Government, Washington D.C.

Kaufman, S. R., and Braun, N., 2004, *Good News for All Creation: Vegetarianism as Christian Stewardship*, Vegetarian Advocates Press, Cleveland.

Knox, P. L. and Marston, S. A., 2007, *Human Geography: Places and Regions in a Global Context*, Pearson Prentice Hall, New Jersey.

Lerner, G., n.d., 'The Creation of Patriarchy: Summarized from *The Creation of Patriarchy*,' Available from: http://faculty.stedwards.edu/bobs/documents/patriarchy_creation.pdf. [4 January 2014].

'Livestock's Long Shadow: Environmental Issues and Options,' 2006, Food and Agriculture Organization of the United Nations, Rome. Available from: http://www.fao.org/docrep/010/a0701e/a0701e00.HTM. [15 June 2014].

Martin, A., 2008, 'One Country's Table Scraps, Another Country's Meal,' *The New York Times*, 18 May. Available from: http://www.nytimes.com/2008/05/18/weekinreview/18martin.html?_&_r=0. [23 February 2013].

'Meat Production Continues to Rise,' 2011, *WorldWatch Institute: Vision for a Sustainable World*. Available from: http://www.worldwatch.org/node/5443. [23 November 2011].

Pimm, S. L. and Raven, P., 2000, 'Extinction by Numbers,' *Nature*, 403, pp. 843–845.

Reynolds, L. and Nierenberg, D., 2012, *Worldwatch Report 188: Innovations in Sustainable Agriculture: Supporting Climate-Friendly Food Production*, Worldwatch Institute, Washington, DC.

Riley, S. S., 2003, 'Ecology Is a Sistah's Issue Too: The Politics of Emergent Afrocentric Ecowomanism,' in *Worldviews, Religion, and the Environment: A Global Anthology*, ed. Foltz R. C., Thompson, Belmont California, pp. 472–481.

Sarkar, S., 2012, *Environmental Philosophy: From Theory to Practice*, Wiley Blackwell, West Sussex.

Schwartz, R. H., 2001, *Judaism and Vegetarianism*, Lantern, New York.

Semon, J. J., 2012, 'If Everyone Lived Liked Americans, How Many Earths Would We Need?,' *SiOWfa12: Science in Our World: Certainty and Controversy* October 24. Available from: http://www.personal.psu.edu/afr3/blogs/siowfa12/2012/10/if-everyone-lived-liked-americans-how-many-earths-would-we-need.html. [22 February 2013].

Simpson, A., 2005, 'Sons and Fathers/Boys to Men in the Time of AIDS: Learning Masculinity in Zambia,' *Journal of Southern African Studies*, vol. 31, no. 3, pp. 569–586. Available from: http://africomnet.org/events/practicum/2007/resources/resource1%20%2810%29.pdf. [26 April 2014].

Spelman, E. V., 1982, 'Woman as Body: Ancient and contemporary views,' *Feminist Studies*, vol. 8, no.1, p.109-131.

'Suze Orman's Intervention with "Octomom" Nadya Suleman,' n.d., *Ophra.com*. Available from: http://www.oprah.com/oprahshow/Suze-Ormans-Intervention-with-the-Octomom_1/4. [18 February 2013].

'10 Billion Lives: North American Tour,' 2012, *FARM*, May. Available from: http://www.10billiontour.org/. [3 February 2013].

'Ten Facts About the Child Tax Credit,' 2011, *IRS* 10 February. Available from: http://www.irs.gov/uac/Ten-Facts-about-the-Child-Tax-Credit. [27 January 2013].

'Vegans Save 198 Animals a Year,' 2010, *PETA*, 13 December. Available from: http://www.peta.org/blog/vegans-save-185-animals-year/. [11 Oct 2014].

'Vegetarian 101,' n.d., *People for the Ethical Treatment of Animals*. Available from: http://www.peta.org/living/vegetarian-living/vegetarian-101.aspx. [12 December 2011].

'Wage Gap Remains Statistically Unchanged,' 2012, *National Committee on Wage Equity*. Available from: http://www.pay-equity.org. [13 February 2012].

Warren, K., 2000, *Ecofeminist Philosophy: A Western Perspective on What It Is and Why It Matters*, Rowman and Littlefield Publishers, Inc., Lanham, Maryland.

'Why are Mothers Glorified?' n.d., *The Childfree Life: A Safe-haven in a Baby-Crazed World*. Available from: http://www.thechildfreelife.com/forum/viewtopic.php?f=4&t=10050. [14 June 2014].

'Women CEOs' *CNN Money*,' n.d. Available from: http://money.cnn.com/magazines/fortune/fortune500/2011/womenceos/. [15 February 2012].

World Bank, 2014, *Energy Use (kg or equivalent per capita)*. Available from: http://data.worldbank.org/indicator/EG.USE.PCAP.KG.OE. [1 May 2014].

'The World at Six Billion,' n.d., *United Nations*. Available from: http://www.un.org/esa/population/publications/sixbillion/sixbilpart1.pdf. [15 June 2014].

'World's Seven Billionth Baby is Born,' 2011, *The Guardian*, October 30. Available from: http://www.theguardian.com/world/2011/oct/31/seven-billionth-baby-born-philippines. [15 June 2014].

'World's Wealthiest 16 Percent Uses 80 Percent of Resources,' 1999, *CNN*, 12 October. Available from: http://www.cnn.com/US/9910/12/population.cosumption/. [11 October 2014].

# SECTION II

# Bringing Change

## Activists and NGOs

**Poem:** Randall Gloege

### An Absurd Design

Once upon a time, there arose
from roiling spirals of carbon
the forms of me and you
and of all animated, self-updating things,
the low, the high, the wet, the dry—
shape-shifting jellyfish, sidewise-scooting crabs,
serene coral snakes, hanging sloths,
path-ravening ants, invisible patient iguanas,
insinuating sand flies, priest-cowled capuchin monkeys,
poised black widows, wobbly swift bears,
syrup-spitting hoppers, sand-pawing bison,
luminous scorpions, light-absorbing ravens,
gilded green beetles, diving foxes,
cannibalizing gophers, cliff-bounding goats,
blind moles and taloned eagles.
On and on and on and on,
a multitude to be sensed
in the squeeze, glide and slidder,
the dither, march, creep, zip and dance,
the leap, soft shoe, whirr and gallop,
the creep, flap, jerk, pounce, scurry,
scramble, twitch and crunch
of the concrete.

# 21

# HUNTING FOR A TROPHY HUNTING ETHIC

## Big Bears and Little Men (with Big Guns)

*Chris Darimont, Chris Genovali, Christina Service, and Paul Paquet*

*On a rainy fall day with only eagles to bear witness, we lost a gentle giant. He was an old friend after whom an entire ecosystem has been named.*

*The river was thick with salmon that day, its banks swollen with rippling water. Riverside branches bounced up and down as they danced a tug-of-war between the flowing water and their well-grounded roots. Leaves of crab apple trees, now in their golden-red hues of autumn, gently shook as if they could already feel the harshness of an impending winter.*

*Knowing too that lean times were coming, bears were lumbering along trails that lead to fish-laden streams. In pursuit of precious calories, they placed massive paws in millennia-old pockets worn into thick moss where generations before them had rhythmically stepped in search of salmon.*

*Just one week before, we had been among these giants, our friend included, bearing witness to autumn feasts. Moist breath hovered around their ever-active nostrils. Beads of water flew as they shook their massive bodies. Huge tongues gently lapped precious salmon eggs freshly deposited in the gravelly bottom of a clear stream. When the wind was right, we could smell grizzly bear—a wild-earth fragrance. Despite the raucous screams of gulls scrapping over prized leftovers, we could hear the bruins fishing—Animals the size of a Volkswagen beetle were sprinting through knee-deep water, swinging and thrashing 50 kg (110 lb) forearms at fish who were just as determined to escape the bears' jaws.*

*This magnificent interaction between natural predator and prey could only be severed by a new and wholly aberrant predator—one who turns bears into trophies. Just days after our experience in this beloved watershed, a group of camouflaged killers struggled up the same busy river. The jet boat could only deliver them so close to their quarry. Their high-powered rifles, lethal from up to 350 meters (400 yards) or more, compensated for what their office-softened bodies could not accomplish. The modern human super-predator's only physical challenge is steadying soft hands rocked by testosterone and bombast.*

*With some effort, the lily-white fingers choose their moment and twitch just enough to pull a mechanical lever. The rest is done for them. A bullet flies through the damp air and tears through muscular flesh. Its victim recoils, flailing in confusion at the sudden advent of excruciating pain, hot, expanding metal tormenting his viscera.*

*Gulls quiet. Men hoot, delighted that this "prize" is theirs.*

*Powerful forearms fail the panicked bear as he tries to run but tumbles into a nearby bush. Blood flows from his nose and mouth. His stomach, ravaged by metal, deposits freshly acquired salmon eggs onto the forest's dark floor. Shrubs snap and branches shatter under his great weight as a once peaceful bear protests newfound suffering.*

*Terrified of the bear's great force, the trophy hunters stand back and wait for the bear to die. Time slows, as it always does when death cannot come quickly enough.*

*Shiny knifes—rarely used except on such occasions—slide under the bear's skin, claiming his body parts as evidence of their manly conquest. They walk away with a hide that connects a severed head to severed paws and claws that were, only minutes before, connected to an individual—a bear—who hunted fish for survival. Even when the men are gone, blood spoils the fragrance of damp earth.*

*The hunters depart. They do not look back at the hauntingly human-like body that they have left, skinless, lying face down on the forest floor. Where the great paws used to be, only bloody bone lay curled into the earth like an old man fallen streamside, waiting to be taken back into the soil.*

## Introduction

For fifteen years, the Raincoast Conservation Foundation has been protecting species and habitats in the river valleys of coastal British Columbia, Canada. In this essay we explain how we came to protect bears—bears as individuals and bears as a species—and their *Great Bear Rainforest*. In the process, we share stories that illustrate who we are and why we work to save bears and forests. Without modesty, we celebrate "our" successes because we work for bears and the forest—our success only has meaning because it is fundamentally their success. Without shame, we share the pain of our failures, knowing that the bears and forests suffer for our shortcomings. In the process, we unveil how landscape and bears are linked—both victims of violent exploitation, both dependent on one another. In concluding, we look to what the future might hold for this fascinating animal and magnificent wilderness, knowing that the fate of the one is the fate of the other, that the future of each bear is tied to the Great Bear Rainforest.

## Raincoast for this Rainforest

Initially, our focus was on species and populations. We focus on grizzly bears, and other "iconic" species. We did not conceive of ourselves as an animal welfare group—at least not at first. Part of our journey, as this essay demonstrates, moved us from thinking about bear populations to thinking about individuals within those populations. This transformation required spending time with individual bears we grew to care about, as well as learning about and applying ethics to our activism.

It was Raincoast that put forward the area's now popular moniker—the *Great Bear Rainforest*—and helped this giant patch of wilderness move from obscurity to renown. In this remarkable and massive forest (60,000 square kilometers, 25,000 square miles), more than 1,000 salmon rivers flow through vast stretches of pristine watersheds and thousand-year-old trees harbor the southern stronghold of North America's remaining coastal grizzly bears. Thanks primarily to hunters and ranchers, as well as hydro development, urban sprawl, and fish-eating humans, there are no longer any coastal grizzly bears south of British Columbia (BC), though they once roamed the areas that are now designated as Washington, Oregon, California, and northern Mexico.

Raincoast's conservation strategy uniquely blends the tenacity and acumen of seasoned campaigners with the rigor and authority of scholars. We are conservationists, professors, students and citizens of this area, all united by our determination to learn about and save this precious coast and all of its wondrous inhabitants. We increasingly weave ethics into our arguments and decision-making process. We practice pragmatic idealism, coupling the best available scientific evidence (our own, carefully vetted by peers in academic journals) with "informed advocacy." In short, we represent bears and salmon as species, and ancient forests in the boardrooms of industry, around policy tables and, importantly, in public discourse. We have learned to go beyond advocacy for populations and habitat. No matter how pragmatic we were, the thought of an *individual* suffering as a result of human activities (in an otherwise "conserved" population) eventually hit home. We now speak for every individual who cannot speak for himself or herself. And when conventional earth and animal advocacy fail us, we go straight to direct action. In these ways we have chartered new territory for environmentalists and animal advocates alike, even as we have taken big risks on behalf of this old forest and its many residents.

## Animal and Forest as One: the Great Bear Rainforest

Fifteen years ago, Raincoast Conservation Foundation's inaugural environmental campaign focused on threats of industrial clear-cut logging. As timber companies destroyed ancient temperate rainforests in south-coastal BC, the great bear and great logging machines came into competition for valued lowland forests. The same enormous cedar trees that protected favorite forage plants for grizzlies also harbored vast quantities of lumber. The streams where wild salmon migrate were the first to suffer; siltation from deforestation choked out salmon life. Bears in this region depend on salmon, and the health of this region simultaneously depends on the bears. The fate of a forest is the fate of an ecosystem, in this case from tiny salmon eggs to massive grizzly bears.

Fortunately, the environmental movement met this threat head on. As logging migrated farther and farther north in search of yet more forests, Raincoast joined others voicing concerns against BC's so-called "war-in-the-woods," (Rossiter, 2004). To put this little-known but massive place on the map, we needed a name as significant as the rainforest itself. No easy task; what could do justice to an area almost completely devoid of modern industrial activity, a place where wildlife trails outnumbers roads by 1,000 to 1, where there were more grizzlies (a species that is sparse in any ecosystem) than businesses? How could we do justice to a place where, upon entering, humans felt like they had stepped back 1,000 years to a time when people were not dangerously overpopulated, and industry had not oppressed every inch of North America?

BC's biggest bears—and a sizeable proportion of the province's total bear population—are salmon-eating grizzlies along the coast. Brown bears hold a magical place in the hearts, minds and imaginations of most humans. Combining the majesty of a forest with our fascination for bears, Raincoast (along with other environmental organizations) settled on the name, *Great Bear Rainforest*. Little did we know the full significance of the name we had chosen.

At the time, one of Nature's most elegant ecological relationships was being meticulously uncovered by scientists. With salmon as intermediaries, Canadian researchers discovered the role of these "great bears" in the rainforest's intricate workings: Bears are the forest's gardeners. Each bear transfers hundreds of pounds of fish "fertilizer" alongside streams and onto the forest floor. Consuming primarily the choice eggs and brains of spawning salmon, bears leave the rest of the fish carcasses along stream edges, where they provide critical nitrogen and phosphorus to otherwise nutrient-limited vegetation (Darimont et al., 2010). In no small part, this seasonal fish subsidy contributes to some of the richest soils and largest trees that have ever graced the Earth.

Indeed, in beautiful reciprocity, the fate of the forests is linked to the fate of bears. As environmentalists, we were initially focused on ecosystems and populations. Meanwhile, Canadian scientists had just uncovered the critical importance of fish-eating bears—black and brown—to the health of the Great Bear Rainforest. Understanding these interconnections brought us to seek an equally interconnected solution, and this new understanding became central to our rainforest campaigns.

## Dead Bears Leaving the Rainforest

We learned early in our campaigns that we were not the only people to venture among these famously tall trees and clear salmon streams in search of bears. A thriving and lethal industry was being fuelled by hunters who killed grizzly bears for pleasure and amusement. We had never imagined such a pastime—how could a flagship species be killed for mere trophies? "How bizarre, sad, and perverse" we thought, not really knowing how to think about such a selfish and wasteful form of entertainment. A couple of appalling encounters with carcasses (like the one left by the hunters in our opening story) motivated us to learn more about this odd fixation with sport-killing.

Internet boasting, complete with macabre photographs and videos, revealed more than we wanted to know. We learned that trophy hunters kill around 300 grizzlies every year in BC. Close to half are killed by "non-resident hunters," including U.S. and European "clients," who pay Canadian guides to take them into British Columbia's wilderness because humans in their own areas have already wiped out all of the local grizzlies. We already knew that we could not separate the fate of bears from the fate of their forests. Therefore we had to fight for an end to grizzly trophy hunting *throughout* BC just as we must fight for the forests themselves. What would we gain if we preserved a forest that was no longer a home for grizzlies and other wild animals who *belong* there? Motivated by the knowledge that most conservation losses are permanent while most gains temporary, a twin Raincoast campaign was born. However, our strategy evolved as time passed, we shifted our focus from numbers and population dynamics to individuals and ethics. We would begin, unconsciously, to cross the imaginary border between earth and animal advocacy.

## Partial and Ephemeral Mathematical Solutions

Our initial strategy was to highlight inherent uncertainty in the provincial government's bear population estimates. Because official estimates were never the result of rigorous scientific method, such speculation as to how many bears existed, which underlay government policies regulating how many could be killed annually, were easy targets for criticism. How could management make key decisions on these critical and fundamental aspects of wildlife management based on such incomplete information? More puzzling, why did their estimates of bear population numbers—a result of so-called "expert-based" modeling—increase constantly, despite decades of logging, declines in salmon, increased human encroachment, and trophy hunting?

Our numbers campaign was bolstered by several high-profile bear biologists. Some of these biologists (including co-author Paul Paquet) had previously served on the province's "independent" grizzly management committee, the members of which, ironically (and revealingly), were fired by the Deputy Minister for "intransigence." Apparently, whistle blowing on laissez-faire management was not embraced by upper level bureaucrats. Their loss was our gain. Sympathetic to our cause and armed with scientific information about bear numbers, reproduction, and the effects of trophy hunting, these biologists added the "informed" content to our "advocacy," creating our trademark approach—"informed advocacy."

We warned a largely uninformed public that the BC government supported killing grizzlies for sport, yet maintained little to no dependable information about bear populations. We posted billboards throughout Vancouver and Victoria, placed ads on buses and transit shelters, debated the issue on local and regional news programs, and penned myriad opinion pieces for every print media venue we could find.

Sparks began to fly. The most fervent and extreme opposition came from the pro-trophy-hunting faction. At the height of the campaign, we began to receive a litany of threatening and nasty emails and letters from "anonymous" bear hunters. We sometimes found the combination of vitriol and illiteracy humorous, like the guy who, after having described in detail the violent things he would do to us if we happened to wander into his hunting camp, ended his missive with the valediction "Fuck You Very Much." (As part of our reply, we provided this charming man with a link to an anger management website.)

One of Raincoast's early turning points came about because of a film we produced with a zealous young videographer. Only five minutes long, this hard-hitting piece juxtaposed the voices of a provincial government biologist and an internationally known scientist and environmentalist Dr David Suzuki in 45 second clips. Suzuki noted: "The idea of a 'harvestable surplus' of grizzly bears is an anachronism in today's world." The well-meaning and otherwise gifted government biologist then stumbled over the province's bear population estimates, revealing an obvious lack of confidence in government calculations.

To mount an effective campaign, we had to gather additional background information on BC grizzlies and the bear trophy hunt, so Raincoast member (and co-author) Chris Genovali, in company with a British television producer, visited a taxidermist-cum-trophy hunter workshop on Vancouver Island. While the producer engaged the taxidermist with (mostly) innocent questions, Genovali wandered across the room to two grizzly bear corpses "in progress." These two once beautiful bears were rearing up on their hind legs—positioned as if they might be a great threat to even the manliest of men armed with high-powered automatic weapons. Building on this absurd bluster, the taxidermist had bared the bruin's teeth in a position of unlikely menace. (In truth, before writhing in pain, their faces likely showed nothing but content as they feasted on salmon.)

We knew the truth, and apparently viewers did, too. Our video showed a similar taxidermied trophy from a local gun shop, and this surreal image proved to be very effective. A wave of support washed across Raincoast, and we learned something unexpected: Presenting the lifeless, taxidermied corpse of a recently vibrant bear created a critical *emotive* element in our film, and in our campaign. The success of this clip suggested to us that we ought to give issues other than numbers more prominence in our campaigns. From this we learned that animal advocacy not only suited our minds, but our hearts and our cause, and that much-maligned emotions were a very real part of being human, and a very effective tool for winning the hearts and minds of the public. The lines between earth and animal advocacy had become blurred, and in the process, we became more effective.

We soon found opportunity to leverage our newfound groundswell of support. Anticipating the upcoming provincial election of spring 2001, we met with the leaders of the New Democratic Party (NDP). Consequently, the grizzly hunt controversy was nudged into a central topic of the upcoming election. To satisfy an increasingly green and bear-aware electorate, the NDP enacted a three-year province-wide moratorium on grizzly bear trophy hunting. These gentle giants were given their first reprieve from blasting guns since the advent of sport hunting.

We could not help but feel a little bittersweet: While grizzlies would be spared, we knew full well that many other animals would be subject to the ravages of trophy hunters. Wolves, cougars, black bears, and many other animals—less "flagship" but every bit as sentient—would continue to be subject to sport hunting, dying by the thousands in upcoming seasons. And for what—wall hangings, rugs, perhaps collars for parkas? Grizzlies won our hearts and minds early on with their charisma and ecological sensitivity, giving them a powerful edge over individuals from other species. Consciously or not, we turned our backs on others, but we were not aware of this at the time. This was the cost of pragmatism. Now, as fully fledged animal and earth advocates, we are more aware: There are no longer any favored species, only suffering individuals, each of whom is critical to ecosystem and Raincoast health.

Against an astonishing 30-year graph of kills, this victory seemed even more bittersweet. Alas, as the Liberal party (read conservative) gained control of the province, our victory was short-lived: Their first order of business—the very first - was to reinstate grizzly killing. Just after the spring hunt was put off, Liberal leaders announced that the fall hunt was back on.

## Cutting off Europe

Despite this defeat, we were on the cusp of a significant gain. Returning to numbers and governmental mismanagement, Raincoast used an international scientific process in the hope of cutting hunting guide clientele from targeting BC grizzlies. Working in concert with other environmental groups (UK's Environmental Investigation Agency and Germany's Pro Wildlife) we petitioned the scientific review panel from the European Union to ban traffic in grizzly remains. Raincoast effectively argued that, although a common practice among government agencies and wildlife managers, intuition or improvised guesswork is no substitute for rigorous science when it comes to management and conservation, particularly given the serious plight of most bears worldwide. In response, the EU Scientific Review Group noted that British Columbia had failed to protect critical grizzly bear habitat and could not demonstrate that their grizzly hunt was sustainable. Consequently, the EU banned the importation of grizzly bear body parts, effective in November of 2001. As a result, European hunters were discouraged from killing BC bears—What trophy hunter would pay large sums of money if they could not return home with a trophy? Very few. Despite persistent government appeals and heavy lobbying by the BC provincial and Canadian federal governments, the EU upheld their ban—twice—on the importation of BC grizzly bear body parts. The EU, in concert with Raincoast, noted that government appeals contained questionable science using numbers "like a drunk uses a lamppost; more for support than illumination."

## A Legal Battle Royale

We wanted more information about their questionable science, their spurious but much-touted numbers. Accordingly, for five years we battled the provincial government over hunter information, urging them to release specific geographic data for the province's roughly 8,000 recorded grizzly kills over the last three decades. Our goal was to commission an independent analysis of grizzly deaths in each region of British Columbia, and to investigate which features (such as roads and access to food) made certain areas particularly dangerous for bears. Even though we had lined up some of the best conservation scholars available to help with a project that had significant management implications, the Ministry of Environment consistently denied us access. Why? We could only imagine two possible reasons: The government did not know these critical details, or the government did not want anyone examining their management. We could not decide which explanation was worse, and we suspected that both motives might apply.

Angry and appalled, we looked for legal assistance. The Sierra Legal Defense Fund (now Ecojustice) took up our plea for the bears in the spring of 2000, battling a very reluctant government. In light of the Freedom of Information and Privacy Act (FOI), our request was granted by the Information Commissioner. But the government, pressured by powerful hunting lobbies, appealed the Commissioner's ruling. Nonetheless, in November of 2002, justice prevailed at a higher level: The Supreme Court of British Columbia rejected the government's specious appeal. Now no amount of threatening phone calls and emails could quell our independent analysis, or our jubilant celebration.

We especially rejoiced in the nature of the ruling. The government had claimed—ironically—that the release of this information would negatively affect grizzly bear conservation. They argued that knowledge of where grizzlies were vulnerable to killing made grizzlies yet more vulnerable to killing. Neither the Information Commissioner nor the courts bought their argument. The defendants had offered no convincing evidence of such risk; the province's actual objection, the courts reasoned, was that this coveted information could be used to criticize the government's grizzly bear management. So much the better.

Even as we were celebrating, the Guide Outfitters Association of British Columbia (GOABC), a special interest lobby that represents well-armed men who lead trophy-tourists to bears, stepped forward to help the fallen government: They threatened lawsuits over the release of *their* "proprietary information." Precisely where clients killed bears on public lands, they argued, was *private*, and such lucrative information need not be shared with greenies who might interrupt hunts, or with competitors who might thereby find the easiest places to bag a "trophy." The outfitters' indignant, pseudo capitalistic posturing failed to overturn the Supreme Court FOI ruling in a BC Court of Appeal. Again, the judges proved to have both intelligence and integrity; the hunters' increasingly desperate pleas fell dead as a coveted "trophy bear."

Nonetheless, the same conservative government that had quickly overturned the short-lived, province-wide moratorium, and that had promised to be the most "open and accountable" government in Canada, continued to suppress information. Delays piled on delays: Database troubles, staffing shortages, and other common administrative challenges seemed to reach epic proportions, preventing even the most simple of tasks—passing on information. When the data finally arrived, our confidence in BC government fell yet further: We received a pile of paper stacked higher than a grizzly's shoulder hump. Did our government not have computers, or was this yet another shameful attempt to stall our analysis?

In any event, with the help of collaborators from both the David Suzuki Foundation and academic colleagues, basic analyses uncovered important results: The government's own quota (that "harvestable surplus") had been repeatedly surpassed in some "grizzly bear management units." Over time, with improved techniques and digital data, we even documented a considerable proportion of *legal* kills in *protected* areas.

Important findings to be sure, but even these statistics were not enough to change policy. Numerical arguments were effectively downplayed by the Ministry of Environment as misinterpretations of "bear allocation" calculations put forward by their biologists. Moreover, the law, and a decades-long legacy of wildlife management, stood squarely on the side of a small population of vocal special interests—trophy hunters. (Have you seen any bears making policies lately?) Our efforts were successful on all counts except the final and most critical count—bringing lasting change for bears. How could such telling numbers be so roundly ignored? We regrouped to ponder our next move.

## Ethics to the Forefront

For every biologist advocating caution, several others legitimize the *status quo*. Hunting lobbies—and indeed government officials—have legions of bullet-blazing "wildlife biologists" ready to justify a grizzly hunt. The province continued to appeal to the EU to reverse the ban on importing grizzly parts, and we wondered how long our victory in Europe could hold up against the onslaught of special human interests backed by wealthy lobbies. Is not a debate about numbers futile when working with a provincial government willing to bend to the pressures of a powerful hunting lobby? Even worse, what if statistics from population and hunter kills showed that there *were* "enough" bears to support a "management" or hunter's vision of "harvestable surplus"? Even if the Great Bear Rainforest's grizzly *population* was able to withstand some trophy hunting, we had, somewhere along the line, committed to a policy of individual protection, of flat-out no-kill for the animals in the Great Bear Rainforest. We wanted to end the trophy hunt permanently and completely. Whether or not the population could sustain such a hunt, individuals could not. Killing grizzlies for trophies was wrong even if these great bears were plentiful.

This notion of a "harvestable surplus," however grotesque, had become a valuable test of who and what we wanted to be *as conservationists*. We were no longer concerned only with some abstract idea of "population health." The individual bears we had come to know and love deserved to live out their lives peacefully in their forest home—even if their death might not affect overall population health. Could we enjoy the knowledge of "conserving" a population (that is, keeping species numerically stable over time) if we knew that individuals within those populations—individuals as capable of feeling pain as you and I—were suffering and dying prematurely at the hands of hunters? Clearly not—not any more.

This shift strained some of our most prized and deep-rooted understandings of who we were: We were a team of earth advocates and scientifically-informed conservationists. We began to think about the seemingly inherent tension between environmentalists—especially conservation scientists—and animal advocates. Some conservation researchers go so far as to rationalize the suffering of individual animals as necessary to larger goals, such as species preservation, or even necessary for personal research goals (Darimont et al., 2008). In other words, the means (sacrificing individual animals) is used to justify the end (species preservation or research aimed at this goal). From an ethical perspective, this is unacceptable. We know this because we do not accept this outlook with regard to human populations and human lives.

A philosophical refutation of the conservationist's species-oriented outlook cannot, however, by itself, clear the tension and unite earth and animal advocacy. For that, we must refocus on deeper, shared concerns: The destructive tendencies of humanity that liquidate our precious stands of old-growth temperate rainforest are supported by the very same outlook that fosters humanity's destructive tendencies toward animals, including trophy hunting. Moreover, what befalls one befalls the other. When forests are removed, untold individual animals suffer (Paquet and Darimont, 2010). And when bears suffer—any bear—the forests as a whole suffer from the damage or loss of an important inhabitant.

Once we recognized this, we looked beyond population assessments and wildlife science. The simple elegance of an ethical argument—using time-honored and rigorous philosophical methods to reach a conclusive answer regarding *right* and *wrong*—seemed very appealing. But exactly how do ethicists determine what is right and what is wrong?

Almost all societal conflicts—from the influence of the state to the workings of legal systems—have been negotiated by an overarching ethical philosophy of *utilitarianism*. Simply stated, decisions are right if they produce the greatest good for the greatest number of all who will be affected: Ethics, therefore, can be derived by estimating the likely outcomes of various actions (Curry, 2006).

We had rephrased our question: Was trophy hunting ethically justified using a utilitarian equation? In the simplest algorithm, the wellbeing gained by the trophy hunter must outweigh the wellbeing of all others who are affected: other citizens, the bear, the bear community, and the many other residents of the bear's ecosystem, and the forest. By way of utilitarian assessments, let us first take the perspective of the trophy hunter. The trophy hunter derives no basic needs (such as food, shelter or security) by killing grizzlies. It is important to note that most EU hunters stopped killing BC bears when they could not bring home trophies. This indicates that most trophy hunters kill largely to gain a trophy. For the hunter, the trophy merely signifies a perceived accomplishment so that viewers of his or her trophy stir a sense of pride (Gunn, 2001).

Next, let us take the perspective of the bear. What does the bear gain from the trophy hunt? He or she indisputably pays the ultimate price for the hunter's trophy:

excruciating pain, suffering, and loss of life. (Death constitutes the greatest loss for any living being (Gunn, 2001)). The bear gains nothing and loses everything. Similarly, the bear community gains nothing but loses everything, as does the bear's ecosystem.

This raw calculation exposes the utilitarian reality of trophy hunting—all trophy hunting. But how do we weigh a bear's loss against human gains? A trump card, played often in ethical debates involving those who are less powerful, distinguishes those who make the rules from those who do not. Where nonhuman animals are concerned, humans overwhelmingly assign greater importance to themselves, thereby trumping any interests other than their own (Taylor, 1981). In contrast, in perhaps the most significant extension of utilitarianism, philosopher Peter Singer famously argued, simply and powerfully, that there is no logical reason why the utilitarian calculation of net good should benefit only humans, or, for that matter, why it should not reach across all sentient species. Anything else, he argued, is "speciesism,"—irrational and unjust human privilege over other beings (Singer, 1975).

The core of Singer's argument, rooted in suffering (basic physiology), was stated by an earlier utilitarian who famously asked, "the question is not, *can they reason?*, nor, *can they talk?*, but *can they suffer?*" (Bentham, 2007). In asking this question of grizzlies, we had fully erased the line that once stood between Raincoast as an environmental organization, and Raincoast as an organization advocating for animals. We all knew that bears suffered when hot shards of metal tore through their vulnerable flesh.

Utilitarianism is now a cornerstone of Raincoast campaigns because we have found this ethical argument posed by animal advocates effective in Raincoast's efforts to protect BC bears. Not only that, but we quickly discovered that this method worked with the public, as well, for at least two reasons. First, wildlife management policies invoke numbers (calculations of "harvestable surpluses" for example), and ethics. For example, most jurisdictions outlaw hunting bears with dogs as an egregious abuse, likely to cause morally unacceptable levels of suffering. Moreover, many provinces ban activities that deviate from "fair chase" principles. For instance, bear baiting, which draws bears to a hunter's gun, is widely outlawed. This means that ethical arguments are already in play, and need only be brought to the fore—and consistently applied.

Second, ethical arguments are important because they resonate with the (voting) public. Simply stated, while most people have not consciously done a utilitarian analysis, most people oppose killing animals for trophies as frivolous bloodlust. Indeed, four province-wide polls over the last decade (conducted by third party pollsters) have shown that 73 to 81% of British Columbians support a ban on trophy hunting grizzlies. This support is *not* based on population dynamics or numerical harvest regulations, but on a visceral repulsion for unnecessary slaughter, for pleasure-killing to feed one's ego. This wide-spread disgust for trophy hunting is rooted in society's sense of how humanity *ought* to treat nonhuman animals, and a reverence for life.

## Talk to Action: "Raincoast Outfitters Ltd."

Policy is slow to catch the ethical gestalt of society. Meanwhile, in the case of trophy hunting, animals continue to suffer and die for no morally justifiable reason. Professional hunting guides in the Great Bear Rainforest (and beyond) continue to help tourist trophy hunters kill hundreds of BC bears every year. While the EU ruling hit them hard (about half their clientele were European) outfitters clung tenaciously to their way of life, taking what tourists they could find, and grizzlies kept dying.

But a serendipitous breakthrough was headed our way. During the fall hunting season of 2004 an outfitter moored his bloody boat, the *Silver Grizzly,* in a small marina on Denny Island, Raincoast's northern outpost. Leaving the remains of grizzlies in the boat's abattoir, the primary operator went ashore to use the nearest payphone. There, by remarkable happenstance, he left his lengthy dossier. By another amazing quirk of good fortune (for the bears), one of Raincoast's friends used the phone next. Greedily falling upon the papers, he found in the outfitter's weathered stack of records a string of profanities directed at provincial bear managers, tightened regulations, the EU ban, and other threats to trophy hunting that he blamed, at least in part, on Raincoast. Only through these documents did Raincoast learn that outfitters were now receiving a host of phone calls accusing them of unethical killing. Most importantly, the manager of the *Silver Grizzly* mentioned financial troubles that put him on the edge of bankruptcy.

As we read his laments, we rejoiced. We found in this happenstance a glorious opportunity: If we could not beat him, why not *buy* him? In a series of bizarre phone calls and emails that slowly shifted from mutual disgust to a crescendo of courtesy, we struck a deal: We were offered the exclusive right to sell hunting licenses to foreign hunters in an area three times the size of Yellowstone National Park (about 13,500 square miles). Thanks to donors as diverse as school children and oil executives, we were able to raise a remarkable $1.3 million to clinch the deal.

As the proud owners of one of the largest exclusive hunting territories in North America, Raincoast now had certain legal obligations. Hunting must go on—or at least the sale of "hunting" licenses. Raincoast must fill government coffers with money from "kill tags" by selling eager clients the right to "hunt" bears. By some people's estimations, Raincoast was famously unsuccessful as an outfitter: Not one animal has been shot since we bought the rights to outfitting in this vast region. There has been, however, a lot of commercial "shooting"—with cameras. Importantly, Raincoast invited local people to benefit along with the animals by partnering with us as wilderness guides. The hunting territory that Raincoast purchased lies within First Nation lands, and their support has been especially important to our continued success.

The forest was the victor, along with the grizzly bears and other local victims of trophy hunting, such as black bears and wolves. Our monumental accomplishment earned Raincoast front-page coverage. We had found an effective form of advocacy—animal *and* earth advocacy. We had also started a trend: Following Raincoast's lead, the Nature Conservancy—in collaboration with local First Nations—bought a large hunting tenure on the Taku River, abutting southeast Alaska. Currently, along with an independent consortium of coastal nations, Raincoast is negotiating the purchase of rights to several other hunting territories. The goal? Shut down commercial trophy hunting along the entire coast of British Columbia.

## Our Day with eBay

Buoyed by success, and an evolving ethical framework, Raincoast next took on a company that makes even the largest outfitters look like a lemonade stand.

It all began when Raincoast received a shocking email from a concerned supporter. He informed us that alongside valuable stamp collections and Nintendo consoles, guided grizzly hunts (in BC and beyond) were auctioned on eBay. Any eBay visitor could bid on the life of a bear (and other nonhuman animal—and nonhuman animal body parts). Raincoast sprang into action. First we tried strong-arming: We threatened scandalous exposure through the very same internet media. For this, we were granted an afternoon with their chief legal counsel, their entire "green team," and various other players. We presented them with the ecological, evolutionary, and ethical implications of trophy hunting. Many attendees choked back tears as they viewed gruesome images of dispatched animals posted to their very own eBay site. We highlighted their hypocrisy in applying eBay rules, which state that items or activities that "glorify or promote violence" are prohibited. At the end of our presentation, we posed what we viewed as a key question: Does eBay's rule relate only to violence towards humans? We closed with a warm "Thank-you," complete with California handshakes, then left with complimentary EBay coffee mugs and stationary. We felt confident we were on the verge of another major victory.

But we were wrong. The corporate world, we learned, is not motivated by ethical arguments. In their reply, they answered our closing question with a resounding, "Yes!" Apparently, eBay only prohibited the sale of items that might involve violence toward humans. They focused on local wildlife hunting policies as a way to pad their bottom line: If a "harvestable surplus" existed for trophy hunters, then eBay ought to join in the fray.

Their answer did not satisfy Raincoast and we are now regrouping for another round. Perhaps with corporations, public image is the rightful target, and strong-arming might be more effective than ethics.

## Conclusion

Changing the government's human-centered wildlife management strategies is extremely difficult—frankly, we found that no amount of painstakingly gathered evidence won our case. So Raincoast found a more effective method of advocacy: We bought the exclusive legal right to guide commercial hunts, then extinguished the commercial hunt altogether. In this way we preserved the forest along with the many citizens who would have been killed by trophy hunters.

But only in the bizarre human world are rights to kill bought and sold. What of the world of bears—black and brown—in their rainforest home? We did not set out to own hunting licenses, but (by chance) this avenue proved a promising way to protect bears and the rainforest from trophy hunters. So we found both the courage and the financial means to try a completely new method of advocacy—radically deviating from planned methods and short-term objectives. Our flexibility and courage paid off, and we were able to reach our long-term goals much more quickly as a result.

But our success also hinged on soul-searching—on facing our own truths. We had to look within and honestly assess our own motivations. In the process, our work took on a moral character. The issues shifted from focusing on wildlife mismanagement, exposed through numbers and legal codes, to an examination of values—right and wrong. When we reorganized our efforts around this fundamental truth, we found commonality with the general population, and were able to move mountains—or at least buy out hunting rights for mountains—on behalf of wild residents and their forest homes.

In the absence of the *Silver Grizzly* and its lethal clientele in this particular guide outfitting territory, watersheds have again come alive. Whereas we once had to "stalk" gun-shy bears to get close enough for a glimpse, bears now trust our presence as we stand at a respectful distance. We watch them go about their important lives, simultaneously enjoying their personal existence while they perform a central role in the rainforest ecosystem: Through their massive deposits bears spread seeds into areas far from parent plants, and distribute nutrients from salmon remains around the forest floor. Do bears shit in the woods? *You bet!* But these days we are as interested in the individual bears as we are in their ecosystem. We have come to see that the bears whom we know enjoy life in their forest home. While Raincoast's massive purchases likely save only a handful of these individuals each year (because grizzly territories are famously large), the lives of each of these individuals matters, and the value of Raincoast's investment—with the help of many generous people—is therefore priceless. While it remains true that these bears are crucial to the ecosystem in which they live, it is no less true that *their* health is also important. As Peter Singer noted in his utilitarian assessment, *their* individual happiness matters, too. How could it not?

What will the future bring for these marvelous bears and their magnificent forest? Voracious clear-cutters haul away trees at a dizzying pace—more toilet paper is needed to wipe an ever-expanding human population; relentless commercial fishers strip the seas; and our diet of animal products and reliance on petroleum continues to threaten the world with climate change. Humans threaten grizzlies and ecosystems with their consumer habits. Bears suffer and die, and rainforests are destroyed because of *our* consumer choices. Additionally, individual bears (and many other nonhuman animals) continue to be killed for pleasure—whether the pleasure of the palate or the pleasure of the hunt.

Contemporary wildlife "management" has failed to protect bears—individuals and species—*and* their ecosystems; environmental groups like Raincoast now offer balance to the human-centered focus of government agencies and vocal and politically powerful (though minute) special interest groups. In the process, Raincoast has learned a valuable lesson: Ethics can trump scientific numbers. As Nelson and Millenbah note in their recent paper, *The Ethics of Hunting*, "inasmuch as the wildlife community begins to take philosophy and ethics more seriously, both as a realm of expertise that can be acquired and as a critical dimension of wildlife conservation, many elements of wildlife conservation and management will change" (2009). Whereas science can provide information, science cannot grant permission to behave unethically. In other words, "harvestable surpluses," and other ghoulish "management" concepts, do not (and *cannot*) justify unethical practices such as trophy hunting.

As we straddle the conventional divide between earth and animal advocacy, our goal is now twofold: Raincoast works to be sure that population-level safeguards are in place for the purposes of species conservation, including assessments of a habitat's "carrying capacity," *and* that any wildlife discussions about policies are informed by something we now recognize as critical to every individual within every population—humanity's "*caring* capacity."

## Discussion Questions

1 Describe the work of the Raincoast Conservation Foundation, their methods, and what they offer the larger environmental movement.

2 What constitutes a dangerous predator? Are you a dangerous predator?

3 What motivates bear hunters? Morally speaking, are these legitimate reasons to kill bears? Are these legitimate reasons to kill humans?

4 What political and economic forces affect the environmental work of Raincoast? What political and economic forces affect environmental activists and issues in your area? Do you feel that political and economic forces make the cause of nonhumans and the environment hopeless? What can you do to mitigate the power of political and economic forces and bring more hope to environmental causes?

## Essay Questions

1 What is the importance of individual bears in relation to and compared with the larger population of bears? What is the importance of individual humans in relation to and compared with the larger population of humans? Do you feel there is a difference? Why? Is this discrepancy reasonable or just?

2 In what ways do your life choices assume human supremacy over other animals and the environment?

3 Create a moral argument in favor of encouraging our species to die out on behalf 3 of the environment.

## Suggested Further Reading

Gunn, A. (2001) 'Environmental Ethics and Trophy Hunting.' *Ethics and the Environment* 6.1: 68–95.

Nelson, M.P. and Millenbah, K. (2009) 'Can We Have Our Animal Rights and Eat Them Too?' *The Wildlife Professional* 3: 33–34.

Paquet, P. and Darimont, C. (2010) 'Wildlife Conservation and Animal Welfare: Two Sides of the Same Coin?' *Animal Welfare* 19: 177–190.

Singer, P. (1975) *Animal Liberation: A New Ethics for Our Treatment of Animals*. New York, NY: Random House.

## References

Bentham, J. (2007) *Introduction to the Principles of Morals and Legislation*, second edition. New York, NY: Dover Publications.

Curry, P. (2006) *Ecological Ethics: An Introduction*. Cambridge, UK: Polity Press.

Darimont, C., Reimchen, T., Bryan, H. and Paquet, P. (2008) 'Faecal-centric Approaches to Wildlife Ecology and Conservation; Methods, Data and Ethics'. *Wildlife Biology in Practice* 4: 73–87.

Darimont, C.T., Bryan, H.M., Carlson, S.M., Hocking, M.D., MacDuffee, M.M., Paquet, P.C., Price, M.H.H., Reimchen, T.E., Reynolds, J.D. and Wilmers, C.C. (2010) 'Salmon for Terrestrial Protected Areas.' *Conservation Letters* 3: 379–389.

Gunn, A. (2001) 'Environmental Ethics and Trophy Hunting' *Ethics and the Environment* 6.1: 68–95.

Nelson, M. P. and Millenbah, K. (2009) 'Can We Have Our Animal Rights and Eat Them Too?' *The Wildlife Professional* 3: 33–34.

Paquet, P. and Darimont, C. (2010) 'Wildlife Conservation and Animal Welfare: Two Sides of the Same Coin?' *Animal Welfare* 19: 177–190.

Rossiter, D. A. (2004) 'The Nature of Protest: Constructing the Spaces of British Columbia's Rainforests.' *Cultural Geographies* 11: 139–164.

Singer, P. (1975) *Animal Liberation: A New Ethics for our Treatment of Animals*. New York, NY: Random House.

Taylor, P. (1981) 'The Ethics of Respect for Nature.' *Environmental Ethics* 3:197–218.

# 22

# CAMP UGANDA

*Haida Bolton*

Confucius once said, "Tell me and I will forget. Show me and I will remember. Let me experience it and I will understand." It is for exactly this reason that I created Camp Uganda.

As a child, I reveled in the empowering effect of camps. Surrounded by nature, living with many other kids, cooperating for chores, singing, hiking, canoeing, and playing games together! Canadian camps offered the happiest, most memorable, and educational times of my childhood. At camp, I felt a sense of belonging both within a community and as part of the natural world. Remembering those days at camp, I can still feel how my self-confidence swelled, how I became a more empowered individual and a more effective citizen. The American Camping Association noted that:

> Camp provides children with a community of caring adults, who nurture experiential education that results in self-respect and appreciation for human value. All of the outcomes—self-identity, self-worth, self-esteem, leadership, and self-respect—build personal competencies. These personal competencies are reflected in the four "C's" of the camp community: compassion, contribution, commitment, and character! For years, campers' parents have reported that when their children return home from camp they are more caring, understand the importance of giving, are more equipped to stand up for what they know is right, and are willing to be more responsible.
>
> (American Camp Association, 2014)

Camp was magic! In fact, camp-induced confidence ultimately led me to work for children's camps, which led me to create Camp Uganda. Camp Uganda is living proof of the power of camps for youth.

## Camp Uganda

In April of 1995, I was invited to go on a nature walk with Jane Goodall and Robert Bateman. Dr Goodall and I talked a fair bit. She asked me to write to her. I was elated. I did write, expressing my two great desires: I wanted to run my own children's camp, and work with wild chimpanzees in Africa. I asked Dr Goodall if I could work at Gombe Stream National Park. She suggested that I create a children's camp at a chimp sanctuary, and I knew I had found my purpose.

Dr Goodall's suggestion simmered for ten years. In 2005 I began to create Camp Uganda, and in January of 2007 Camp Uganda opened at the Uganda Wildlife Education Centre, where dormitories, food services, wildlife (most importantly, chimpanzees) and education staff were already available. We began with 22 campers from two different schools, inviting students who had shown dedication to their school's wildlife club. We also invited three teachers to attend our four-day camp. That first camp was a success, and we continued to grow. By 2010 we expanded to thirty campers and seven teachers from three different schools, and we try to offer a seven-day camp every year. We hope we can eventually provide more than one camp per year.

Children's camps provide the inner tools that young people need in order to take effective action to preserve forest habitats, which simultaneously preserve thousands of individual animals and plants. These tools, including greater self-esteem, confidence, independence, caring, knowledge, and experience, build a healthy foundation for positive leadership.

At Camp Uganda, we design activities to connect humans to nonhumans, and enable them to help to build a caring bond between campers and the other animals, who are their neighbors. Children at Camp Uganda help with chores, feed chimpanzees, lions, antelopes, zebras, rhinoceroses, and warthogs; they touch a rhino and a python, and they spend a good deal of time observing nonhuman animals, especially chimpanzees, at the Uganda Wildlife Education Centre. We discuss the similarities that campers witness between humans and chimpanzees. Ugandan campers also watch chimpanzees playing, and they never fail to notice how similar we look and act to our primate relatives. They *see* that chimps require food and sometimes get sick, just like we do. We visit a vet clinic, and they see that the chimps sometimes need to see a doctor, too. We also use creative methods to connect children to other animals, such as nature art, eco-games, skits, singing and dancing. When campers look into the eyes and hear the sounds of these close cousins—who share 98.6% of our genetic material—they form an emotional connection with their relatives, chimpanzees (Jane Goodall Institute of Canada, n.d.). This connection leads to a deeper understanding, and greater caring, which motivates young people to help preserve forest habitats.

The Uganda Wildlife Education Centre was nothing more than a zoo when it first opened. Luckily, the Centre's interests shifted, and Uganda Wildlife Education Centre now cares for confiscated, injured, orphaned or otherwise rescued wildlife. Some monkeys and snakes, crested cranes, peacocks, squirrels, and a few warthogs, roam the grounds freely. The animals are no longer exploited for human entertainment, but rather, are tended by a professional staff. And the Uganda Wildlife Education Centre has also come to specialize in education.

One of the greatest ecological problems in Uganda is deforestation, made worse by locals collecting fire wood for archaic cooking stoves. Over time, this has devastated much of the forest land, and much of the habitat that is necessary for Uganda's wildlife. A simple change of cooking methods is perhaps both the simplest and the most effective way for individual citizens to slow deforestation in Uganda. For example, solar cookers have been tested by the New Nature Foundation in hopes of reducing the use of wood. Unfortunately, these particular stoves did not work consistently in cloudy weather in forested regions, and were therefore not considered a workable replacement for wood. Research and testing continues.

Thanks to a generous grant from the International Primate Protection League, Camp Uganda added a new activity—teaching children how to build fuel-efficient rocket stoves to reduce their families' need for firewood. This is a particularly important issue in Uganda, where most people still cook over open fires. As part of this process, children engage in a science experiment that actually compares a newer rocket stove with a traditional three stone method of cooking over an open fire. Children readily see that the newer stoves are more fuel-efficient—using less wood, creating less smoke, burning more safely, requiring fewer trips to the forest—and therefore causing less deforestation, which means more homes for chimpanzees and other wildlife. Ugandans collect wood from nearby forests (New Nature Foundation, n.d.).

A narrated stroll through the pristine jungle at the Botanical Gardens awakens young minds to how their country might have looked centuries ago, when the forests were lush and full, reaching for hundreds of miles across the landscape. They learn methods to combat deforestation—they learn to plant two or more trees whenever they cut down a tree. Naturally this means that one of our activities is tree planting, to be sure that children know how to plant and tend young trees. At our annual 2010 camp we planted seven seedlings, and each of our campers and teachers took a seedling home to plant in their communities. To help inspire students, campers travel from their dorms each morning to the Elder Tree with the simple instruction: "Do your best and enjoy every step!" We hope children might remember this in their daily lives back home.

At Camp Uganda, we try to help people to realize that nonhumans and humans share a similar fate. After just six days, Camp Uganda visitors understand that trees are important to both chimpanzees and humans. They learn that trees help create oxygen and control rainfall; trees hold ground water, and remove carbon dioxide from the atmosphere. Campers see that forests are essential for chimps if they are to find food, shelter, and security. In the ways of small children who are just learning to express themselves on paper—in a second language at that—children remind us that our time at Camp Uganda is well spent:

- Clare: "If you cut at least one tree please try to plant more trees in order to avoid deforestation, soil erosion, and trees help in the formation of rainfall."
- Gerald: "A new experience I had was that sick animals are also treated, like human beings."
- Grace: "When I go home I will tell my friends that I learned not to cut trees because trees are home for wild animals."
- Christopher: "To stop deforestation is good for chimpanzees to get food."

Camper comments also suggest that campers have built a connection with other animals, especially chimpanzees:

- Muse: "I had look chimpanzee face to face. I had look lion face to face."
- Agnes: "When I go back home, I will tell people that chimpanzees can walk on two legs like we people."
- Aivan: "I want Camp Uganda to continue because I learned how to conserve forests and animals, I saw chimpanzee which I will never see it, I saw bush buck and water buck which I will never see it."
- Badru: "[Chimpanzees do many things that we do], eat like people, clap their hands."

To explore the effectiveness of Camp Uganda, we interviewed campers. Through our surveys, we were told that only 44% of the children cared about helping wild animals before attending camp, but after camp, a whopping 92% of those who attended Camp Uganda stated that they cared a great deal about helping animals. By the end of their camp experience, 96% of campers reported that they cared very much about protecting chimpanzees. While questionnaires are by no means scientific data, they suggest that we are headed in the right direction.

After Camp Uganda, a camper named Rose commented that chimpanzees were very much like humans: They feed only during the day, and they look after their young with great care. She also noted that lions—carnivores and therefore less like human beings—eat only every other day. Her comments indicated that she had seen both similarities and differences, and recognized core, shared similarities across species: "a new experience was the animals eat like human beings." Campers' comments, like those of Rose, help us to see our work through the window of a child's mind.

How many of us know that children are unlikely to realize that animals need food, and that their source of food is the forest and other natural environments, unless or until they actually *see* wild animals eating in the forest? If they do not know this, they cannot know the importance of preserving forests, lakes, streams, and meadows. Two years after attending Camp Uganda, thirteen-year-old Ronald told me that he had convinced many neighbors, as well as his own family, to stop beating monkeys with sticks when they come to the village in search of food. He told me that he was now an advocate for monkeys because he understood that "the monkeys need food to live too." He understood that Uganda was once covered with forests, but now the forests have been cut back, and villages built, so that monkeys have less homeland, and must live close to humanity. He was clearly very proud of himself for carrying new information back to his village to bring a positive change. Once they understand the connection between wildlife and habitat, children have a foundation for environmental sensitivity and activism, and the likelihood that they will act on behalf of the environment and nonhuman animals is much greater.

Camp Uganda is not only an experience for children, but also for adults. Through the wondering eyes of campers, through their exclamations of amazement, *we* learn. One Camp Uganda teacher, reflecting on the process of "connecting the dots," wrote to the Board of Directors: "I am glad to inform you that our kids can now keep the environment clean, which was not the case before." If educators do not understand what children do not know, and how our ignorance leads to environmental destruction, then we cannot bring positive change.

Oftentimes, what is missing from a child's education is actual experience in the natural world, connecting with forests and wild animals. If we can bridge this gap, if we can offer children the knowledge of shared life and beauty in the natural world, we can help people to understand the importance of making environmentally sound choices.

## Conclusion

If children are to take action, we must give them information, skills, and a direct experience with the environment that will bring both a deeper understanding of their homeland, and a heartfelt connection with the natural world. Even this is not the sum total of the challenge. For a child to bring change, he or she must have personal tools such as self-esteem, self-confidence, and a measure of independence. Children who bring change need to be empowered, to have the courage to suggest ways that are different from those of their parents and grandparents.

If youth living on the edge of the Ugandan forests are only taught about wildlife and trees in the classroom, they are less likely to make meaningful connection with the natural world. When we take children outdoors, they are more likely to see, feel, and remember what we teach. Children's camps strive to offer an experience that will change lives, which can change our collective future. If we are to build a generation of citizens who care about nonhuman animals and the environment, then we need to promote children's educational camps internationally. Children living next to forests can play an important role in what happens to local forests and wildlife. Attitudes, knowledge and personal tools, formed at a young age, can inspire and empower youth to take action on behalf of the natural world.

## Discussion Questions

1 Given the writings of authors in this anthology regarding hierarchy, power, and dominion (see Chapters 2–4, for example), what must be considered when Canadians, Americans, or Europeans, for instance, travel to other continents and countries with intent to bring change?

2 Perhaps the best way to bring life-affirming changes to other nations is by way of example and empowerment. Does your nation provide a worthy example? Does your nation invest in support for other nations with regard to changes on behalf of the environment? Do you? In what ways? Where do you find room for improvement?

3 Is there more of a tendency for humans to be disconnected from nature in industrialized societies? Why? What are the likely implications of this shift away from the natural world and animals? Should we make an effort to disrupt this tendency, and if so, how?

4 Is it important to promote a love of nature among youth? If so, how do we do this without exploiting the natural environment and nonhumans for our ends?

5 Do you think children can change the way adults behave toward animals and the environment? Have you ever seen this happen?

6 What are the advantages of investing your activist dollars overseas? What do you see as the disadvantages of this choice?

## Essay Questions

1 Describe a memorable encounter with the natural world from your childhood. Why do you suppose you remember this incident? How has it shaped your relationship with the natural world? What is the value of this memory for your present relationship with the natural environment?

2 On the internet, find and learn about a handful of environmental organizations located in Asia or Africa. Which of these seems the most effective, and why? Which holds the least appeal to you, and why? What are three ways that you can support the organization of your choice aside from sending money?

3 If you were starting an NGO to work with people in other nations on behalf of the environment and animals, what sort of an organization would you create? Describe the problems you will tackle and target goals. What will likely be your biggest obstacles?

4 If you have a choice between investing $20 in an organization overseas or locally, which would you choose and why?

## Suggested Further Reading

American Camp Association, 2014. *Camping Magazine*. [Online] Available at: http://www.acacamps.org/camping-magazine

Camp Uganda, 2014. *Chimps and Children*. [Online] Available at: http://www.campuganda.org/index.php?option=com_content&view=article&id=6&Itemid=6

Louv, R., 2005. *Last Child in the Woods*. New York: Algonquin Books.

## References

Jane Goodall Institute of Canada, n.d. *About Chimpanzees*. [Online] Available at: http://janegoodall.ca/about-chimps.php [Accessed 22 May 2014].

New Nature Foundation, n.d. *Building Stoves*. [Online] Available at: http://newnaturefoundation.org/?page_id=29 [Accessed 22 May 2014].

# 23

# PRESERVING EARTH IS PROTECTING ANIMALS

*Phaik Kee Lim*

Sahabat Alam Malaysia (SAM) was established in Malaysia by Mr S. M. Mohd Idris more than thirty years ago. SAM focuses on a variety of issues affecting the environment and nonhuman animals, such as land use, deforestation, pollution, fisheries, energy use, water quality, dams, agricultural techniques, indigenous people's rights, the effects of tourism, and the protection of wildlife and their habitat. Before SAM, there was not much concern for animal rights or the environment in Malaysia. With offices in two cities in Malaysia, Penang and Sarawak, SAM works on behalf of both the environment and nonhuman animals because we recognize that they are inextricably connected.

## Destroying Lives, Damaging Ecosystems—Wild Animals in Captivity

The Royal Indian circus came to my hometown when I was a small child and my father bought tickets. The most renowned act for this particular circus was a pair of counting dogs—small-breed dogs, who could reportedly count and total numbers. The audience was asked to provide two numbers for the dogs to add, which were written on a whiteboard. The dogs were supposed to pick the correct answers, from one to ten, from a variety of cards. One of the dogs picked a wrong number, and the trainer savagely slapped the dog across the face. I was very young, but I knew then that I was done with circuses.

Malaysian theme parks and resorts similarly mistreat "exotic" animals, even though there is nothing entertaining (or educational) about wildlife performing silly human tricks. In 2005, I was shocked to find that orangutans were forced to perform tricks at Malaysia's Kuala Lumpur Bird Park. One of the young orangutans was slow to perform, and I noticed that the trainer pinched the tender skin under her arm. Outraged, I decided to bring this to the attention of the authorities, hoping that the plight of these youngsters might change.

Orangutans are an endangered species, listed under CITES Appendix I. Consequently, it is illegal to transport orangutans across international borders, and I was able to use this against the Kuala Lumpur Bird Park. I questioned the source of these orangutans and called on officials at Malaysia's Trade Record Analysis for Flora and Fauna in Commerce (TRAFFIC) to do their job—to make sure that these orangutans had not been smuggled across international lines. As a result, the Wildlife Department conducted DNA tests to determine if they were captive bred or wild-caught and illegally transported across international lines. Ultimately, at least some of the orangutans were returned to Indonesia, and the orangutan "show" at the Kuala Lumpur Bird Park was closed due to letters of complaint from NGOs and the larger public.

I also noticed that A'Famosa Resort (in Malacca) was running an orangutan show, where orangutans were dressed in bikinis and forced to play golf. I wondered where these orangutans had come from, and whether or not their parents had been shot in order to kidnap their young for sale to the entertainment industry. While the A'Famosa show has since shut down, "wildlife" shows persist because of human indifference and ignorance. Those who pay to stare at these tortured animals fail to recognize the suffering and misery they pay to support. Most visitors have never given a thought as to how dogs, tigers, bears, and elephants are acquired, trained, or kept when they are not "performing." Few circus visitors are aware that protective adults are often shot so that their young can be kidnapped and sold into the entertainment industry. Not many realize how many of these animals perish once taken from their wild homes, or how this trade affects ecosystems. Few are aware that circus animals are chained, caged, or tethered for most of their lives in heavily barred and barren cages, or that they are trucked across many miles in these small cages for the sake of human "entertainment." The public does not generally know that trainers cause excruciating physical pain to wild animals to break their spirits and to force them do unnatural "tricks." Such acts are neither entertaining nor educational, they devalue both nonhuman animals and ignorant viewers who fail to recognize the wrongness of what they have paid to see. To protect ecosystems and individual animals, we must not visit or stay in places that exploit nonhuman animals, and when we see such exploitation, it is best to explain what we have seen to an organization like SAM, which might be able to help these unfortunate animals.

Trade in "exotic pets"—even without frivolous tricks—harms individuals and ecosystems. For example, reptile parks are a serious animal welfare and environmental concern. In February of 2008, a python park opened in Malaysia's UluBendul Kuala Pilah Recreational Park (near Seremban). This facility displayed several imported species of snakes, including anacondas, albino Burmese pythons, Indian pythons, and South American boas. Though snakes can live for decades—even in captivity—almost all of the 70 snakes died within eight months of their arrival in UluBendul ('Pythons' 2008). Where did these snakes come from? What if these imported species escape or are released and take root in Malaysia?

There is also a brisk and damaging trade in "exotic birds." Birds are supposed to be hatched into a world of green space, but for the sake of human interests, many are put into cages so small that they cannot spread their wings. In Asia, captive parrots and macaws are chained by one leg, sold for a song, and spend their lives perched on a tiny stand. Each year, hundreds of exported birds die en route to pet shops in foreign countries. Most people who buy these birds do not know how many birds die along the way, or how they are stolen from their wild ecosystem. On an individual level, a bird in a cage is cruel and therefore depressing. In the larger picture, trade in "exotic birds" devastates ecosystems and endangers species. Many birds sold into the pet trade are illegally caught in the wild, stolen from local ecosystems in order to satisfy consumer demand at home and abroad. Some species have been driven to the brink of extinction by this trade in "exotic pets." Do not buy caged birds or "exotic" species as "pets." If you would like to adopt a bird (or dog or cat or horse) from a rescue, there are many sanctuaries around the world—especially in Western Europe and North America—almost all of which are filled to capacity with animals waiting hopefully to be placed with responsible caretakers.

While people sometimes care about the suffering of dogs or elephants, the suffering of fish is almost universally overlooked. Nonetheless, the current trade in aquarium fish raises both environmental and animal rights concerns. Aquarium fish are often wild-caught, causing significant declines in specific fish populations, while damaging coral reefs. These fragile oceanic beings, meant to live in a moving and living sea, suffer miserably when forced to spend their lives enclosed in glass aquariums.

In Malaysia, it is now a fad to keep fish as part of *fengshui*, an ancient Chinese school of thought that teaches the art of balancing energies of a given space to bring health and good fortune. *Feng* means wind and *shui* means water ("wind-water"), both of which are associated with good health. Many *fengshui* "symbols" come from nature, such as Mandarin ducks (for love and marriage), tortoises (protection and stability), or koi (carp) or gold fish (abundance). These "symbols" are living, breathing individuals, yet many wind up in a state of serious neglect. *Fengshui* fish are usually kept in dirty water with pumps that do not function properly. People who buy koi or a turtle simply plop them into a tank, hoping to bring good fortune to their homes, then quickly forget and neglect their *fengshui* fish or turtle. Because fish (and tortoises) cannot cry out, or in any way draw attention to their plight, they usually die a slow and silent death. But SAM does not need to hear their screams; we know that it is wrong to treat fish or any other animals as a means to our ends. Wildlife does not belong in human homes, and is not merely an object for our purposes. Tortoises, ducks, and fish are part of wild ecosystems, where they have a life of their own, independent of human beings.

Most "zoo animals" are also miserable in confinement, and Malaysian zoos are notoriously inadequate. As SAM's officer, I visit local zoos, documenting unskilled animal care and poor zoo management, both of which I find frequently in Malaysia's zoos. I also note abnormal behaviors, such as rocking from side to side (elephants) or continuous pacing (tigers) in bored, stressed zoo captives. While zoos announce with pride the birth of a lion or a chimpanzee, they do not report to the public when animals die of diseases, accidents, negligence, or depression. Nor do they explain where baby animals will go when they have grown up—most people have no idea how zoos deal with "surplus" animals. Comparatively safe when they are young and cute, animals in zoos are often sold to the highest bidder when they reach maturity. This is true in nations around the world. Zoos are also disastrous for ecosystems and species. Because zoos buy wildlife from animal dealers, poachers pilfer wildlife in order to sell these animals to zoos, as well as into the pet trade (Kemmerer 2010). If we are to protect ecosystems and endangered species, we must shut down trade in "exotic" species, including any trade to or from zoos and pet shops.

## Domestic Animals, Animal Rights, and the Environment

SAM also works on behalf of cats, dogs, kittens, puppies, and other domesticated animals. In the pet trade, these vibrant beings are kept in small cages for long periods of time, with no room to run or play—alongside "exotics" in pet shops. When they are in their "cute" phase, they are more likely to sell, and consequently, so many kittens and puppies are snatched from their mothers before they are weaned. Forced to feed on dried food (when they should have their mother's milk), many kittens and puppies sold in pet stores are malnourished. Stuck in tiny cages with a big price tag, unwitting buyers perpetuate the system.

Farmed animal abuse and environmental degradation caused by animal agriculture are also of concern at SAM. Malaysia has learned how to factory farm from the West, we now debeak, dehorn, and castrate farmed animals—without anesthesia. Poultry, cattle, chickens, and pigs are transported to slaughter across many miles under the hot Malaysian sun without shade or water. When these large trucks have accidents, farmed animals are strewn across the roadways, trapped in the wreckage, and those still alive are in shock and often injured. In short, farmed animals in Malaysia are treated just like farmed animals in the U.S.

Animal agriculture is the largest single cause of greenhouse gasses and dead zones (Kemmerer 2014). Consumption of animal products in North America rapidly depletes fresh water supplies and destroys rainforests abroad. Farmed animals consume monoculture crops sprayed with pesticides, as well as antibiotics and hormones that are flushed into the water system, leaving us with piles of manure that also wash into and pollute the water system. Whether one is concerned about animals or the planet, animal agriculture is a serious concern.

## Bringing change for earth and animals

SAM has lobbied to bring the Malaysian government to reform two key documents for the protection of animals, the Wildlife Protection Act of 1972 (amended in 1976) and the 1953 Animal Act. After years of patient but persistent negotiations, the former was upgraded, and is now called The Wildlife Conservation Act 2010 (Abdullah 2010). The new Act was put into place to improve wildlife management for facilities such as zoos and circuses.

With recent amendments, the Act now has a wider scope protecting more species (including derivatives, hybrids, invasive, and nonnative species) and covering more activities that affect wildlife. The Act also specifically addresses issues of animal welfare and cruelty, including stringent penalties for such infringements as poaching, which should help to deter crimes against wildlife. Penalties now include fines of up to RM 500,000 (about U.S. $150,000), with a jail term of not more than five years and a minimum penalty for several offences of not less than RM 5,000 (roughly U.S. $1,500). Jail time is now mandatory for offences that involve protected wildlife, including up to five years and a fine of not less than RM 100,000 (U.S. $30,000), and not more than RM 500,000.

Additionally, the Act now includes stiffer rules protecting animal welfare, effective as of 2012. For example, zoos, commercial captive breeders, circuses, researchers, and wildlife exhibitors can only operate legally if they hold a government approved permit. In order to secure this operating license/permit, zoos must comply with guidelines set by the Natural Resources and Environment Ministry; zoo operators who fail to comply will lose their license and their animals will be confiscated by the Malaysian Wildlife Department. Malaysian zoos will be held to a higher standard, especially in light of a new Conservation Act (December 2011), and will necessarily be much more humane. As a case in point, Saleng Zoo (in Johore) was recently closed due to poor animal care when they failed to meet national, regional, and international welfare standards.

In other areas SAM has had less success in bringing tighter reforms through the 2010 Act. For instance, the Act offers little change with regard to hunting, largely because hunting licenses provide a major source of revenue for the Wildlife Department. Similarly, "Special Permits," which allow certain individuals to keep endangered wildlife, came under scrutiny but remain. These permits regulate both indigenous and foreign wildlife that are "owned" and kept by private parties, theme parks, and zoos (Chiew 2005). Though not successful as of yet, SAM continues to argue that wildlife does not belong in cages, whether in private cages, public spaces, or for personal profit, and that "Special Permits" should therefore not be granted.

SAM has had even less success in bringing change to the archaic and woefully inadequate 1953 Animal Act, which covers prevention of cruelty to animals. Only one word has been changed to update this 1953 document, and that word was not in the text, but merely an inconsequential part of the title. Nonetheless, we remain hopeful. There is now a greater need and more pressure to update the Animal Act (proposed Animal Welfare Bill) due to the horrendous treatment of animals that has recently been highlighted by the media. Under proposed amendments, anyone charged and found guilty of violating the Animal Act will be fined up to RM 50,000 (roughly U.S. $15,000) or jailed for a year, or both. This is a great improvement over the current maximum fine of just RM 200 (U.S. $60), or a jail term of not more than six months, or both.

Many activists and NGOs lobby for more humane methods of animal control, noting that local council members, charged with collecting and housing strays, regularly, flagrantly, and openly engage in animal abuse. SAM is also hopeful that proposed legislation controlling pet shops, which have come to harbor backyard breeders, will successfully pass into law. Then again, laws and amendments are of no use unless the Attorney General recognizes the importance of prosecuting animal abuse cases. If authorities are unwilling to take the matter of animal welfare seriously, the recently amended Act and proposed Animal Welfare Bill are of little use.

SAM also works internationally. In June of 2001, SAM reacted to a feature story in the *Guardian Weekly* that was reporting that prominent former members of the United States Government (current members of Safari Club International—SFI), were undermining the Botswana Government's efforts to protect lions by pressuring the Botswana Government to lift a ban on lion trophy hunting (McGreal 2001). Africa's large animals have been targeted by white safari hunters since the early 19th century ('White hunter' 2014). For short-sighted hunters, the pleasure of the kill, and whatever joy they gain by hanging the head and or skin of a deceased animal on their wall, is more important than the life of another individual, or the protection of endangered species. Consequently, many of Africa's large animals—rhinos, elephants, leopards, jaguars, tigers, and of course lions are at risk. SAM sent letters to the Government of Botswana, encouraging them to continue to protect their few precious lions. The Honorary Consul of the Republic of Botswana sent a thank-you letter for supporting their wildlife protection policies against powerful U.S. officials.

In 2002, when SAM learned (from International Primate Protection League—IPPL) that four gorillas had been imported to Malaysia's Taiping Zoo, we worked with African officials on behalf of this endangered species (Lak 2009). Gorillas are protected, and therefore must not be bought or sold nor imported or exported without strictly regulated paperwork. Both SAM and IPPL were suspicious as to how these primates, dubbed the "Taiping Four," had been acquired. It was likely that they had been illegally caught from Cameroon, then smuggled across the border into Nigeria before being sold overseas. The mystery was solved when border officials caught smugglers transporting primates—two baby chimpanzees—across the border between Cameroon and Nigeria. SAM, along with other organizations, wrote letters of protest demanding the return of the Taiping Four to their country of origin. International cooperation between various groups working on behalf of animals, species, and ecosystems brought the Taiping Four home and the gorillas returned to Cameroon in 2007 (Paula 2007).

SAM also mobilized on behalf of twenty Rockhopper penguins, shipped across borders, when they arrived at Malaysia's Langkawi Underwater World. This story began in 2003, when Captive Animals' Protection Society (CAPS) announced that a notorious South African wildlife dealer, John Visser, had 146 Rockhopper penguins in his possession, destined for zoos around the world ('Penguins' 2003). Exported from Nightingale Island (part of the UK's Overseas Territory of Tristan da Cunha in the South Atlantic), at least four penguins had already died en route, or shortly after their arrival at Visser's notorious and disgusting shark aquarium in Jeffrey's Bay, South Africa. Though permission for capture, sale, and transport had been granted by the Administrator of Tristan and the British Foreign Office, SAM did not want these penguins to be sent to zoos—particularly zoos in Asia and Africa, which are ill-suited to the health and welfare of wildlife—particularly wildlife equipped for cold-weather conditions. SAM protested to the British Foreign Office and the Administrator of Tristan, noting how hard (and miserable) a change to captivity would be for these wild-caught penguins. We also noted that robbing so many individuals from one ecosystem alters and upsets the balance of nature. While we were unable to gain the release of these kidnapped birds, we hope that we made enough of an impression that there will be no more penguins captured from Nightingale Island. We do not need penguins in local zoos—we can see Rockhoppers in action online. People learn more about Rockhoppers—and all animals—when we understand that wild animals are best left in the wild.

Canada's seal hunt is the largest marine mammal hunt in the world, currently with a three-year kill quota of almost one million. Last year 365,971 seals were reportedly killed; 96.6 percent of those were less than three months old. More than a million seals were killed over the previous three years, the largest kill in more than half a century ('Canada' 2005). This infamous "hunt" destroys more than 300,000 baby seals, some as young as 12 days old, for fur. Most recently the Canadian government had given seal hunters the green light to club up to 319,500 baby seals.

In March of 2005, SAM first learned of the Canadian government's three-year seal hunt quota, and rallied on behalf of Canadian seals. Working with International Friends for Animal Welfare (IFAW), SAM sent out an action alert criticizing the hunt as brutal and barbaric, demanding an end to seal slaughter in Canada. Although Canada replied that the hunt would continue as planned, we soon received news to the contrary: A worldwide outcry had temporarily halted the seal massacre. Again, earth and animal activists had effectively united to protect individual animals, species, and ecosystems. We hope that this Canadian killing will one day be called off permanently.

Also in 2005, the Thai government arranged for the Kenyan Government to capture 300 wild African animals for Thailand's Chiang Mai Night Safari Zoo ('Local' 2006). We sent action alerts to other activists—both earth and animal organizations—calling for letters of objection to the Kenyan and Thai Embassies. We also sent letters to Thailand's Prime Minister, and Kenya's President and the Minister for Environment, joining an international outcry against the capture, transportation, and permanent confinement of Kenyan wildlife. We urged Kenya to keep their wildlife at home and reminded officials that wildlife is central to Kenya's rich national heritage. The High Court in Nairobi eventually ruled that this transfer of wildlife was illegal and Kenya's wildlife was allowed to remain in their homeland, free and wild ('Kenyan' 2005).

In 2003 and 2004, Thai and Indonesian government inspectors discovered 115 orangutans, many of whom were not registered with the authorities, constrained in squalid, cramped conditions at Thailand's Safari World. Safari World claimed that the orangutans had been produced through a breeding program, but a disproportionately high number of young orangutans made this claim implausible. SAM suspected that Thailand's Safari World orangutans had been illegally imported from Indonesian Borneo, and we joined forces with the global earth and animal activist community, demanding repatriation of these primates for rehabilitation and release back into the wilds of Indonesia. We sent letters to the Thai Prime Minister, and their Director General of National Parks, and to The President of Indonesia and their Minister of Environment, urging them to do what was just and right—to make sure these illegally captured orangutans were returned to Indonesia. We also encouraged Indonesian authorities to pressure the CITES Secretariat to impose sanctions on Thailand—as well as Cambodia, Malaysia, and Saudi Arabia—for failing to prevent smuggled orangutans from crossing their borders. As a result of activist efforts, most notably a vigorous campaign from the UK's Nature Alert, about 75 orangutans were repatriated.

In 2007, SAM pressured the Indian government (through petitions posted to the prime minister) to halt their 560 million dollar Sethusamudram Ship Canal Project. This project entailed cutting a 90 kilometer channel, 12 meters deep and 300 meters wide, through a chain of small islands (known as Adam's Bridge) between India and Sri Lanka. This project was designed to prevent large ships that are transporting goods between India's western and eastern coasts from having to circumnavigate Sri Lanka. But dredging this channel involves collecting, moving, and dumping sediment further out to sea, which harms marine ecosystems, including small coral and large whales. At least six whales died as a result of digging under the sea to create a shipping route ('Indian' 2007). Though only 43 whale deaths have occurred in this area over the last century, somewhere between six and ten whales (whose navigation sensors are affected by loud noises that result both from dredging and from sonar used for marine surveys) have beached since dredging began. These whales, stranded in the shallow waters near Rameswaram (an Indian island that lies between Sri Lanka and India) either lost their sense of direction because of loud noises, or purposefully stranded to escape the pain caused by such intense sound waves. Environmentalists and whale enthusiasts around the world voiced concern along with SAM, and India's Supreme Court halted the Sethusamudram canal project in 2010, ordering a "full and comprehensive" Environmental Impact Analysis and an exploration of the feasibility of an alternative route.

In 2009 SAM called attention to orangutans illegally held captive in the homes of Indonesian Government officials, and in the homes of those holding high offices in the Indonesian Ministry of Forestry. Because these matters were being ignored in their home nations, SAM brought these "pets" to the attention of Indonesian Ambassadors in Australia, New Zealand, and the UK, requesting that they urge their governments to take action against officials illegally keeping orangutans as "pets." Unfortunately, every primate rescue center in Indonesia is full—and has a waiting list. With few protesters involved in this issue, and nowhere for rescued orangutans to be relocated, the Indonesian government ignores our pleas.

In 2010, SAM contacted the Prime Minister of India and several departments responsible for India's roads and railways concerning the frequency of elephant deaths along India's railways. As many as 150 elephants have been crushed by speeding trains since 1987 ('PETA' 2010). SAM urged the Indian government to establish a committee to develop an action plan to prevent further deaths. To the credit of the Indian government, we recently received a reply noting that they had implemented all that SAM had suggested.

In 2012 SAM appealed to a variety of environmental and animal welfare organizations to voice disgust and disappointment with Malaysian authorities for the early release of Malaysia's notorious wildlife smuggler Anson Wong Keng Liang, whose jail sentence was overturned because of shortcomings and errors in the investigation and prosecution of his case. As a result, the Court of Appeals reduced his wildlife trafficking sentence from five years to time served—just 17 months ('Wildlife' 2012). His early release—despite an outpouring of worthy letters—forecasts doom and gloom for the world's rare and endangered species: he is a repeat offender, and there is every probability that he will offend again. And why not? His illegal activity has made him rich, while Malaysia finds it almost impossible to catch him, prosecute him properly, or keep him in prison.

In March of 2014, the wife of wildlife trader Anson Wong, Cheah Bing Swee, appeared in a Session Court. She and two others are jointly charged for allegedly possessing five elongated tortoises, a protected species, along with Syarikat Rona Wildlife Enterprise and its director Muthukomar ('Anson's' 2014). Cheah, believed to be the manager of Syarikat Rona Wildlife, is also alleged to have had two female elongated tortoises, the possession of which is prohibited under Section 70 (1) of the Wildlife Conservation Act 2010. The Deputy Public Prosecutor DPP Charanjit Singh has since requested more documents from the prosecution's headquarters in Putrajaya.

## Conclusion

SAM cannot change the world single-handedly, but we can work with other dedicated activists to bring about meaningful change on behalf of the earth and nonhuman animals. Perhaps the most important aspect of networking is informing the larger public. In Malaysia, as elsewhere in the world, most people think very little about birds, fish, reptiles, or even nonhuman primates, let alone insects—even though these animals are critical to innumerable ecosystems. Nor does the public think about the problems inherent in circuses, zoos, and the trade in "exotic" species. But when citizens of conscience learn that farmed animals are mistreated, that forests are plundered, or that endangered animals are captured and sold into a caged life, they often respond by taking a stand against these practices. And when they respond, governments and businesses must change their ways. Together, through education and outreach—*only* with the power of many voices—we can protect animals and the planet.

I am motivated to work for SAM both by love for animals and by love for the natural world. Overall, human beings are uninformed with regard to the interconnected nature of animal concerns and environmental concerns. Animals are part of any natural environment—just as the environment is essential for all living beings. Without wilderness, there can be no wildlife; without wildlife, there can be no real wilderness. Animal agriculture destroys not only billions of individual animals every year, but also millions of acres of rainforest, while plundering water sources and fouling the air. If people could learn to think of and care for individual animals—including insects, fish, farmed animals, and reptiles—we would be able to protect and restore the environment.

## Discussion Questions

1 How does Lim help you to see circuses and other forms of animals-as-entertainment differently?

2 In what ways does it seem that animals in Asia are viewed and treated differently from where you live, and in what ways do they seem to be viewed and treated similarly?

3 What environmental effects and ethical implications come into play when zoos purchase wildlife from poachers and animal dealers?

4 What is the government's role in protecting the natural world and nonhuman animals? What are the laws in your nation/state with regard to importing "exotic" animals," and what would you suggest to your legislators with regard to these laws?

## Essay Questions

1 Ota Benga was an African pygmy captured from Africa and kept in a zoo in New York. Read *Ota Benga*, by Phillips Verner Bradford and Harvey Blume (see suggested further reading below). How is treatment of Ota Benga similar/different from our exploitation of other living beings in zoos today?

2 What are the environmental and ethical problems likely to be involved when humans keep animals in zoos? Which do you think is the most egregious problem, and why?

3 Write a letter to your senators and/or representatives calling attention to local laws in need of change. Explain why the current law is inadequate, how the law ought to be changed, and why.

4 Write a creative essay to zoo visitors from the perspective of a large mammal (not unlike yourself or Ota Benga) trapped in a zoo enclosure. Explain what you want, and why.

## Suggested Further Reading

Bradford, P.V. & Blume, H 1992, *Ota Benga: The pygmy in the zoo*, St. Martin's Press, New York.

Kemmerer, L. (ed.) 2012, *Primate people: Saving nonhuman primates through education, advocacy, and sanctuary,* University of Utah Press, Salt Lake City.

Phelps, N. 2007, *The longest struggle: Animal advocacy from Pythagoras to Peta,* Lantern Books, New York.

## References

Abdullah, M.T. 2010, *Wildlife Conservation Act 2010. Act 716.*, Government of Malaysia, Available from: http://www.academia.edu/3734442/WILDLIFE_CONSERVATION_ACT_2010._Act_716._Malaysia. [11 June 2014].

'Anson's wife in court over possession of tortoises' 2014, *The Star Online* 7 March. Available from: http://www.thestar.com.my/News/Nation/2014/03/07/Ansons-wife-in-court-over-possession-of-tortoises/. [11 June 2014].

'Canada kicks off annual hunt for baby seals' 2005, *International Fund for Animal Welfare* 29 March. Available from: http://www.ifaw.org/canada/node/22931. [11 June 2014].

Chiew, H. 2005, 'Wildlife endangered,' *The Star Online* 22 November. Available from: http://www.thestar.com.my/story.aspx/?file=%2f2005%2f11%2f22%2flifefocus%2f12573232&sec=lifefocus. [11 June 2014].

'Indian ship project killing whales' 2007, *DAWN.com* 4 January. Available from: http://www.dawn.com/news/226106/indian-ship-project-killing-whales. [11 June 2014].

Kemmerer, L. 2010, 'Nooz,' in *Metamorphoses of the zoo: Animal encounters after Noah,* ed. R. Acampora, Rowman & Littlefield, New York pp. 37–56.

Kemmerer, L. 2014, *Eating earth: Dietary choice and planetary health*, Oxford University Press, New York.

'Kenyan judge halts wildlife gift to Thailand' 2005, *Wildlife1.org* 21 December. Available from: http://www.wildlife1.org/news/300-kenyan-judge-halts-wildlife-gift-to-thailand. [11 June 2014].

Lak, C.A. 2009, 'Monkey business,' *The Star Online* 27 March. Available from: http://www.thestar.com.my/Travel/Malaysia/2009/03/27/Monkey-business.aspx/. [11 June 2014].

'Local news from all over' 2006, *Earth Island Journal* Summer. Available from: http://www.earthisland.org/journal/index.php/eij/article/local_news_from_all_over6/. [11 June 2014].

McGreal, C. 2001, 'Lions face new threat: they're rich, American and they've got guns,' *The Guardian* 27 April. Available from: http://www.theguardian.com/world/2001/apr/27/usa.highereducation. [11 June 2014].

Paula 2007, 'The return of the Taiping Four gorillas', *Limbe wildlife center newsletter* 9 December. Available from: http://limbewildlifecentre.wildlifedirect.org/2007/12/09/the-return-of-the-taiping-four-gorillas/. [11 June 2014].

'Penguins captured from wild for zoos' 2003, Available from: http://lists.envirolink.org/pipermail/ar-news/Week-of-Mon-20031215/013535.html. [11 June 2014].

'PETA moots speed guns against trains that kill tuskers' 2010, *Thaindian News* 5 October. Available from: http://www.thaindian.com/newsportal/enviornment/peta-moots-speed-guns-against-trains-that-kill-tuskers_100439338.html. [11 June 2014].

'Pythons at park dying off' 2008, *The Star Online* 4 August. Available from: http://www.thestar.com.my/story.aspx/?file=%2f2008%2f8%2f4%2fnation%2f21996685&sec=nation. [6 June 2014].

'Re-mention of wildlife case against Wong's wife on May 8' 2014, *The Star Online* 9 April. Available from: http://www.thestar.com.my/News/Nation/2014/04/09/Remention-of-wildlife-case-against-Wongs-wife-on-May-8/. [11 June 2014].

'White hunter' 2014. Available from: http://en.wikipedia.org/wiki/White_hunter. [11 June 2014].

'Wildlife trader Anson Wong free' 2012, *The Malaysian Insider,* 22 February. Available from http://www.themalaysianinsider.com/malaysia/article/wildlife-trader-anson-wong-freed. [11 June 2014].

# SECTION III

# Bringing Change

## Personal Encounters and Reflections

**Poem:** Tami Haaland

### Goldeye, Vole

I say sweep of prairie
or curve or sandstone,

but it doesn't come close
to this language of dry wind

and deer prints, blue racer
and sage, its punctuation

white quartz and bone.
I learned mounds of

mayflowers, needle grass
on ankles, the occasional

sweet pea before I knew
words like perspective or

travesty or the permanency
of loss. My tongue spoke

obsidian, red agate,
arrowhead. I stepped

through tipi rings, leaped
buffalo grass and puff ball

to petrified clam.
Jawbone of fox, flint,

blue lichen, gayfeather,
goldeye, vole—speak to me

my prairie darling, sing me
that song you know.

"Goldeye, Vole" is included in *Breath in Every Room*, a poetry collection from Story Line Press, 2001.

# 24

# KEEPING CAGED ANIMALS, KEEPING SECRETS

## Working as a Keeper at the Zoo

*Bethany Dopp*

Even now I feel mixed about my experiences at the zoo. It was sad and fun, interesting, and depressing, educational, and disturbing. I now *know* that zoo keepers can be very kind people who genuinely care about animals. They usually believe that they are helping animals, and as such, are well employed and doing the right things. Still, as the saying goes, the road to hell is paved with good intentions.

When I was a little girl I went to the Portland Zoo with my family. I made it to the penguin exhibit before I was stopped in my tracks, too fascinated to move on to the next pen. I spent the rest of my time at the zoo with the penguins, absolutely rooted to the spot. They were *so* fascinating! They waddled across their human-designed landscape, dove into their pool, and swam gracefully past the observation windows in the pale green water. From that moment on, I loved penguins. Their antics intrigue me to this day.

Ideally, I think this is how zoos are supposed to work—they instill a love of wildlife in "citified" people through a visual encounter. Though our society is now painfully isolated from the natural world, zoological institutions can provide a bridge, a window into the world of our ancestors, through which individuals can recreate the awe and wonder behind the myths of old, behind human respect for nonhuman animals. Ideally, exposure to captive animals cultivates a desire, in our hearts, and motivation to treasure and protect wildlife. *Ideally*.

Now fully grown and still enthusiastic about nonhuman animals, I was working towards a bachelors of arts degree in environmental studies. My course of study required an internship, and since I love animals, I decided to work for a summer as an intern at the regional zoo. But I was no longer a child, and I had reservations about working there. I pondered, and pondered ... was it right to work in a place that entrapped nonhuman animals for human purposes? Would the resident animals seem happy and well cared for? Or would they be clearly sad and pathetic? As I pondered these concerns my ever-thoughtful husband noted that working hands-on in a zoo would allow me to answer these questions for myself, to decide whether or not they are ethical institutions, and to thereby discern how I feel about the practice of keeping animals in captivity. So, I signed up for an internship as a keeper for the summer of 2007. The account which follows is based on my personal experience and reflections.

Being a keeper is a very special job at the zoo. Keepers engage in close, frequent interaction with captive animals. It is, quite literally, a hands-on experience. Keepers measure, cut, prepare, and deliver food to the cage. They also periodically clean leftover food debris, animal waste, dead plant material, old bones (and other items given to the animals for entertainment and enrichment) from holding pens and exhibits. Additionally, they plan and execute enrichments—special treats or entertaining objects put into pens to arouse curiosity and stir activity in an attempt to reduce the inevitable boredom of captive animals. Finally, keepers groom and exercise animals as needed, and they call in a vet when zoo residents seem amiss. In short, they are the lifeline for captive animals— everything an animal needs and does is orchestrated by a keeper.

During my time in this role I never saw any obvious, in-your-face cruelty, such as one would see in a slaughterhouse or at a cockfight. I witnessed a quieter kind of suffering: eagles who cannot fly; an owl forever imprisoned in a small, indoor cage; predators who will never hunt; and mature adults who cannot mix with their own kind.

When friends and acquaintances learn that I have been a keeper-intern at the zoo they are always excited, even envious, about my seemingly marvelous good fortune. Generally speaking they understand zoos to be fun places and often ask me to tell stories about my experiences—about my interactions with noble, marvelous, fascinating wild animals. Confronted with their requests, I find myself bereft of charming stories about lions, tigers, and bears. While it was entertaining to feed red pandas grapes with my bare hands and to stroke the sweet calf in the petting barn, it was heartbreaking to watch animals die every week—and animals *did* die every week. It was gruesome and horrifying to prepare food for carnivores. While it was enlightening to see how fast the wolverine could smell out food that was hidden for her, and I learned quite a bit about the care of animals in the process, I became more and more aware that visitors knew almost nothing of these things.

At the end of the day though, I seriously doubt that the institution where I was working had much of an educational impact on visitors. When asked by these well-intentioned friends, I invariably launch into a long, confused rant, venting my frustrations about the trapped "wildlife" and fabricated "food chain." Even now, eight years later, I feel powerfully mixed emotions—working at the zoo was simply unnatural in more ways than I can begin to express.

Most of the time, when people ask about my time working at the zoo, my mind turns first to Emilio and Tokata, the resident bald eagles. As the story was told to me, Emilio came to live there in December 1998 and Tokata in June of 1999. Their flight was stolen from them by human beings when they were still living free in the wild. Emilio was shot in Alaska and Tokata was likely injured when she flew into a powerline on the C.M. Russell National Wildlife Refuge in Montana. In the wild both Emilio and Tokata would have died, but humans intervened. They survived. The zoo provides Emilio and Tokata with a cage, food, and medical care in exchange for their presence. At first this seemed to me a rather noble gesture—save and tend unfortunate animals who have been debilitated by humans. Isn't this our rightful responsibility when we cause damage to other beings? But as I gazed into their enclosure, day after day, I began to wonder what life was like for an eagle who cannot soar on the winds high above the sandstone cliffs and sage-strewn prairies of Montana, who can never pluck a salmon from a rushing, Alaskan stream, or a rabbit from among the dried yellow prairie grasses. What is it like for an eagle who can never gather nesting materials to create a true wonder of architecture high in a tall cottonwood tree—a nest that will be built onto over the course of a lifetime, where generations of young are carefully tended?

Unless one of them dies prematurely, bald eagles typically pair for life. Tokata was a juvenile when she hit the power lines—she likely never had a chance to find a mate or raise chicks. But I wonder how long Emilio's mate waited for him to return before accepting that he was never going to come back to their nest.

I watched Emilio and Tokata perch or hop about on strategically placed branches in their small pen which did not need a tall fence—or even a roof —as these are "flightless eagles." I observed the bit of grass and dirt beneath the perches and a small wooden shelter at the rear—protection against Montana's harsh weather. Occasionally I noted that the eagles hopped about and dropped to the grass, but for the most part they just sat on a dead cottonwood branch all day long, watching the world go by without them.

Emilio is now almost twenty-three years old; Tokata is about eighteen. They have spent the majority of their lives sitting on dead branches in a cage. Tokata has been an earth-trapped eagle for fully three-quarters of her life. Emilio has spent nearly three times as long on a perch as he did on the wind. In the wild, eagles live an average of twenty to thirty years. In captivity, however, animals tend to consistently outlive their wild counterparts (Anon 2014). Knowing this I was left to wonder just how long those two magnificent creatures would languish in their small pen. It could be decades more yet. Both eagles have grown acclimatized to life in captivity and are utterly dependent on humans. They chirp loudly when zoo staff drive by in little electric carts, which portend food: The zoo's very own Pavlov's Dogs, conditioned to respond to certain stimuli (Cox and Baker 2002). They seem to me to represent a twisted parody of one of the greatest symbols of our national freedom.

As these thoughts played about my mind, I began to wonder—why exactly did we "save" Emilio and Tokata? Was it for their welfare or our own? I pondered whether or not these eagles have a life worth living and in the process came to see that we saved Emilio and Tokata for our own sake. Ultimately, watching these giant, land-ridden birds day after day I came to see a flightless eagle as a tragedy. It is clear to me that eagles are born for the air.

Even if I had come to see a flightless eagle as having an acceptable quality of life my experiences revealed yet more to complicate the equation. As noted, Emilio and Tokata cannot hunt, we must hunt for them. And what is a carnivore to eat in captivity? I ask this question because I am a vegan: As a biological omnivore I have a choice as to whether or not I eat other animals or animal products, such as meat, milk, and eggs. Out of compassion, I choose not to eat animal products. Eagles, and other true carnivores, do not have this option—by definition, a true carnivore requires and can only eat flesh (Anon 2005). This leaves a person to consider the very important question: how does a zoo feed eagles, wolves, and other carnivorous residents who have no biological choice but to eat flesh?

Tucked away behind the scenes where only staff travel is a run-down, unobtrusive building—not too small, not large, with dirt floors—which holds a line of dismal cages. These cages hold the source of cheap food—rabbits—for larger carnivorous "exhibits." Not too far away is another behind-the-scenes baby-mill, this one supplying food—mice—for smaller resident carnivores.

In the zoo's bunny breeding facility a dozen or so nameless, white female rabbits run around with a couple of nameless, white male rabbits. They are allowed to mate as often as they choose. After a female gives birth she and her offspring (called kittens) are put into small, wire cages that are mounted on the sides of the shed. Here they remain until their babies are old enough to be prematurely and forcibly weaned at which point the doe (female rabbit) is put back with the other rabbits to be re-impregnated. This process is repeated over and over in order to produce as many "food" bunnies as possible. The kittens are born—through human maneuverings—only to be fed to others. The doe will never see her babies again. The kittens will never again feel the comforting warmth of their mother. When a nameless mother rabbit ages and is no longer "productive" she is killed, fed to carnivores, and replaced with a younger female. On reflection, I could see that this procedure was eerily similar to the egg, dairy, and flesh industries, where hens lay eggs and cows are "milked" for profits until they become slightly less productive, and sows are bred for piglets until their "production" slows down—then they are discarded into the slaughterhouse to become food for human beings. Though rabbits, like factory farmed animals, are still young when they are "spent," they are replaced by even younger females, ripe for exploitation.

It was in this small breeding facility that I witnessed rabbits kept in wire cages from immediately after birth until they reach "food size," as one staff member told me, at about four or five pounds, before they are then gassed and fed to other zoo captives. The gas chamber is actually a modified cooler—a red cooler with an opening for the gas hose and a window on top so the executioner can tell when the deed is done, when the animals are dead. Fulltime keepers called this "the red box," or sometimes more bluntly, "the red box of death." Killing mice did not require a cooler. Mice, who are cost effectively killed in batches, are gassed in a large, plastic jar (with a hole in the top for a gas hose). As the mice start to die—or more accurately, realize they are being killed—their small bodies writhe with what I can only describe as sheer panic as they try to crawl over each other in a vain attempt to escape the jar and their sad fates. Slowly they all grow still and lifeless until they are just tangled lumps of fur and tails at the bottom of the jar—mice miraculously transformed into snake food.

Sometimes zoo keepers gas newborn rabbits, too. These vulnerable, pink babies are so small that they fit easily into the palm of a hand. Another keeper explained to me that some of the smaller carnivores need correspondingly smaller "pinkie rabbits"—as they were euphemistically called. I also learned about "pinkie mice." The keeper I was shadowing told me it wasn't worth the money and hassle to gas such small mice, so he demonstrated how he simply pinched their tiny little heads between his fingers, crushing their skulls and brains, rendering them lifeless. This procedure was so brutal that I could scarcely believe it was standard zoo practice at the institution where I was employed. But it was.

It was clear to me that the lives of these tiny, defenseless newborns, still naked and blind—barely out of the womb—were insignificant to keepers. This baffled me because the keepers all professed to love animals very much. I soon learned that they love cows and pigs and mice and rabbits a whole lot less than they love marmosets and tigers and bears. They gassed bunnies and mice in the morning, contentedly ate cows and pigs for lunch, then crooned over a pigmy marmoset in the afternoon. I found their inconsistencies baffling.

I, too, profess to love animals, but not just a few animals, not just ones I think are handsome, or noble, or cute, or somehow better than other animals. That is the primary reason I am a vegan. I don't want *any* innocent, sentient creature to suffer or die because of how I choose to eat, drink or be entertained, no matter what their species. You can perhaps imagine my shock when watching keepers—responsible for the welfare of nonhuman animals, crushing the skulls of mice with their bare hands and gassing baby bunnies in batches. It does not matter, to my mind, *whose* babies these helpless beings are. They are helpless, and they are sentient. That is all that matters. They, too, have lives and it seems reasonable to assume they would prefer to live than to be killed. I know that carnivores must eat flesh in the natural order of things, but "pinkie production," is not the natural order of things.

While mouse and bunny breeding facilities were morally problematic on many levels I have not yet mentioned what may be the most twisted aspect of the breeder facilities. Like many similar institutions, the zoo where I worked maintains a petting zoo where visitors—particularly children—can pet, hold, and cuddle young, domestic animals. Alongside the calf, horses, goats, ducks, and chickens in the "petting barn," there are half dozen or so cuddly bunnies. These bunnies are especially popular with little kids, and zoos around the world understand that families will pay entrance fees to bring small children specifically because children want to interact with these cute, snuggly critters. What these children—and their parents—do not know is that these adorable bunnies are fed to zoo carnivores as soon as these little cuddlers reach the magic "age" of four-to-five-pounds. A new batch of fuzzy white babies maintains the illusion that all is well in the petting barn. This is why the zoo likes to use only white rabbits—the kids can return every week and still never know that the bunnies they hugged last week have been gassed, cut up, and their bones spit out on the dirt floor of a nearby cage. Because the bunnies pretty much look the same, toddlers never ask their Mommies or Daddies, "What happened to the little brown and white spotted bunny I petted last week?" The bunny they petted last week is long gone through the contrived zoological food chain. It was my experience that the zoo, and perhaps most similar organizations across the nation, hide these ugly practices from youngest visitors—and from all visitors, if possible. But these institutions know exactly what they are doing and why.

Carnivores such as eagles and lions need to eat flesh. This is a fact of life. I am not in any way suggesting that zoos ought to veganize carnivores in their care. Instead, I am questioning the ethics of keeping carnivores and concealing the reality of what this means for keepers and "food" animals kept at zoos. It is disturbing that all of this carnivorous killing occurs behind the scenes, *purposefully* enacted in obscure buildings closed to visitors. Why isn't this facility open to everyone? Why don't visitors see carnivores consuming baby rabbits? *Dead bunnies are the reality of keeping carnivores in captivity.* Animals are killed every day at zoos across the country—around the world—to feed resident omnivores and carnivores. If these places are really educational institutions, as they claim to be, why do they hide the realities of what it entails to feed carnivores in zoos? It is an utter contradiction. For me it seems a grave injustice to "protect" visitors from the realities of caged carnivores. Zoos are places where previously wild animals cannot hunt and where even fully capable residents are fed on "gassed" bunnies and mice. Why not be honest?

There was only one exception to this "hidden death" policy in which I was involved. Fish were a favored enrichment for the zoo's river otters. They were fed live goldfish as a special treat for both otters and zoo guests who were delighted to watch the otters swim quickly after fish, snatching them from what had become their watery tomb. The most frequent complaint I heard from zoo visitors was that the animals never "do" anything, and this is one simple way to put on a show, as it were, to entertain an easily bored humanity. Fish guaranteed that the otters would "do something." Because this activity was popular with both otters and visitors keepers fed goldfish regularly.

I never saw anyone appear disturbed and I never heard anyone complain about feeding live fish to river otters. I found this surprising, especially since goldfish are a common first pet for many North American children, but perhaps this stands as evidence of our early disconnect from nonhuman animals. On further reflection, I am not really surprised that I did not hear even one complaint. To be honest, it really was amazing to watch the otters dash after those goldfish, to see how quickly they could catch them in their watery environment and how many they could hold in their mouths at one time. I knew that watching otters devour goldfish offered a more realistic vision of an otter's inherent, wild nature. This was as close as it got to "real" in this highly orchestrated, carefully constructed world. Most of the time, seeing animals in cages does not allow us to have much of a sense of what these animals are actually like: free and wild, beautiful yet bloodstained.

The only reason I could ever decipher for the double standard that the zoo maintained between bunnies and goldfish was that people do not tend to be as disturbed when they watch fish die—even little kids. I know from my own circle of friends that many people give up eating chicken and cattle and pigs, but continue to eat shrimp, tuna, salmon, and cod, becoming "pescetarians" (Anon 2012). Most of us are more disturbed when we see a "cute," fuzzy, large-eyed mammal die, than a shimmering, fan-tailed, bubble-eyed goldfish. If watching goldfish die in the mouths of otters had been unpopular with zoo guests I am sure that otters would have lost their occasional hunting privileges and they would be quietly fed flesh only during off hours, like the rest of the carnivores at the zoo.

With the exception of enrichment foods, which are offered throughout the day (to entertain visitors as much as to entertain resident captives), all of the carnivores are fed before the zoo opens in order to spare zoo patrons the sight of such carnage. Hiding the deaths that are necessary to caged carnivores in this way contradicts the zoo's stated educational mission: displaying animals as close to their wild settings as possible. By definition, predators hunt. They must eat—and they must eat flesh. If the honest mission of a zoo is to educate the public, and to keep animals as naturally as possible, then this is not only a glaring omission, but downright dishonest and deceitful.

I also feel that the zoo gave mixed messages. For example, on the one hand, both through staff commentary and via the rare educational cage signs, the zoo reminds visitors that the animals therein are wild, not tame—certainly not pets. On the other hand, all of the caged animals have names, some of which are displayed on these same signs, and repeated by the same staff. Just like family dogs—Andy, Prince, Taylor, Lucky, Rocket, and Dolly—each animal was named.

Two wolves, a brother and sister, fell partially under my care while working as a keeper. Resident wolves are perhaps the clearest example of these mixed messages. Their diet consists primarily of "meatballs." The first time that I fed these wolves they kept their distance (or so the keeper told me)—I was a stranger. Even still, they were only about ten feet away! When the wolves realized that I was their source of food, they approached me to catch the "meatballs," which I tossed into their open, waiting, snapping mouths. The wolves were always anxious to be fed in the morning so they loped over to us expectantly as soon as we appeared at the entrance to their exhibit.

Cain seemed to hold himself higher than Phor, who crouched submissively before her brother. They both anxiously shifted positions and pawed at the earth as they waited for "prey" to land in their mouths. I was told that the male, Cain, always had to be fed first because he was the top dog in their little two-wolf "pack." Phor, the female, was to be fed second. At first I was a little fearful of getting the two wolves confused—they looked identical to me, and I was unsure what Cain would do if he was mistakenly fed second. Anyway, I didn't really want to find out: There was nothing between us but air. I was armed with a baton, but this did little to reassure me.

The wolves' food consisted of fortified ground meat, which I shaped into small balls, then underhand-tossed into one open mouth, and then the other. They snapped their pointy teeth closed as the ball came near, ripping it out of the air, then smacking their tongues, never taking their intense eyes off of me. Despite their domesticated lives, their eyes remained remarkably sharp and aware; holding eye contact with a wolf is strange and intimidating, even when the wolves are bizarrely tame. When the bowl of meat was empty, I slowly backed out of the exhibit, relieved. Standing so close to a wolf was strange and unnerving. It was more than a little peculiar to toss them wads of meat in the same way that I used to toss treats to my cocker spaniel, Herbie. That was not how I imagined wolves—not even in a zoo. The wolves at the zoo seemed more like domestic dogs than those I had seen in Yellowstone National Park. Their unnatural behavior seemed an injustice both to the wolves and to visitors who pay to receive an entirely inaccurate depiction of "wild" animals.

Several times I entered cage exhibits with other keepers and wandered around their bleak living space to spray against ticks and fleas, to check fences, and pick up rotting bones. When we entered the wolves' pen, they followed us around from a distance of ten to twenty feet, pausing when we paused, walking when we walked. As I struggled with this surreal experience, I felt angry: I was part of presenting to the public what was perhaps the worst message imaginable with regard to wildlife—that these animals were somehow both wild yet tame. Small wonder that zoo visitors later approach bison or grizzlies in Yellowstone National Park for a photo, to offer food, or even to pet these wild animals, only to be surprised when these great beasts are aggressive. Watching animal "professionals" and "educators" at the zoo, I could see how it is possible for zoo visitors to confuse the behavioral differences between wild animals and domestic pets.

My experiences working at the zoo reminded me of my childhood visit to the Portland Zoo, which fostered the illusion that captive penguins are wild. This allowed my childish brain to forget that I was in Portland, to imagine these birds splashing through ocean waves, diving, and waddling across ice-flows. I wonder what sort of impact the Portland Zoo would have had on me, as a child, if I had seen keepers mingling with the penguins, walking alongside, touching, and swimming with them in their small pool. Would I have been as mesmerized? Would such an experience have instilled a deep appreciation for the magnificence of these unusual birds? Or would the keeper's presence have changed my experience entirely?

Children who watched me walk with the wolves were super-excited about a human right in among these mythic animals. They would point, laugh, shout, and of course, offer their best imitation of howling. But I don't think they were so much fascinated with wolves as they were envious of me—jealous that I could walk among the wolves while they had to remain behind a fence. They wanted to do what I was doing—to come inside, to walk with wild animals. Instead of imagining these large canines wandering the hills, or resting together on a warm sandstone rock, it appeared to me that the visitors imagined *themselves* mingling with wolves. Is this the education that the zoo intends to offer young people—the hope that they, too, can mingle with caged wild animals? This unnatural relationship with wildlife should not be the message that children carry away from a zoo, but I fear that it is.

Zoological institutions purport to teach children (as well as adults) about wildlife. Zoos usually claim to instill a deeper appreciation of animals and to foster our natural excitement at the marvels of the animal kingdom—the graceful way a tiger lifts her nose to sniff the wind, the methodical way that the sika deer stretches his long, slender legs when uncurling from a nap. They can only do this if cages offer some semblance of a natural setting. When visitors see a keeper inside an exhibit with wildlife visitors lose this sense of the real wild. Instead, they are presented with a lie.

The animals locked up in captivity are placed in a setting where they unwittingly pretend to be something they are not. When I was a kid I loved the tale of Little Red Riding Hood. In this story, the villain is a wolf pretending to be someone he is not (a loveable grandmother) in order to secure food—Red Riding Hood. Zoos are similar with regard to deception. For most visitors, zoos are a fun place to visit, a place where they enjoy themselves while gaining an education by watching caged wildlife. As a keeper, I unwittingly pulled back the covers, and peeked underneath. I now see zoos as akin to the wolf in Grandma's bed: Dressed in the trappings of civilization wild animals seem friendly, but they are still wild animals—some are carnivores with killer teeth and killer instincts.

Over several months as zoo keeper I realized that the zoo is designed not to provide a worthy education, nor for wildlife, but to attract visitors and dollars. The animals trapped in zoo cages are neither performers nor entertainers, they are neither tame nor volunteers. They are living, breathing, independent creatures who deserve a life of freedom—something which all living beings seem to prefer. Zoos, by their very nature, can never provide the reality they profess to teach—they can never offer the public a "wild" animal which by definition entails freedom and autonomy. Zoo practices do not match zoo ideals.

In Little Red Riding Hood, the grandmother is rescued from death in the belly of the wolf by a woodsman. Our vision of the wild and dangerous wolf is fostered and the tough man saves the helpless grandmother. Our treasured stereotypes are upheld. In truth, all of the unfortunate creatures trapped in zoos around the world are waiting for someone to rescue them. Collectively, it is time for us to burst into the little cabin in the woods—to dismantle our idealized vision of zoos so that we can see what is really happening in these capitalistic enterprises—and provide some form of release for these imprisoned individuals.

I am still torn up on the inside about what I saw as a zoo keeper. My time at the zoo was both an amazing opportunity and a tragic discovery. My experiences with the animals were simultaneously magical and devastating. Despite my emotional confusion, I am absolutely certain about one thing: Neither education about the natural world, nor animals of any kind, belong on concrete floors, behind thick glass, nets, chains, and metal bars.

I now understand why wildlife conservation and education should not be rooted in deception and half-truths—why conservation and education must not cater to urban humans at the expense of reality. Like the bears, snakes, and wolves themselves, exposure to these wild animals *belongs* in the wild. Zoos run counter to wildlife conservation because they exist under the illusion that formerly wild animals kept in environments engineered by humans can stand in place of wild animals in their natural environment. We safeguard these creatures in zoos—just in case we wipe them out in the wild—which allows us to continue our careless, selfish, destructive ways. Having worked at even just this one zoo, I can say that we are wrong about this: It is not possible to maintain the true wonder of a species outside of a natural environment and community. When the last *wild* tiger disappears, then the last *tiger* disappears.

True, some of us will never have the chance to see many different animals in the wild—I will likely never travel to the antarctic to watch the penguins waddle, or hike into a jungle to see a sloth or gibbon move about in the trees. Still, it seems to me that it would be better never to witness these individuals in the flesh than to see them in the kinds of mock wildlife experiences that zoos provide to satisfy our paltry desires to be fascinated by other beings. It would be better never to see a wolf or an eagle than to enslave these individuals within the restrictive confines of capitalistic enterprises that we call zoos. When we visit zoos we are not "seeing" wildlife; we are being shown a projection of what we imagine wildlife to be. As an environmentalist concerned about wildlife conservation, as well as the quality of life for each individual animal, I learned at the zoo that we can only experience the true nature of wild animals in the natural world. Only in the wilds, should we be so lucky, can a wildlife experience be honest, fair, and therefore truly meaningful.

## Discussion Questions

1 Is it possible to learn anything of significance about wildlife from visiting a zoo? Ideally, what will zoo visitors learn about wild animals from seeing them caged for viewing? How else might we gain this same information?

2 What does Haaland's poem, "Goldeye, Vole" (page 311)indicate about our experiences with nature in relation to our larger lives? How is this the more powerful given that the author grew up to become an accomplished poet and an academic—who continues to write about blue racers, quartz, and needle grass?

3 What moral argument might be posed for holding wild animals in cages for educational purposes? Is there a morally relevant distinction between holding carnivores and holding herbivores? What about caging reptiles (generally carnivores) versus elephants (herbivores)?

4 Utilitarianism suggests that the few might legitimately be sacrificed for the benefit of the many in certain situations. Zoos turn this equation on its head, sacrificing many for the few. Is it possible to pose a moral argument to justify killing the many for the few? If you do so, is it reasonable to expect that humans can always be spared?

5 What is the difference between a zoo and a sanctuary? Is it sometimes okay for a sanctuary to buy an animal? If so, when? Why might others view this as inherently problematic?

## Essay Questions

1 What are the moral problems entailed when we keep wildlife in zoos?

2 Will you ever visit a zoo again? Why or why not?

3 Research the lives of Ota Benga and Saartjie Baartman, and Social Darwinism. What is the relationship between Social Darwinism and the exploitation of others, whether human or nonhuman? What aspects of your community reflect Social Darwinism? How are zoos a reflection of this way of thinking?

4 If humans destroy the last tiger or rhino in the wild, is it a legitimate enterprise to attempt to preserve tigers and rhinos in zoos? Why or why not?

5 Read Acampora's *Metamorphoses of the Zoo* (see suggested further reading below), then describe what you find to be the parameters of a morally acceptable zoo.

## Suggested Further Reading

Acampora, R.R. (ed.). 2010. *Metamorphoses of the Zoo: Animal Encounters After Noah*, Lanham, MD: Rowman and Littlefield.

Bekoff, M. and Goodall, J. 2007. *The Emotional Lives of Animals: A Leading Scientist Explores Animal Joy, Sorrow, and Empathy – and Why They Matter*, Novato, CA: New World Library.

Malamud, R. 1998. *Reading Zoos: Representations of Animals and Captivity*, New York, NY: NYU Press.

## Note

Special thanks to the editor for going the extra mile to create a polished piece from a rough draft.

## References

Anon. 2005. 'carnivore.' *Collins Dictionary Of Biology*.

Anon. 2012. 'pescatarian' or 'pescetarian.' *Merriam-Webster's Collegiate(R) Dictionary*.

Anon. 2014. Bald Eagle. *National Wildlife Federation*. Available from: http://www.nwf.org/wildlife/wildlife-library/birds/bald-eagle.aspx [Accessed May 29, 2014].

Cox, C. and Baker. L. 2002. 'Pavlov, Ivan Petrovich' in *Biographical Dictionary of Psychology*, Routledge, London.

# 25

# INTERSECTIONS

## Activism and the Natural Sciences

*Xylem. T. Galadhon*

### Unfolding

I have always cared about animals. From my early youth, I remember feeling a keen sensitivity to the emotions and modalities of the nonhuman world. Though I was never allowed to keep a dog, no matter how desperately I wanted one, I was always drawn to these beings at friends' homes, or simply on the street. I have a clear memory from a family trip to India, when I was 10 years old, revealing my early orientation toward the Animal Kingdom: By commandeering my 2-year-old cousin's baby bottle, I was secretly feeding stray dogs who wandered the neighborhood, using the bottle to squirt milk towards the dogs' mouths. When my uncle caught me in the act, I received quite a scolding—an early indication that my family did not value animals in the same way that they valued humans. I was to learn later that my family merely reflected a general lack of concern for animals held by our larger society. But to be fair to my family, they imbued me with a strong sensitivity to humans that later spilled over to nonhuman animals.

My uncle's scolding did not deter me, and I proceeded to help animals throughout my childhood. On another memorable occasion, I adopted a guinea pig from another junior high student, whose parents wouldn't let him keep the guinea pig any longer. Not liking the little fellow to be caged all day, I decided to give him freedom to run around in our backyard. When I checked in on him later, he had been bitten by a predator, perhaps a neighborhood cat traipsing through our yard. I brought him in, but he started to tremble, then spasm, and just a little while later, died. I still don't know the medical reason for his passing, but my guilt and sadness—my sense of responsibility—was intense and profound, permanently affecting my sense of responsibility for other beings.

Yet another indelible (and even more unpleasant) childhood memory was killing snails in our garden. Encouraged by my parents, who did the same, I took the task to heart, killing with morbid efficiency. Looking back, I still feel slightly queasy when I remember how much death and destruction I doled out to Class Gastropoda. Yet no one in my family, nor among my friends, batted an eyelash at such methodical killing. Perhaps this is because cruelty to small animals is so endemic and rampant among young boys as to be hardly noticed. In fact, one of my earliest exposures to Christianity was when a local church invited us to help pay for group ski trips by turning in garden snails (who were shipped off to a restaurant). I am quite certain that this "discovery"—that Jesus didn't mind if we sent snails to their Maker a little prematurely—"greased the skids" for further snail massacres in our backyard. (It is important to note that, contrary to what I was exposed to when young, many religious individuals are intrinsically concerned with animals and recognize that sacred texts provide clear moral guidance for rethinking our exploitative approach to nonhuman animals and the Earth; e.g. Regan 2004; Kemmerer 2012).

Such generally accepted religious nonchalance did nothing to encourage my inborn sensitivity to animals. In fact, society's tendency to ignore cruelty towards other species did little to help me on my path—even though we now know that animal cruelty is an urgent warning sign of a dangerous personality. I, at least, felt remorse as I grew older, as I looked back on the unfortunate snails whose souls I dispatched so triflingly. In retrospect, my thoughtless cruelty resulted in considerable guilt, and perhaps ultimately helped to make me a more decent person.

## Adding in Equations

My love for science, like my sensitivity to life and suffering, came from some deep part of my being. Initially, I was captivated by space exploration in science fiction. Later, in junior high school, I clearly remember reading popular science pieces in *Scientific American*. The articles were far above my comprehension level, but I found their adventurous and exploratory spirit wonderfully exciting.

My trajectory was more strongly confirmed in high school, when I took my first course in physics. I was amazed to learn that I could calculate exactly how high I was standing when I dropped a rock from a bridge into a river: All I had to do was count the number of seconds it took for the rock to hit the water, then plug this number into a very simple algebra equation. This algebra equation is introduced in "classical mechanics"—the branch of physics that examines the movement of bodies in space and which has long been and remains the first physics course required for aspiring physicists and engineers.

Connecting space and time through such a simple equation was amazing to me then, as it is now. The fact that little scribbles on a piece of paper—discovered by infinitesimally small mortals on a tiny planet in the corner of an unremarkable galaxy in our vast Universe—apply to *all* bodies in this Universe—as far as we can see out in space, and back in time—holds a touch of mystery for me. Indeed, some of our greatest scientists have felt humbled and amazed by this realization in their more self-reflective moments. Perhaps this is best encapsulated by physicist Eugene Wigner in a well-known article entitled "The Unreasonable Effectiveness of Mathematics in the Natural Sciences" (Wigner 1960). In the same vein, Einstein noted "the eternal mystery of the world is its comprehensibility… The fact that it is comprehensible is a miracle" (Einstein 1956).

## Many Twists and Turns: An Adventurous Trajectory

My interest in animals and physics seemed quite distinct and unrelated to most onlookers, and this bifurcation deepened under the power of Western education. But it was this seemingly disparate combination that led me, ultimately, to understand the critical connection between animal and Earth liberation.

I first delved more deeply into science—and this is when I truly and deeply fell in love with physics. I took an advanced version of classical mechanics, electromagnetism, quantum mechanics, and special and general relativity—starting with what the Greeks had to teach us, and progressing all the way up to the magic of the weird but wonderful world of twentieth century quantum physics. I marveled at the elegance of equations, the beauty of experiments devised to glimpse the wondrous secrets of Nature, and found them all rather profound and divine. I reveled in science. It was glorious.

Through events intruding from outside the ivory tower, I became steadily more aware that there *was* a world outside of this beauteous scientific elegance, a world full of pain and unnecessary suffering, a world of horrible, terrible things that human beings did to each other and to nonhuman animals. There was no way that a young person, strongly attuned to animals since childhood, could simply ignore this very real world of agony. So I took up activism with the same rigor and intensity with which I approached physics.

My first activism centered on animal rights. I remember the impetus quite clearly: a breaking scandal in 1986 at the Head Injury Laboratory at the University of Pennsylvania. The Animal Liberation Front had broken into this lab and stolen videotapes, created by experimenters, of their grim research. Designed to mimic human car crashes, in order to explore the effects of head injuries, baboons were strapped into sliding sledges that rapidly accelerated, and then smashed into brick walls at high velocity, head-first. One can guess the results without much difficulty—the baboons did not do so well after having their skulls crushed. The videotapes also exposed researchers mocking the injured baboons, demonstrating a derisive, unfeeling response (Newkirk 1986).

Watching this footage hit me like a body blow to the gut. At the time, I knew little of the Universe, of the continual and massive suffering that lies just beyond our comfortable daily environs. These videos provided my first exposure to the dreadful moral cost of exploiting nonhuman animals for science. Indeed, exploiting animals for science in the hope of enhancing human knowledge, or with the hope of improving human health, is fraught with complex ethical questions. Yet my immediate reaction was very simple: I was shocked and appalled—as any decent person would be—on seeing how these baboons were treated, and what experimenters had done to them. I was no longer able to sit idly by on the sidelines. I had seen the horrible violence that continues at every moment both far from our homes, and too often, so very near.

Just as childhood instincts impelled me to feed India's stray dogs, I was now driven to assist others to extricate animals from horrendous human-caused suffering through misguided and calloused animal experimentation. I joined the student animal rights group at my university, and plunged into a maelstrom of activity. I helped to expose the abuses and horror of the Head Injury Lab at the University of Pennsylvania and I opposed an underground animal experimentation facility planned at my own university. Over the next two years, I battled cruel and unjustified forms of animal experimentation at universities, and the fur industry, rodeos, circuses, factory farms, and on to the confrontational (though still *non-violent*) activism of trophy hunt sabotage.

Throughout high school, with the tacit consent of my communities, I had maintained a cool air of cynicism and insouciance toward the suffering of others. As a fully engaged activist, I could not ignore suffering, which led me to further round out my education, taking courses on environmental ethics, the U.S. role in Central America, the nuclear weapons race, the military-industrial complex, etc. I then took action based on what I had learned. In the years that followed, I repeatedly discovered connections between various forms of injustice, and I expanded my activism accordingly. Activism carried me over a winding and circuitous path: I traveled to Nicaragua for solidarity work, I organized against the first Gulf War, and I wrote articles connecting the war in the Middle East with fossil fuel consumption, all the while engaging in many other collective activities aimed at helping reduce unnecessary suffering—aimed at making the world a better place for all animals, including humans.

## Coming Down to Earth

The two sides of my life—science and social justice activism (most notably animal activism)—could more probably be the stories of two separate individuals. Yet both are central to who I am. This led to a measure of curiosity about intersections and interactions between the two. Ultimately, this combination led me to environmental activism.

As I worked for social justice, I felt myself shifting toward environmental causes, partly due to my training in the natural sciences. Additionally, I had always been an "outdoors person": my parents took me to national parks when I was young, which fostered a strong appreciation and interest in Nature. Indeed, positive childhood exposure to Nature is critical to one's later relationship with the outdoors.

I was drawn to forest preservation, at least in part because this subject allowed me to *be in forests*, which brought me profound happiness. When outdoors, the science part of my mind ponders everything around me—molecules, photons, electromagnetic fields, neutrinos from solar fusion, gravitational waves—in terms of equations. Another part, perhaps the part that inspires my animal-connectedness, relaxes and extends outward into the wider natural Universe, in a way that I cannot fully relax or extend when surrounded by the trappings of society and humans.

While it is common to believe that these spheres of human experience are diametrically opposed, this is not at all the case. Rather, the knowledge of the specific mechanics of how things work deepens my appreciation for the beauty and wonder of Nature, as expressed by renowned US physicist Richard Feynman:

> Poets say science takes away from the beauty of the stars—mere globs of gas atoms. Nothing is "mere." I too can see the stars on a desert night, and feel them. But do I see less or more? The vastness of the heavens stretches my imagination—stuck on this carousel my little eye can catch one-million-year-old light. A vast pattern—of which I am a part ... What is the pattern or the meaning or the why? It does not do harm to the mystery to know a little more about it. For far more marvelous is the truth than any artists of the past imagined it.
>
> (Feynman n.d.)

Science and wonder are both central to our deepest spiritual impulses. All forms of social justice advocacy are intertwined with religion and spirituality, but the connection between environmental awareness and issues and spirituality is particularly important. Indeed, William James noted: "religious awe is the same organic thrill which we feel in a forest at twilight, or in a mountain gorge" (James 1902). It is sad that a church, which encouraged me to kill snails, did not understand this basic truth.

## Absolute Truth

Being a physicist who was interested in Earth and animal liberation created considerable tension in my life—in my experience, the world of scientific inquiry and the world of social change do not currently mix well. This made me yet more curious about the juxtaposition of these seemingly disparate elements of my life—science and social justice advocacy (with leanings toward animals and environment).

I have been fortunate to find noteworthy role models to help me connect elements of my life—great scientists who have combined their zest for scientific truths with their interest in moral truths, and who have therefore worked for social change, such as Albert Einstein (1956) and Carl Sagan (1997). Through the writings of such scientists, I came to understand how physics supports both Earth and animal liberation, and I have seen that their many excellent writings continue to foster social justice action in young scientists.

Over time I found one clear strand that runs through both sides of my life: a rigorous concern for truth. The word "truth" is loaded, and I use the term carefully, with a clear sense of context and definition. Let me be clear: All natural science that humans have discovered that holds true for our Universe—Newtonian gravity, quantum physics, statistical mechanics, cosmological evolution, and so on – are absolutely true *in their domain of applicability*. In other words, within measurement precision limits, information gathered by people in the natural sciences has *repeatedly* and dependably reached the same conclusions.

There are other overarching scientific paradigms and theories beyond those we have discovered, but any new information we gather must work within the confines of the physical theories we already know, if they are to be accepted. Furthermore, mathematical descriptions of these new theories must conform in the relevant sub-domains with extant equations. Consider this example: Einstein's description of gravity must prove applicable to any gravitational forces we discover, such as particle orbits in the immediate vicinity of a black hole, which constitutes an extremely strong-gravity environment. Furthermore, Einstein equations *must* reduce down to, and be consistent with, the much simpler and earlier Newtonian equations for "weak gravity" domains, such as the motion of bodies on Earth, of planetary and satellite orbits in the solar system, of stars in galaxies, and those between galaxies themselves (i.e. all the domains that scientists were familiar with early in the twentieth century). In short, motion displays a certain identical "truth" in all contexts, and we can demonstrate this commonality through mathematical equations. The necessity that scientific equations be consistent with earlier discoveries creates weighty constraints, which Einstein had to operate within as he extended Newtonian theories of motion. While I am most familiar with the sciences, it seems that such constraints are central to academics more broadly, and to advances within every field of study.

There is also an "absolute truth" in the world of ethics, summed up concisely and emphatically by Kurt Vonnegut: "God damn it, you've got to be kind" (Vonnegut 1965). Some formulation of this basic moral truth is found in every religious tradition; Christians know this moral truth as The Golden Rule, which was reformulated (more or less) in Western analytic philosophy by Immanuel Kant in his Categorical Imperative, which essentially states that we should act only in a way in which we could, at the same time, will that our behavior become a universal law (Kant 1785). Indeed, if all humans lived by this rule, our world would be *much* more humane. But many of us rationalize, justify, and squirm our way out of honoring this basic moral truth—we are unwilling to accept that this rule applies to *each* of us, *all* the time, in *all* places.

Indeed, integrity is often lacking among politicians and in human relations more generally, but scientists are used to working in a black and white world of mathematical certainty, a world where it is more difficult to twist truths in order to force them into one's own contorted framework. (Not impossible, just more difficult—humans are not necessarily rational.) For example, it requires remarkable self-deception for a scientist to deny that other phylogenetically advanced beings are sentient, but it has been done. And it is difficult to defend the notion that we are entitled to more wealth than other human beings simply because we were born into wealth, or that we ought to cause excruciating pain and premature death to other animals in the hope of learning something more about head injuries, however minor or clinically inapplicable that information might be, but both have been done. Sometimes this scientific demand for consistency and honesty is reflected in the scientist's tendency to be politically progressive—many natural scientists vote as liberal progressives, though this does not necessarily mean that they engage in social activism. Activism seems to require an extra degree of integrity, commitment, and humility than I have found in many scientists, who often lay claim to some measure of intellectual superiority above their fellow citizens.

## This Great Green Earth

I have come to see that social justice issues are all connected, and if we do not maintain a supportive and functioning natural environment, *all* is lost, including animals (both human and nonhuman). If we eat factory-farmed meat, milk, and eggs, we harm the environment, visit inordinate and unnecessary cruelty on animals, and degrade employees who work at feedlots and slaughterhouses for slave wages—they are all connected. If we are cruel to humans, it is likely that we will be cruel to nonhumans, and vice versa. If we don't care about turkeys or voles, it is more likely that we won't care about the hungry children of the Sudan.

As a scientist, I have come to see things as connected. All things are connected through energetic and material interactions over time and space—including the particles that form our bodies and the bacteria that swirl inside and outside of our bodies, including the water and air in which we evolved and on which we now depend, including the stars and supernovae that breathed out the heavy elements of which we are made. The diversity of life and inanimate matter are interconnected. It behooves us to work to protect this grandeur. If we are to achieve this end, a holistic vision is essential. One cannot love and protect Earth without protecting *everything* therein.

## Conclusion

We are privileged to live in an absolutely beautiful Universe. I have come to understand this universe much more profoundly through physics. Most importantly, I have come to recognize that this planet is populated with marvelous creatures, both human and nonhuman, which my scientific quest for truth compels me to protect, holding fast to fundamental truths—both scientific and moral.

We must contend with the ups and downs of life, with the difficulty of finding balance. Ultimately, we must look at ourselves in the mirror in the morning and find purpose to continue. I personally continue to work for change with regard to poverty, animal exploitation, war, and the forests, but I currently focus on environmental concerns. This planet and its many creatures offer so much beauty and wonder and joy, all of which balance the darkness, dreck, and dross of so much human indifference, destruction, and cruelty. Knowing this, with communities of like-minded others, I forge onwards, determined to protect throughout my life, as well as I am able, the entirety of this grand and great green Earth on which I feel privileged to exist.

The arc of the moral universe seems very long, and so there is likely to be much unnecessary suffering when each of us passes from this tiny little pale blue dot of a globe. Nonetheless, it is incumbent on *each* of us to do what we can during our lives and our time to work towards a more perfect world.

## Acknowledgements

This essay benefited from close readings and comments by V. Chalcraft, L. Kemmerer, A. McKim, L. Pease, and G. Reinking, and their assistance is greatly appreciated.

## Discussion Questions

1 What might be a reasonable way of responding to a church that asked children to collect/kill snails? What might you say to a child who explained this fund-raising endeavor to you?

2 Does scientific knowledge of the world come with moral responsibility for the environment?

3 How are Western science and Asian religions similar and different in their understandings of the world as interconnected? Which outlook do you prefer and why?

4 Is the point of view of natural scientists privileged with regard to environmental issues? In what ways might the training of Western science hinder our ability to assess our place in the universe and heal our damaged world? Does science leave room for the role of faith or emotions such as empathy and caring?

## Essay Questions

1 Describe the ideal occupation for bringing change on behalf of Earth and animals, and explain why this job would be ideal. What other job options might a social justice activist choose if interested in combining income with advocacy?
2 Design a university course (complete with readings, field trips, and assignments) exploring the interconnected nature of social justice causes.
3 List five criteria to assess whether or not a social justice cause is sensitive to other causes. Create questions assessing organizational commitment to these criteria. Meet with the leader of a local environmental/animal activist organization and use your questions to assess the organization's sensitivity to other social justice causes. Write up your findings.

## Suggested Further Reading

Einstein, Albert (1956). *Out of My Later Years*. Citadel, Secaucus, NJ
Sagan, Carl. (1997) *The Demon-Haunted World*. Ballantine Books
Wynne-Tyson, John. (1989) *The Extended Circle: A commonplace book of animal rights*, Paragon House

## References

Einstein, Albert. (1956). *Out of My Later Years*. Secaucus, NJ: Citadel.
Feynman, Richard. (n.d.). "Science and the Admiration of Nature." http://www.fotuva.org/online/poem_elliot.html, Accessed May 27, 2014.
James, William. (1902). *The Varieties of Religious Experience*. Gifford Lectures on Natural Religion. http://csp.org/experience/james-varieties/james-varieties2.html , Accessed May 27, 2014.
Kant, Immanuel. (1785). Translated by James W. Ellington (1993). *Grounding for the Metaphysics of Morals*, 3rd ed., Hackett. pp. 30.
Kemmerer, Lisa. (2012). *Animals and World Religions*. Oxford: Oxford UP.
Newkirk, Ingrid. (1986). "Unnecessary Fuss." http://www.youtube.com/watch?v=B6W_hL_gyJo, Accessed May 27, 2012.
Regan, Tom. (2004). *The Case for Animal Rights*. First Edition, updated. University of California Press.
Sagan, Carl. (1997). *The Demon-Haunted World*, Ballantine Books.
Vonnegut, Kurt. (1965). *God Bless You, Mr. Rosewater*. http://www.goodreads.com/quotes/show/1889. Accessed, May 27, 2014.
Wigner, Eugene. (1960). "The Unreasonable Effectiveness of Mathematics in the Natural Sciences." *Communications in Pure and Applied Mathematics* 13:1, http://www.dartmouth.edu/~matc/MathDrama/reading/Wigner.html, Accessed, May 27, 2014.

# 26

# ECOLOGY, FOOD, AND HOLISTIC POLITICS

*Deric Shannon*

## Class Society and Radical Politics

Like many people, my entrance into radical politics was a direct result of personal interest. When I was a teenager, my father left us, and, like many of my contemporaries, I was raised by a single mom. As part of my United States indoctrination, I learned that those who work the hardest are those who become wealthy. If someone is poor, it's their own damn fault—but this didn't jive with my personal experience. My mother worked hard—tirelessly, sometimes as much as fifty or sixty hours a week. Yet she could barely make rent and keep food on the table. I paid for school lunches with a stamp, visibly demonstrating to everyone around that we required state aid to make ends meet. At some point I realized that those who made more money than my mother actually tended to work fewer hours, had better working conditions, and had more empowering jobs than she did. It occurred to me that the vast majority of the wealthiest members of our society seemed to be complete strangers to work—born into wealth, simply owning property as a living.

Oddly enough, it was punk music that set the stage for my radicalization. Bands like the Dead Kennedys, Crass, and Black Flag spoke a language I understood. They spoke about structural conditions: Poor people aren't poor because they don't work hard enough. Rather, poverty exists because capitalism creates a class society—one in which the wealthy benefit from things produced by workers. Workers, ironically enough, are responsible for creating everything, yet often have almost nothing. These bands talked about the *systemic* cause of poverty amid wealth. With their help, I came to understand that capitalism guarantees a split between haves and have-nots.

I was around nineteen or twenty years old when I started developing a taste for social theory (e.g. Marx 1961; Kropotkin 1968).[1] Karl Marx explained how exploitation allowed some people to own without labor, while others labored to make money, producing goods that they could not afford. Workers eke out a living off of whatever the owners are forced to share in the form of wages. But Marx's "dictatorship of the proletariat" didn't exactly sound like a worker's paradise to me. Further, watching the USSR and China's governments made me distrust Marxism—or at least their interpretations of Marxism. Indeed, reading works by leading Marxist revolutionaries like Lenin (1929) made me even more suspicious. They seemed to lean toward creating a state that forced a "better" world on everyone, whether they liked it or not, through a vanguard Party, not with the effort of (extra)ordinary working people like my mother. And if capitalism were smashed, but not the state, wouldn't we just trade economic bosses for political bosses? This is what seems to have happened in the USSR, China, the Eastern Bloc countries, and Cuba. I had suffered at the hands of the state's domestic military—the police—as had a number of my friends. It seemed that we ought to be able to create something better than the state as an organizing power.

Anarchist thinkers and writers seemed to agree. Mikhail Bakunin (1916), for example, argued that our ideas about God, the state, and capitalism had created in humanity a need to be led—to be ruled. And Peter Kropotkin (1968), a Russian prince who renounced his title and his wealth to throw his lot in with the workers, explained quite elegantly how a cooperative society could be organized without the need for hierarchical directives from bosses—whether political or economic. But it was Murray Bookchin (1971) who sealed the deal for me, with his fiery attack on Marxists who, as he saw it, did not understand that we cannot use hierarchical methods to bring about a non-hierarchical future. That is, we must live and organize in ways that mirror the sort of society we would like to create, inasmuch as that is possible. If a dominating institution like the state brings about social changes, then we end up living in a society dominated by the state.

That might have been the end of my story. I had come to accept that "hierarchy" referred to the state, organized religion, and capitalism. This certainly suited my understanding, and made the solution to the world's problems seem simple: We workers needed to organize ourselves, form non-hierarchical associations, and replace capitalism and the state with social relations of mutual aid. Of course, as an anarchist I was also opposed to racism, sexism, and homophobia, but it seemed that these "secondary" matters could be addressed after proper material conditions (i.e. equality) had been established.

## Feminist Encounters

I had moved far beyond the United States' rhetoric of work hard and succeed. I began thinking like a radical instead of a liberal. In the years that followed, I experimented with various political practices, trying to test idealistic visions. I refused to work, drank too much, and bummed around the country in order to avoid the inherent exploitation of capitalism by scamming my way through life. When I grew weary of this challenging and risky existence, I worked just enough to pay for food, drink, and rent—and complained loudly while doing so, in order to remain clear about my ideals. I tried organizing fellow workers with little success (most likely because I was arrogant and unable to listen to those who had more experience). Undaunted, emulating my earliest musical teachers, I played in bands that sang about society's problems, and hoped I might move others to question the status quo. I joined local anti-racists: I went to meetings and protested police abuse and racism. And of course I swore I'd never go back to school, since those privileged and shallow university yuppies were a big part of the problem.

Some of my youthful endeavors were less embarrassing than others, but they all proved important to where I was headed. To make a long and sordid story short, I eventually grew tired of life on the fringe and decided to take the best scam at hand—I enrolled in college.

I was very surprised to find radicals at Ball State University, where I enrolled. There were even anarchists—though certainly not amid the faculty. Radical professors were typically Marxists. In retrospect, this makes sense, since Marxism (as developed by Lenin) gave power to professional intellectuals. In any case, as an undergraduate I defended and advocated for anarchism, while doing as little as possible to earn decent grades. I didn't think much about the future.

After graduation it made more sense to go to graduate school than to work another "straight" job. Indeed, school was a pretty sweet scam for a working class kid from Indiana. Yet more amazing, I received a stipend for my graduate studies—something I still find unbelievable (if we are going to keep universities around, they should be open to everyone, not a luxury for a minority of wealthy students).

One of the course offerings in my doctoral program was a class on feminist theory. I couldn't really pass up this chance to get paid to read and talk about politics. After all, I'd read about women's subordination—I was an anarchist! I didn't believe in *any* hierarchies, so this course was a natural fit—or so I thought. I enrolled, and this turned out to be the best enrollment decision I've ever made (though I've generally learned more valuable lessons *outside* of university classrooms than *in* them).

Feminist theory challenged my core assumptions about politics. Learning about feminism began a long process of political re-alignment that continues to this day, and, with any luck, will continue for the rest of my life. The writings of bell hooks (1984) and Patricia Hill Collins (1990) taught me that we need not always prioritize some struggles over others, and that class struggle can't be fully understood, or fought, without analyzing exploitation and oppression more broadly, recognizing both as overlapping phenomenon in complex ways. Further, I learned that it is unrealistic to fight against only one form of hierarchy and oppression, expecting the rest to somehow naturally fall into place. I discovered that class struggle, as it is typically conceived, often fails to address certain hierarchies.

Feminism demanded that I take into account the ways that men oppress women in the here and now—as the anarchist women's group Mujeres Libres noted so many decades ago; other hierarchies (patriarchy, white supremacy, heteronormativity, etc.) could not be resolved "on the morrow of the revolution" (Ackelsberg 2004, p. 38). Supplemented with readings in post-colonialism and critical race theory (Fanon 1963; Said 1994; Delgado 1995) as well as a good dose of queer theory—a body of work that became very near and dear to me (Jagose 1996; Butler 2004; Wilchins 2004), my studies slowly widened my view of organizing beyond the narrow concerns of class, giving critical support to struggles against *all* hierarchies—or so I thought.

## The Non-Human World and "Personhood"

It should be obvious that reading has been critical to my politics. Theory informs my political practice, which in turn informs my theory, *ad infinitum*. I began to work with a small group of radicals while I was working on my doctorate, and because reading was so important to us politically, we started a small radical reading group to study and learn together.

One of the members of the group suggested that we do some readings in animal liberation. I had always been supportive of vegetarians, vegans, "locavores," people who grew their own food, urban farmers, and guerilla gardeners, but I also thought that these sorts of interventions were pointless exercises in the politics of consumption. I believed that meaningful change came from struggle, not from personal changes in one's own lifestyle and more specifically, certainly not from building different relationships with food. I knew that some folks protested animal abuse in a variety of ways, but I conveniently reduced all politics of animal advocacy to diet, and disassociated animal advocacy from its deep and sure connection with ecology and food writ large. I can't remember what we read for that particular meeting. I do remember the conversation we had afterwards, and my overwhelming sensation that I had clearly missed the elephant—or shall we say cow—standing in the middle of the room.

I again turned to books, starting with Peter Singer's (1975) tome on animal liberation, in which he suggests that, if our ethics are rooted in our capacity to experience pain, then our ethics apply to *all* animals. Indeed, the exploitation and suffering of animals in modern societies is absolutely unethical and, worse still, unnecessary. However, Singer's views on reproductive freedom (he argues that a woman's freedom to make choices about her own body should be limited) turned me to other sources. I next read Best and Nocella's (2004) volume on animal liberation, which led me to their work on earth liberation (2006). Both of these books connect struggles for earth, animal, and human liberation, and root these links in what has since become familiar to me—feminist theory.

Best and Nocella's collections call our society's construction of "personhood" into question, a conception that I had uncritically swallowed somewhere long ago, along with the idea that those who work hard become wealthy. And this is, perhaps, the most valuable lesson I took away from these readings—personhood can be far more inclusive than I had understood.

I always believed that my organizing and way of life reflected a commitment to the liberation of *all* persons. But I had never investigated what this meant. Is a non-human animal *not* a "person"?

At that time, as always, I loved and cared for companion animals, and certainly recognized them as endowed with personhood and worthy of ethical consideration. Why did I not extend this understanding to include other nonhuman animals? And if it is problematic to assume that non-human animals are here simply for human use, why not apply this analysis to streams, trees, and mountaintops as well? How had this false sense of separation from all other life—from all else that exists—become so entrenched in Western thought—in *my* thought? How do we begin repairing the damage that this separation has done to countless living things, including human beings? Where might *I* start on this next journey?

Links between the ways that we exploit the non-human world and our own oppression became obvious. If we destroy the environment, how will humans have clean air, water, and a livable environment? Considering the apocalyptic and criminal proportions of our current environmental destruction (see Kolbert 2014; White and Heckenberg 2014), human liberation is meaningless if we don't also move towards ecological sustainability—the consequences of failing to do so seem like mass suicide. Finally, I wondered how a movement dedicated to the liberation of workers could routinely ignore the dehumanizing effects of working in slaughterhouses and on factory farms. A cursory glance at animal industries reveals callousness and inhumanity required of and ingrained in workers, who are often poor immigrants, exploited by wealthy, white capitalists. What good would an anti-capitalist movement be without addressing these fundamental problems?

## Applying Lessons Learned: By Way Of Conclusion

Radical politics do not posit any simple solution to the problem of hierarchy. However, most radical theories note that hierarchies do not exist in a vacuum. Relations of ruling run throughout our institutional fabric, and are rooted in our daily lives. These hierarchies support each other, naturalize domination, and reinforce assumptions that underlie the coercive organization of our communities and our world. Given this, we might create understandings of the world that are geared toward dismantling *all* hierarchies.

If history demonstrates anything, it demonstrates that regulatory tinkering does not effectively address domination (though such tinkering may reduce or limit its effects in some contexts). If we are to meaningfully alter our social relationships, a holistic understanding of relations is essential, an understanding rooted in our willingness to struggle to overturn the existing social order, and to undo the damage we humans have done to each other and the entirety of the non-human world.

For this to happen, we might begin by recognizing that the earth and non-human animals do not exist for our use, that they have inherent value, independent of human beings. Thousands of years of human domination over the non-human world must come to a close—not just for the benefit of humanity (after all, we won't survive on the planet if we continue despoiling and destroying everything in our path as we greedily grab at energy, profits, comfort, and convenience), but also for the benefit of the biosphere and all that dwell therein.

Whether people refer to themselves as "anarchist" or "feminist" or "liberationist" or "queer" matters little if our conception of revolutionary struggle is limited to human beings. We would do well to struggle to alter deeper aspects of our understanding and our culture: diet matters, daily life in general matters—not merely as patterns of *consumption*, but as ways that we relate to each other, ourselves, and the natural world. We might also empower ourselves to realize and actualize our own worth and collective power. In the meantime, we might also liberate spaces, places, and persons. What good are radical ideas divorced from practice (or vice versa)?

Acting directly on our own behalf, and with others, is critical. Rather than waiting for our legislators to protect us, and those who are most vulnerable, we might take a cue from those engaged in direct action: Rescue exploited animals; intervene against institutions that bring *us* miserable conditions. Rather than allow capital to continued pillaging nature, we might put ourselves physically in the way of such destructive actions—occupying vulnerable spaces treated as the nightmarish playgrounds for capital, a collection of "things" to exploit for profit. Actions need only be limited by our collective imagination. With a diversity of tactics to match our expanded diversity of concerns, we have the ability to fight for nothing less than *total* liberation—a *process*, not an *event*.

The radical imagination equips us for creating a holistic movement, but it took me years to connect the dots. We live in an indoctrinated and fragmented society, in which the idea that fighting for human liberation is *synonymous* with fighting for earth and animal liberation is most often not even recognized, or is only represented as hopelessly utopian. *Fighting for the earth and animals must be a piece of any liberatory political practice.* Together, we can construct alliances based on mutual aid, and those of us who are able, can show others how and why liberation movements are interconnected.

Ecology, food, non-human animals—these are often seen as the separate and specialized concerns of either animal rights activists or environmentalists, disconnected from the politics of human liberation. My own travels have shown me that they are not separate issues, and that we don't have the luxury of treating them as such. Marx suggested over a century ago that it's not enough to think about the world and interpret it. Rather, our task is to change the world. If we can agree to these things, we can liberate ourselves. And in the process, we will liberate both earth and all animals.

## Discussion Questions

1. Is there a relationship between education and militancy? If so, what is the nature of this relationship? How does the status quo seek to mitigate these effects in early education?

2. How do relations of domination like white supremacy, capitalism, patriarchy, or heteronormativity overlap with environmental devastation? What challenges do we face in trying to communicate these relations to others?

3 In what ways can the plight of working people under capitalism reasonably be compared to the suffering of animals in factory farms? What might be the limits of such a comparison?

4 What does the author mean by total liberation, and how do you envision this? Is total liberation desirable or realistic?

## Essay Questions

1 Shannon notes that in his quest for a liberation lifestyle there is always yet one more oppression to examine, what might be the oppression of the future, as yet currently undetected?

2 If you were teaching others about interconnected relations of domination, how would you do so? What analogy might you develop? (Be careful not to use analogies except from your own experiences so as to avoid offending others.) What projects or exercises might you create for use in a college classroom?

3 Create a visual prop for teaching about interconnected relations of domination.

4 What forms of institutionalized domination affect or have affected you personally? What three changes are most important if we are to avoid the effects of such domination in the future?

5 What are the first three steps you would make to move toward total liberation in your own community?

## Suggested Further Reading

Ackelsberg, M. 2004, *Free Women of Spain: Anarchism and the Struggle for the Emancipation of Women*, AK Press, Oakland.
Bookchin, M. 1971, *Listen, Marxist!*, Times Change Press, New York City.
hooks, b. 1984, *Feminist Theory from Margin to Center*, South End Press, Boston, MA.
Kolbert, E. 2014, *The Sixth Extinction: An Unnatural History*, Henry Holt and Co., New York City, New York.

## Note

1 As a side note, a number of my comrades-in-arms have a tendency to decry theory as irrelevant, "intellectual masturbation." I disagree—I will never let anyone convince me that ideas don't matter—that we need more "action" and less "thought," as if there were a wall between the two. Ideas and theory matter—they have been critical for me, personally, and countless movements throughout history. For a few stand out examples, see Best and Nocella (2006), Bookchin (1971), Fanon (1963), and hooks (1984).

## References

Ackelsberg, M. 2004, *Free Women of Spain: Anarchism and the Struggle for the Emancipation of Women*, AK Press, Oakland.

Bakunin, M. 1916, *God and the State*, Mother Earth Publishing.

Best, S. and Nocella II, A. (eds) 2004, *Terrorists or Freedom Fighters: Reflections on the Liberation of Animals*, Lantern Books, Brooklyn, New York.

Best, S. and Nocella II, A. (eds) 2006, *Igniting a Revolution: Voices in Defense of the Earth*, AK Press, Oakland, CA.

Bookchin, M. 1971, *Listen, Marxist!*, Times Change Press, New York.

Butler, J. 2004, *Undoing Gender*, Routledge, New York.

Collins, P. H. 1990, *Black Feminist Thought: Knowledge, Consciousness, and the Politics of Empowerment*, Unwin Hyman, Boston, MA.

Delgado, R. (ed.) 1995, *Critical Race Theory: The Cutting Edge*, Temple University Press, Philadelphia, PA.

Fanon, F. 1963, *The Wretched of the Earth*, Grove Press, New York.

Hooks, B. 1984, *Feminist Theory from Margin to Center*, South End Press, Boston, MA.

Jagose, A. 1996, *Queer Theory: An Introduction*, New York University Press, Washington Square, NY.

Kolbert, E. 2014, *The Sixth Extinction: An Unnatural History*, Henry Holt and Co., New York.

Kropotkin, P. 1968, *Fields, Factories, and Workshops; or, Industry Combined with Agriculture and Brain Work with Manual Work*, B. Blom, New York.

Lenin, V. I. 1929, *What is to Be Done? Burning Questions of Our Movement*, International Publishers, New York.

Marx, K. 1961, *Economic and Philosophic Manuscripts of 1844*, Foreign Languages Publishing House, Moscow.

Said, E. 1994, *Orientalism*, Vintage Books, New York.

Singer, P. 1975, *Animal Liberation: A New Ethics for our Treatment of Animals*, New York Review, New York.

White, R. and Heckenberg D. 2014, *Green Criminology: An Introduction to the Study of Environmental Harm*, Routledge, New York.

Wilchins, R. A. 2004, *Queer Theory, Gender Theory: An Instant Primer*, Alyson Books, Los Angeles, CA.

# INDEX

For Product Safety Concerns and Information please contact our EU
representative GPSR@taylorandfrancis.com
Taylor & Francis Verlag GmbH, Kaufingerstraße 24, 80331 München, Germany

www.ingramcontent.com/pod-product-compliance
Ingram Content Group UK Ltd.
Pitfield, Milton Keynes, MK11 3LW, UK
UKHW021504240926
475702UK00035B/359

* 9 7 8 1 1 3 8 8 2 5 8 7 1 *